AF388678

Industrial Applications of
Power Electronics

Industrial Applications of Power Electronics

Editors

Eduardo M. G. Rodrigues
Edris Pouresmaeil
Radu Godina

MDPI • Basel • Beijing • Wuhan • Barcelona • Belgrade • Manchester • Tokyo • Cluj • Tianjin

Editors
Eduardo M. G. Rodrigues
Management and Production
Technologies of Northern
Aveiro—ESAN
Portugal

Edris Pouresmaeil
Aalto University
Finland

Radu Godina
Universidade NOVA de Lisboa
Portugal

Editorial Office
MDPI
St. Alban-Anlage 66
4052 Basel, Switzerland

This is a reprint of articles from the Special Issue published online in the open access journal *Electronics* (ISSN 2079-9292) (available at: https://www.mdpi.com/journal/electronics/special_issues/IndustrialApplications_PE).

For citation purposes, cite each article independently as indicated on the article page online and as indicated below:

LastName, A.A.; LastName, B.B.; LastName, C.C. Article Title. *Journal Name* **Year**, *Article Number*, Page Range.

ISBN 978-3-03943-483-1 (Hbk)
ISBN 978-3-03943-484-8 (PDF)

Contents

About the Editors

Eduardo M. G. Rodrigues received his degree in Electromechanical Engineering from the University of Beira Interior (UBI), Portugal. Before becoming a PhD Researcher with UBI under the EU FP7 funded Project "Smart and Sustainable Insular Electricity Grids Under Large-Scale Renewable Integration", he gained significant industrial experience in different industrial sectors, namely in control valve manufacturing, the automotive industry, and in steel casting foundries, covering different technical areas such as quality management systems, product development, electrical and mechanical assistance in the heavy mechanical sector, automation development, and technical support in MW-size resonant inverter-controlled induction furnaces. As part of the researching consortium and during the first stage of his PhD studies, he collaborated onsite with the team responsible for energy storage technology at Department of Future Technology Execution, Alstom Power (Switzerland). He collaborated as Postdoctoral Researcher in the Compete 2020 – Portugal funded project "Enhancing Smart GRIDs for Sustainability" and as task leader at University of Beira Interior. Currently, he is invited Assistant Professor at University of Aveiro – School of Design, Management and Production Technologies, Aveiro North. In addition, he has a managing role in an engineering firm, being one its founders, where he has acquired the function of hardware and firmware engineer, providing industrial machine retrofitting services, automation-based control solutions, and industrial electronics products designed in-house. He has authored or co-authored more than 70 indexed international journals, book chapters and conference proceedings papers. He is Editor for journals *Electronics, Automation,* and *Ad Hoc & Sensor Wireless Networks* and has also served as a member of more than 40 international conference committees. His research interests are in industrial electronics, advanced industrial power electronics applications, instrumentation and signal acquisition, digital signal processing, and the implementation of advanced control techniques.

Edris Pouresmaeil received his PhD degree in Electrical Engineering (with honors) from the Technical University of Catalonia BarcelonaTech (CITCEA-UPC), Barcelona, Spain, in 2012. After completing his PhD, he joined the Department of Electrical and Computer Engineering, at the University of Waterloo, ON, Canada, as a Postdoctoral Research Fellow. Following his postdoctoral appointment, he was appointed Associate Professor at the Centre for Energy Informatics, University of Southern Denmark (SDU), Odense, Denmark. He is currently Associate Professor with the Department of Electrical Engineering and Automation (EEA) at Aalto University, Espoo, Finland. His main research fields are operation and control of power electronics in power systems, integration of large-scale renewable energy sources into the low-inertia power grid, ICT-based power networks, smart grids, microgrid operation and control, and simulator design and software development for smart grid applications. Dr. Pouresmaeil is Senior Member of the IEEE.

Radu Godina received his PhD degree from University of Beira Interior (UBI), Covilha, Portugal, and since 2019, has served at the Department of Mechanical and Industrial Engineering as invited Assistant Professor at NOVA School of Science and Technology (FCT), Universidade NOVA de Lisboa (NOVA University Lisbon), Portugal. He has authored or co-authored more than 110 indexed international journals, book chapters, and conference proceedings papers. He served as a member of 50 international scientific conferences related to such topics as energy, sustainability, and industrial engineering. Radu Godina is Editor of the Swiss journal *Applied Sciences* and of the *European Journal* *of Sustainable Development Research.* His main research fields are industrial symbiosis, sustainability, lean manufacturing, electric vehicles, circular economy, model predictive control, and quality control.

Editorial

Industrial Applications of Power Electronics

Eduardo M. G. Rodrigues [1,*]**, Radu Godina** [2,*] **and Edris Pouresmaeil** [3,*]

1   Management and Production Technologies of Northern Aveiro—ESAN, Estrada do Cercal 449,
     Santiago de Riba-Ul, 3720-509 Oliveira de Azeméis, Portugal
2   UNIDEMI, Department of Mechanical and Industrial Engineering, Faculty of Science and Technology (FCT),
     Universidade NOVA de Lisboa, 2829-516 Caparica, Portugal
3   Department of Electrical Engineering and Automation, Aalto University, Office 3563, Maarintie 8,
     02150 Espoo, Finland
*   Correspondence: emgrodrigues@ua.pt (E.M.G.R.); r.godina@fct.unl.pt (R.G.);
     edris.pouresmaeil@aalto.fi (E.P.)

Received: 16 September 2020; Accepted: 18 September 2020; Published: 19 September 2020

Abstract: Electronic applications use a wide variety of materials, knowledge, and devices, which pave the road to creative design, development, and the creation of countless electronic circuits with the purpose of incorporating them in electronic products. Therefore, power electronics have been fully introduced in industry, in applications such as power supplies, converters, inverters, battery chargers, temperature control, variable speed motors, by studying the effects and the adaptation of electronic power systems to industrial processes. Recently, the role of power electronics has been gaining special significance regarding energy conservation and environmental control. The reality is that the demand for electrical energy grows in a directly proportional manner with the improvement in quality of life. Consequently, the design, development, and optimization of power electronics and controller devices are essential to face forthcoming challenges. In this Special Issue, 19 selected and peer-reviewed papers discussing a wide range of topics contribute to addressing a wide variety of themes, such as motor drives, AC-DC and DC-DC converters, electromagnetic compatibility and multilevel converters.

Keywords: power converters; electrical machines; power grid stability analysis; inverters; power supplies; power quality; multilevel converters; motor drives; power semiconductor devices; electromagnetic compatibility

---

1. Introduction

In recent years, power electronics have been intensely contributing to the development and evolution of new structures for the processing of energy. It is becoming very common to generate electrical energy in different ways and convert it into another form in order to be able to use it—for instance, renewable sources, battery banks, and the transmission of electric power in direct current (DC), which makes the voltage of the network available in different levels, in detriment to the supplied voltage from the grid [1]. The main users of these signals are pieces of electronic equipment that use voltages at levels different from that which is available from the grid; the drives of electrical machines, which modify the voltage of the electrical network (amplitude and frequency) to control the machines and finally, in electrical systems, DC power transmission and frequency conversion [2].

Two leading trends are currently noticeable in the power systems field of study. The first trend is the increasingly and prevalent employment of renewable energy resources. The second trend is decentralized energy generation. This scenario raises many challenges. Therefore, the design, development, and optimization of power electronics and controller devices are required in order to face such challenges. New microprocessor control units (MCUs) could be utilized for power production control and for remote control operation, while power electronic converters are and could be utilized to control the power flow [3].

Nevertheless, power electronics can be used for a wide range of applications, from power systems and electrical machines to electric vehicles and robot arm drives [4]. In conjunction with the evolution of microprocessors and advanced control theories, power electronics are playing an increasingly essential role in our society [5].

Thus, in order cope with the obstacles lying ahead, original studies and modelling methods can be developed and proposed that could overcome the physical and technical boundary conditions and at the same time, consider technical, economic, and environmental aspects. The objective of this Special Issue was to present studies in the field of electrical energy conditioning and control using circuits and electronic devices, with an emphasis on power applications and industrial control. Therefore, researchers contributed their manuscripts to this Special Issue, and contribute models, proposals, reviews, and studies. In this Special Issue, 19 selected and peer-reviewed papers contribute to a wide range of topics, by addressing a wide variety of themes, such as motor drives, AC-DC and DC-DC Converters, multilevel converters and electromagnetic compatibility, among others.

A significant portion of the currently produced electricity worldwide is mostly generated by centralized systems, based on conventional fossil fuel plants or nuclear power [6,7]. The barriers that policy makers, researchers, and engineers have to overcome when it comes to the operation and control of conventional power plants, and the development of low voltage power generation systems, have paved the way for diverse opportunities of energy generation, closer to the load, by the customers themselves, also known as distributed generation (DG) [8]. Thus, concerning this topic, several papers are published in this Issue.

In [9], an efficient H7 single-phase photovoltaic grid-connected inverter for common mode current conceptualization and mitigation is proposed. Specifically, an extended H6 transformerless inverter that operates with an additional power switch (H7) is utilised for improving the common mode leakage current mitigation in a single-phase utility grid. A new control for a modular multilevel converter (MMC) based static synchronous compensator (STATCOM) is proposed in [10] as an effective interface between energy sources and the power grid. This study showed that the proposed control method led to an effective reduction in capacitor voltage fluctuation and losses. The protection of sensitive loads against voltage drop is a challenge for the power system, especially in face of the rising use of DG. Thus, in order to address this obstacle, a compound current limiter and circuit breaker is proposed in [11] and validated through experimental and simulation results. The authors argue that in this study that the proposed compound current limiter is able to limit the fault current and fast break in order to adjust voltage sags at the protected buses. A data-driven based voltage control strategy for DC-DC converters, which are increasingly used to integrate renewable energy resources, with the aim of applying them to DC microgrids is given in [12]. Because these converters can be used for so many applications, suitable modelling and control methods are necessary for their voltage regulation. Simulations performed in this study show a satisfying performance of the data-driven control strategy. Since DG will most likely cause a higher occurrence of fault current levels, in [13], a multi-inductor H bridge fault current limiter is proposed in order to reduce the frequency of occurrence of such types of problems. Positive results are obtained through experimental and simulated tests.

As for research studies revolving around converters, a full-bridge converter (FBC) for bidirectional power transfer is presented in [14]. The proposed FBC is an isolated DC-DC bidirectional converter, linked to a double voltage source—a voltage bus on one side, and a stack of super-capacitors (SOSC) on the other side and real prototype, compliant with automotive applications. In [15], a single DC source multilevel inverter with changeable gains and levels for low-power loads is proposed. The validation of this inverter was conducted through simulation and experimental tests using nine different modulations. A p-type module with virtual DC links to increase levels in multilevel inverters is proposed in [16], which are able to produce higher voltage levels with a lower number of components, making them appropriate for a wide range of applications. In another study [17], Janina Rząsa proposes an alternative carrier-based implementation of space vector modulation to eliminate common mode voltage in a multilevel matrix converter and evidences that part of the high-frequency output voltage distortion component is eliminated. The proposed modulation method is validated though simulation and experimental results. A comprehensive comparative analysis of impedance-source based DC-DC and DC-AC converters regarding passive component count and size, range of input voltage variation, and semiconductor stress is proposed in [18]. The authors analyzed the main impedance-source converters with or without inductor coupling and with or without a transformer; useing simulations and experiments to validate this.

Converters and inverters are used for several applications, as gathered from the above-mentioned studies, and as so, they are also used for permanent-magnet synchronous machines (PMSM). An improved model predictive torque control combined with discrete space vector modulation for a two-level inverter fed interior PMSM is proposed in [19] and the authors establish a cost function involving the excitation torque and reluctance torque. Simulation and experimental results are used to validate this study, in which the torque ripple and current ripple is reduced. The nonlinear effects, such as voltage and command voltage deviation, of a three-level neutral-point clamped inverter on speed sensorless control of an induction motor are studied in [20]. In this study a new voltage deviation compensation measure based on the volt-second balance principle is proposed and validated through experimental results.

Extensive research is also made in the area of motor drives. A new effective use and operation of fuzzy-logic controller-based two-quadrant operation of a permanent magnet brushless DC motor drive system for multipass hot-steel rolling processes is proposed in [21], with validation through simulation and experimental tests provided. Another study [22], proposed a field weakening control method that employs interpolation error compensation of the look-up table based PMSM method. As in the majority of these studies, the improvement reached by using the proposed method is validated through experimental and simulation tests.

Non-linear ceramic resistors such as metal oxide ZnO-based varistors are mainly utilised to protect electronic and electric circuits from overvoltage. The ZnO varistor, also known as metal oxide varistor, is the most well-known type of varistor, and, as such, it is a topic that attracts attention from the research community. This Special Issue is no exception. An experimental study on the effect of multiple lightning waveform parameters on the aging characteristics of ZnO varistors is proposed in [23], in which the aging rate and surface temperature rise of ZnO varistor under the impact of multi-pulse current was examined. Another experimental study was made in [24] by focusing on the failure mode of ZnO varistors under multiple lightning strokes, in which alterations were observed in the performance of these ZnO varistors after multiple lightning impulses. These changes were analyzed from micro and macro perspectives.

In the semiconductor area of research, a robust electrostatic discharge reliability design of an ultra-high voltage 300-V power n-channel lateral diffused MOSFETs, with elliptical cylinder super-junctions in the drain side, is given in [25]. The authors have concluded that this is a decent strategy and the human-body model capability of these ultra-high voltage n-channel lateral diffused MOSFETs could be successfully improved without altering the basic electrical properties or adding any extra cell area.

In this area of electromagnetism, a novel AC magnetic transmitter current source circuit is proposed in [26] for the application of frequency domain electromagnetic method prospecting. The proposed current source circuit is able to generate high frequency and high constant amplitude currents, which are the main technical problems for the frequency domain electromagnetic method. The results obtained through simulation and experimental tests show that the proposed circuit achieves the linearity of the rising/falling edge, a short reversal time, a low power loss, and a constant amplitude. Since the penetration of electric vehicles is constantly growing in the market, in another study [27], a new topology-based approach to improve vehicle-level electromagnetic radiation is proposed due to the fact that electric vehicles suffer from various electromagnetic interferences. The efficacy of this method was demonstrated through experimental tests that compared the predicted vehicle-radiated emissions at low frequency with the obtained experimental results.

Improved control methods, better energy efficiency and problem mitigation can be achieved at any level and in almost any system, as can be seen in the contributions to this Special Issue. Even though numerous challenges still remain, research and technology are vital tools for overcoming the challenges that arise in power electronics, specially by heading towards responsible and careful use of the environment. Power electronics plays a key role in the development of renewable energy systems and, therefore, in reducing greenhouse gases. Therefore, through small incremental steps, the objective is to strengthen the role of innovation, with the aim of facing the challenges that lay ahead [28] with efficient responses that, additionally, can ensure an economical, reliable and sustainable electrical supply, on which we have grown to become so dependent.

Author Contributions: E.M.G.R., R.G. and E.P.; writing—original draft, E.M.G.R., R.G. and E.P.; writing—Review and editing, E.M.G.R., R.G. and E.P. All authors have read and agreed to the published version of the manuscript.

Funding: This research received no external funding.

Acknowledgments: This Special Issue is the result of the contributions of various talented authors, experienced reviewers, and the devoted editorial team of *Electronics*. Thanks are due to all authors and peer reviewers. Finally, the editorial team of *Electronics* "Power Electronics" Section are to be congratulated for the success of this project, and in particular, thanks are due to Judy Jia, Eric Lin, Xenia Xie and Josephine Yang, and especially to Michelle Zhou, Assistant Editor from MDPI Branch Office, Beijing.

Conflicts of Interest: The authors declare no conflict of interest.

References

1. Rehg, J.A.; Sartori, G.J. *Industrial Electronics*, 1st ed.; Pearson Prentice Hall: Upper Saddle River, NJ, USA, 2006; ISBN 978-0-13-206418-7.
2. Rashid, M.H. (Ed.) *Power Electronics Handbook*, 3rd ed.; Butterworth-Heinemann: Oxford, UK, 2011.
3. Perreault, D.J.; Afridi, K.K.; Khan, I.A. 32-Automotive Applications of Power Electronics. In *Power Electronics Handbook*, 4th ed.; Rashid, M.H., Ed.; Butterworth-Heinemann: Oxford, UK, 2018; pp. 1067–1090. ISBN 978-0-12-811407-0.
4. Pollefliet, J. 15-Applications of Power Electronics. In *Power Electronics*; Pollefliet, J., Ed.; Academic Press: Cambridge, MA, USA, 2018; pp. 15.1–15.44. ISBN 978-0-12-814643-9.
5. Siwakoti, Y.P.; Forouzesh, M.; Ha Pham, N. Chapter 1-Power Electronics Converters—An Overview. In *Control of Power Electronic Converters and Systems*; Blaabjerg, F., Ed.; Academic Press: Cambridge, MA, USA, 2018; pp. 3–29. ISBN 978-0-12-805245-7.
6. Funcke, S.; Ruppert-Winkel, C. Storylines of (de)centralisation: Exploring infrastructure dimensions in the German electricity system. *Renew. Sustain. Energy Rev.* **2020**, *121*, 109652. [CrossRef]
7. Agnew, S.; Smith, C.; Dargusch, P. Understanding transformational complexity in centralized electricity supply systems: Modelling residential solar and battery adoption dynamics. *Renew. Sustain. Energy Rev.* **2019**, *116*, 109437. [CrossRef]
8. Blaabjerg, F.; Chen, Z.; Kjaer, S.B. Power electronics as efficient interface in dispersed power generation systems. *IEEE Trans. Power Electron.* **2004**, *19*, 1184–1194. [CrossRef]
9. Mahmoudian, M.; Rodrigues, E.M.G.; Pouresmaeil, E. An Efficient H7 Single-Phase Photovoltaic Grid Connected Inverter for CMC Conceptualization and Mitigation Method. *Electronics* **2020**, *9*, 1440. [CrossRef]
10. Shahnazian, F.; Adabi, E.; Adabi, J.; Pouresmaeil, E.; Rouzbehi, K.; Rodrigues, E.M.G.; Catalão, J.P.S. Control of MMC-Based STATCOM as an Effective Interface between Energy Sources and the Power Grid. *Electronics* **2019**, *8*, 1264. [CrossRef]
11. Heidary, A.; Radmanesh, H.; Bakhshi, A.; Rouzbehi, K.; Pouresmaeil, E. A Compound Current Limiter and Circuit Breaker. *Electronics* **2019**, *8*, 551. [CrossRef]
12. Rouzbehi, K.; Miranian, A.; Escaño, J.M.; Rakhshani, E.; Shariati, N.; Pouresmaeil, E. A Data-Driven Based Voltage Control Strategy for DC-DC Converters: Application to DC Microgrid. *Electronics* **2019**, *8*, 493. [CrossRef]

13. Heidary, A.; Radmanesh, H.; Moghim, A.; Ghorbanyan, K.; Rouzbehi, K.; Rodrigues, E.M.G.; Pouresmaeil, E. A Multi-Inductor H Bridge Fault Current Limiter. *Electronics* **2019**, *8*, 795. [CrossRef]
14. Pellitteri, F.; Miceli, R.; Schettino, G.; Viola, F.; Schirone, L. Design and Realization of a Bidirectional Full Bridge Converter with Improved Modulation Strategies. *Electronics* **2020**, *9*, 724. [CrossRef]
15. Samadaei, E.; Salehi, A.; Iranian, M.; Pouresmaeil, E. Single DC Source Multilevel Inverter with Changeable Gains and Levels for Low-Power Loads. *Electronics* **2020**, *9*, 937. [CrossRef]
16. Samadaei, E.; Kaviani, M.; Iranian, M.; Pouresmaeil, E. The P-Type Module with Virtual DC Links to Increase Levels in Multilevel Inverters. *Electronics* **2019**, *8*, 1460. [CrossRef]
17. Rząsa, J. An Alternative Carrier-Based Implementation of Space Vector Modulation to Eliminate Common Mode Voltage in a Multilevel Matrix Converter. *Electronics* **2019**, *8*, 190. [CrossRef]
18. Husev, O.; Shults, T.; Vinnikov, D.; Roncero-Clemente, C.; Romero-Cadaval, E.; Chub, A. Comprehensive Comparative Analysis of Impedance-Source Networks for DC and AC Application. *Electronics* **2019**, *8*, 405. [CrossRef]
19. Zhang, G.; Chen, C.; Gu, X.; Wang, Z.; Li, X. An Improved Model Predictive Torque Control for a Two-Level Inverter Fed Interior Permanent Magnet Synchronous Motor. *Electronics* **2019**, *8*, 769. [CrossRef]
20. Li, P.; Zhang, L.; Ouyang, B.; Liu, Y. Nonlinear Effects of Three-Level Neutral-Point Clamped Inverter on Speed Sensorless Control of Induction Motor. *Electronics* **2019**, *8*, 402. [CrossRef]
21. Nandakumar, M.; Ramalingam, S.; Nallusamy, S.; Rangarajan, S.S. Novel Efficacious Utilization of Fuzzy-Logic Controller-Based Two-Quadrant Operation of PMBLDC Motor Drive Systems for Multipass Hot-Steel Rolling Processes. *Electronics* **2020**, *9*, 1008. [CrossRef]
22. Ji, Y.-B.; Lee, J.-H. Feedforward Interpolation Error Compensation Method for Field Weakening Operation Region of PMSM Drive. *Electronics* **2019**, *8*, 1052. [CrossRef]
23. Zhang, C.; Li, C.; Lv, D.; Zhu, H.; Xing, H. An Experimental Study on the Effect of Multiple Lightning Waveform Parameters on the Aging Characteristics of ZnO Varistors. *Electronics* **2020**, *9*, 930. [CrossRef]
24. Zhang, C.; Xing, H.; Li, P.; Li, C.; Lv, D.; Yang, S. An Experimental Study of the Failure Mode of ZnO Varistors Under Multiple Lightning Strokes. *Electronics* **2019**, *8*, 172. [CrossRef]
25. Chen, S.-L.; Wu, P.-L.; Chen, Y.-J. Robust ESD-Reliability Design of 300-V Power N-Channel LDMOSs with the Elliptical Cylinder Super-Junctions in the Drain Side. *Electronics* **2020**, *9*, 730. [CrossRef]
26. Wang, X.; Fu, Z.; Wang, Y.; Wang, W.; Liu, W.; Zhao, J. A Wide-Frequency Constant-Amplitude Transmitting Circuit for Frequency Domain Electromagnetic Detection Transmitter. *Electronics* **2019**, *8*, 640. [CrossRef]
27. Wu, C.; Gao, F.; Dai, H.; Wang, Z. A Topology-Based Approach to Improve Vehicle-Level Electromagnetic Radiation. *Electronics* **2019**, *8*, 364. [CrossRef]
28. Mendes, T.D.P.; Godina, R.; Rodrigues, E.M.G.; Matias, J.C.O.; Catalão, J.P.S. Smart and energy-efficient home implementation: Wireless communication technologies role. In Proceedings of the 2015 IEEE 5th International Conference on Power Engineering, Energy and Electrical Drives (POWERENG), Riga, Latvia, 11–13 May 2015; pp. 377–382.

 *electronics*

Article

Comprehensive Comparative Analysis of Impedance-Source Networks for DC and AC Application

Oleksandr Husev [1], Tatiana Shults [2], Dmitri Vinnikov [1], Carlos Roncero-Clemente [3], Enrique Romero-Cadaval [3,*] and Andrii Chub [1]

[1] Department of Electrical Power Engineering and Mechatronics, School of Engineering, Tallinn University of Technology, Ehitajate tee 5, 19086 Tallinn, Estonia; oleksandr.husev@ieee.org (O.H.); dmitri.vinnikov@ttu.ee (D.V.); andrusha.chub@gmail.com (A.C.)
[2] Power Electronics Institute, Novosibirsk State Technical University, 630073 Novosibirsk, Russia; tanuta91@mail.ru
[3] Power Electrical and Electronics System Research Group, University of Extremadura, 06006 Badajoz, Spain; croncero@peandes.net
* Correspondence: eromero@unex.es; Tel.: +34-924-289-300 (ext. 86787)

Received: 19 March 2019; Accepted: 2 April 2019; Published: 5 April 2019

Abstract: This paper presents a comprehensive analytical comparison of the impedance-source-based dc-dc and dc-ac converters in terms of the passive component count and size, semiconductor stress, and range of input voltage variation. The conventional solution with a boost converter was considered as a reference value. The main criterion of the comprehensive comparison was the energy stored in the passive elements, which was considered both under a constant and predefined high frequency current ripple in the inductors and the voltage ripple across the capacitors. Main impedance-source converters with or without a transformer and with or without inductor coupling were analyzed. Dc-dc and dc-ac applications were considered. Selective simulation results along with experimental verification are shown. The conclusions provide a selection guide of impedance-source networks for different applications taking into account its advantages and disadvantages.

Keywords: impedance-source inverter; shoot-through; dc-dc converter; dc-ac converter

1. Introduction

The Z-source inverter (ZSI) was introduced in 2003 [1]. It was claimed that the converter overcomes the conceptual and theoretical barriers and limitations of the traditional voltage-source inverter (VSI) and current-source (CS) inverter and provides a novel power conversion concept. ZSI utilizes the shoot-through (ST) cross-conduction states to boost the input dc-voltage by switching on both the top and bottom switches of at least one inverter leg. ZSI can buck-boost voltage, minimize component count, increase efficiency, and reduce the cost. This topology also has no forbidden switching states, which improves converter reliability significantly.

This solution is intended for various fields of application: dc-dc, ac-dc, ac-ac, and dc-ac. In particular, it is suitable for grid integration or electric drive control [1–4]. Further, the quasi-Z-source (qZS) network was proposed in [2]. As compared to the Z-source (ZS) network, it has a continuous conduction mode (CCM) of the input current. Also, the volume of the capacitors may be lower in the qZS network. Due to these features, ZS and qZS networks have been named as the most suitable solution for renewable energy applications. Many papers have studied in detail possibilities for such applications [5–7]. Steady-state analysis and dynamic analysis along with different control strategies have been described in detail [8–25].

Since then, many derivative topologies of impedance source (IS) networks have been presented. The main justification of the presented solutions is that they overcome drawbacks of the ZS and qZS networks. Mostly, the benefit is in the low dc-link utilization under constant boost control [15–25].

References [26–32] present a good overview of existing solutions. Almost all the presented novel solutions are based on the magnetically coupled inductors. The main hypothesis in those papers addresses possible reductions of size and volumes of passive components due to the increase in the turns ratio in magnetically coupled elements. At the same time, the overall conclusions and predictions seem to be incomplete. These papers are devoted to very general issues related to the gain factor, and the number of active and passive components.

After an in-depth overview of the comparative papers, some contradictory results can be underlined. As a result of a direct comparison between novel and conventional solutions, the conclusions reached contain high uncertainty [33–39]. Several papers reveal only superior performance of the IS-based converters over conventional solutions [33–35]. At the same time, other papers reveal opposite results. In conclusion, despite many studies devoted to the IS derived converters, there are still many open questions.

This research work focuses on the comparison between each other by overall dimensions and voltage stress on semiconductors. A comprehensive comparative analysis is provided for all key types of IS networks with conventional solutions for the dc-dc and the dc-ac application. Conventional solutions based on the boost dc-dc converter will be considered as well (Figure 1a).

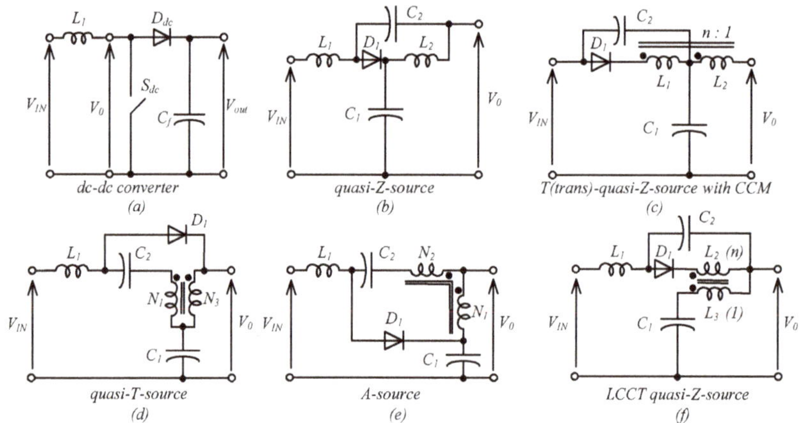

Figure 1. Basic impedance source (IS) networks along with the boost dc-dc converter: (**a**) Boost dc-dc converter, (**b**) quasi-Z-source, (**c**) trans (T)-quasi-Z-source, (**d**) quasi-T-source, (**e**) A-source, and (**f**) LCCT quasi-Z-source.

The paper is organized as follows. Section 2 presents a brief overview of IS derived converters. Section 3 describes a comparative analysis approach while Section 4 represents results of comparison. Section 5 is devoted to the pros and cons discussions of IS networks application in galvanically isolated converters. Simulation and experimental study is summarized in Section 6. Finally, conclusions are presented and discussed in Section 7.

2. Brief Overview of IS Derived Converters

Over 20 different types of IS networks subdivided into subgroups have been presented. They may have separated inductors, magnetically coupled inductors, or transformers. Implementation of magnetically coupled inductors or transformers in the IS network can result in a higher voltage boost factor due to the turns ratio. All IS networks can also be divided into those with discontinuous input current and continuous input current (CIC). At the same time, it has been shown that networks with discontinuous input current have no advantage over topologies that have CIC [30].

Figure 1 shows basic IS networks. The conventional boost converter is presented in Figure 1a as a reference solution that provides the same functionality. Another reason of including the conventional solution here is that some research results claim that novel IS networks have unconditional advantages over the boost converter [33–35].

The first IS network under consideration was the qZS network (Figure 1b). As mentioned above, it has a CIC and the same size and volume of passive components as for ZSI. The other four selected topologies belonged to the magnetically coupled IS networks [29–31]. Since any magnetically coupled inductor can be represented as a combination of leakage inductance, magnetizing inductance, and ideal transformer, any other IS network with a magnetically coupled inductor can be considered a derivation of the above solutions. This was very well demonstrated in [38].

The trans (T)-quasi-Z-source network (Figure 1c) with a CIC was the first network with coupled inductors under consideration [40–47]. As it was shown in [47], despite an additional capacitor, the overall size of capacitors is lower. The quasi-T-source network (Figure 1d) has a CIC and a slightly different configuration.

LCCT networks (Figure 1f) have quite similar features [48]. At the same time, many derivative circuits and types of converters are proposed in [49–53]. Finally, the A-source network (Figure 1e) was one of the latest solutions proposed and will be compared as well [54,55].

The main advantage of all magnetically coupled derived IS networks is the high boost due to the high turns ratio of the transformer. As a result, the ST duty cycle is shorter. At the same time, no studies have demonstrated the impact of the shorter ST duty cycle on the size, volume, and probable cost of the converter.

3. Comparative Analysis Approach

The two possible applications of any IS network are shown in Figure 2. In the dc-dc application, the dc-link voltage is close to the peak voltage across the IS network (Figure 2a). At the same time, the dc-ac converter based on the IS network has no direct dc-link. When the load terminals are shorted through both the upper and lower semiconductor devices of any one phase leg, the energy is accumulating in the inductors of the IS network.

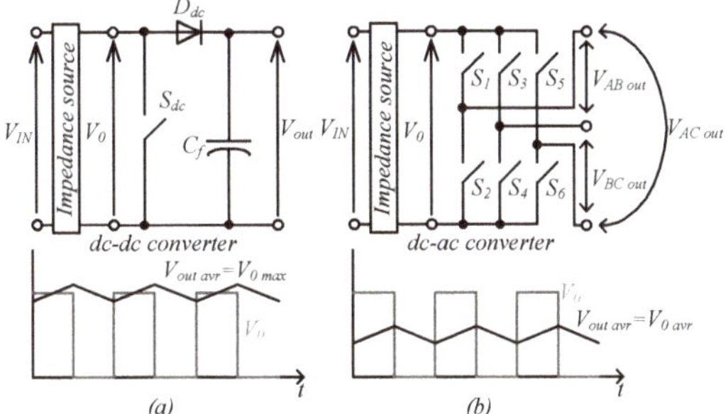

Figure 2. Difference between the output voltage of (**a**) the impedance source dc-dc converter and (**b**) the dc-ac converter.

This ST zero state provides the unique buck-boost feature to the inverter. At the same time, average voltage applied to the ac load is lower than the peak voltage across the IS network. The ST duty cycle inserted in the switching states reduces the effective dc-link voltage. In other words, the peak voltage generated from the IS network across an inverter should be higher than in the conventional VSI to compensate the zero ST states (Figure 2b). It has to be taken into account in the design of the converter. In particular, voltage stress across semiconductors is increasing.

Maximum boost control (MBC) is a well-known approach in the implementation of the ST states without degradation of dc-link voltage utilization, but this approach cannot be considered as an equivalent alternative because of low frequency input current generation. Such a ripple can be mitigated by increasing the size and cost of passive components, which is not a competitive solution. The modulation techniques with equal ST states distribution are considered for a comparative analysis in this work [22].

Usually, parameters such as the amount of semiconductors, size, and volume of the passive components along with overall power losses define the feasibility of a power electronics converter. In order to include such criteria in the comparative analysis, several assumptions were considered.

The first assumption is that the volume of the magnetic components is proportional to the maximum stored energy E_L:

$$E_L = \frac{L \cdot I_{MAX}^2}{2},\tag{1}$$

which is estimated by means of the inductance L and the maximum inductor current I_{MAX}. For convenience and generalization of the analysis, the total magnetics energy in relative units is introduced as:

$$E_{LW} = \sum_{i=1}^{N_L} \frac{L_i \cdot I_{MAXi}^2}{2},\tag{2}$$

where N_L is the number of inductances.

A similar parameter can be introduced for the capacitors:

$$E_{CW} = \sum_{i=1}^{N_C} \frac{C_i \cdot V_{MAXi}^2}{2},\tag{3}$$

where E_{CW} is the total maximum energy stored in the capacitors and N_C is the number of capacitors.

It is well known that size, volume, and cost of the capacitors depend on the maximum voltage and capacitance. It should be mentioned that volume and cost also depend on the technology of manufacturing of the magnetic components and capacitors and some corrective coefficients can be introduced [23].

In order to estimate the contribution of the semiconductors to the discussed topologies, their number and blocking voltage are also taken into account.

$$D_W = \sum_{i=1}^{N_D} V_{BDi},\tag{4}$$

$$T_W = \sum_{i=1}^{N_T} V_{BTi},\tag{5}$$

The blocking voltage of semiconductors is one of the objective parameters that depends on the topology and can be clearly estimated. The current stress could be taken into account, but it is more difficult to estimate it; moreover, it depends on the passive components.

The size of the passive elements depends on the material and switching frequency as well, but these parameters are assumed to be the same for all topologies. Thereby, the above presented approach will provide results that depend only on the topology itself. The power level, input current ripple, as well as the dc-link voltage ripple are considered equal for all cases under comparison.

Figure 3 shows equivalent circuits of the qZS network. It shows two main states of the dc-dc or dc-ac converter that are based on the qZS network. ST state (Figure 3a) corresponds to the accumulating energy time period in which the duty cycle is usually denoted as D_S.

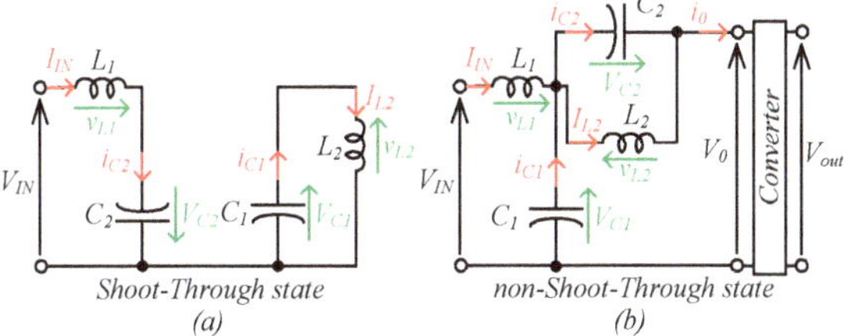

Figure 3. Equivalent circuits of the quasi-Z-source network: (**a**) The shoot-through (ST) state and (**b**) the non-ST state.

During the ST time interval, the current in the inductor is increasing and energy is accumulating in the inductors. An non-ST equivalent circuit state is depicted in Figure 3b. It corresponds to the time interval when energy is provided to the load and charges the capacitors while the current is decreasing. The average current value remains the same in the steady state condition. The inductance value is usually selected to limit this current ripple. This value of inductance along with the peak current value defines the size, volume, and finally the cost of the inductors.

The calculation of the inductance value is exemplified in many papers [1,8,14,26,33]. The input power level, input voltage range, and the predefined input current ripple are completely sufficient to define the parameters of the magnetic components.

Similar processes are happening in the capacitors. The value of capacitance is selected in order to limit the voltage ripple that is proportional to the power level.

After the steady state analysis presented in the above papers, the summarized equations for selected topologies were derived and are shown in Table 1.

Table 1. Summarized equations for selected topologies [1].

	V_C	I_L	C	L	V_0 dc-dc	V_0 dc-ac
a	$V_{C1} = V_0 = \dfrac{V_{in}}{1-D_S}$	$I_{L1} = I_{in}$	$C_1 = \dfrac{D_S(1-D_S)^2 I_{in}}{V_{in}k_{C1}f_S}$	$L_1 = \dfrac{V_{in}D_S}{I_{in}k_{L1}f_S}$	$\dfrac{V_{in}}{1-D_S}$	$\dfrac{V_{in}}{1-D_S}$
b	$V_{C1} = \dfrac{(1-D_S)V_{in}}{1-2D_S}$ $V_{C2} = \dfrac{D_S V_{in}}{1-2D_S}$	$I_{L1} = I_{in}$ $I_{L2} = I_{in}$	$C_1 = \dfrac{D_S(1-2D_S)I_{in}}{V_{in}(1-D_S)k_{C1}f_S}$ $C_2 = \dfrac{(1-2D_S)I_{in}}{V_{in}k_{C2}f_S}$	$L_1 = \dfrac{V_{in}D_S(1-D_S)}{I_{in}(1-2D_S)k_{L1}f_S}$ $L_2 = \dfrac{V_{in}D_S(1-D_S)}{I_{in}(1-2D_S)k_{L2}f_S}$	$\dfrac{V_{in}}{1-2D_S}$	$\dfrac{V_{in}(1-D_S)}{1-2D_S}$
c	$V_{C1} = \dfrac{(1-D_S)V_{in}}{1-(n+1)D_S}$ $V_{C2} = \dfrac{nD_S V_{in}}{1-(n+1)D_S}$	$I_{Lm} = \dfrac{(n+1)I_{in}}{n}$	$C_1 = \dfrac{nD_S(1-(n+1)D_S)I_{in}}{V_{in}(1-D_S)k_{C1}f_S}$ $C_2 = \dfrac{(1-D_S(n+1))I_{in}}{V_{in}nk_{C1}f_S}$	$L_m = \dfrac{V_{in}D_S(1-D_S)(n)^2}{I_{in}(1-(n+1)D_S)(n+1)k_m f_S}$	$\dfrac{V_{in}}{1-(n+1)D_S}$	$\dfrac{V_{in}(1-D_S)}{1-(1+n)D_S}$
d	$V_{C1} = \dfrac{(1-D_S)V_{in}}{1-nD_S}$ $V_{C2} = \dfrac{(n-1)D_S V_{in}}{1-nD_S}$	$I_{L1} = I_{in}$	$C_1 = \dfrac{D_S(n-1)(1-nD_S)I_{in}}{V_{in}(1-D_S)k_{C1}f_S}$ $C_2 = \dfrac{(1-D_S n)I_{in}}{V_{in}(n-1)k_{C2}f_S}$	$L_1 = \dfrac{V_{in}D_S(1-D_S)n}{I_{in}(1-nD_S)k_{L1}f_S}$	$\dfrac{V_{in}}{1-nD_S}$	$\dfrac{V_{in}(1-D_S)}{1-nD_S}$
e	$V_{C1} = \dfrac{(1-D_S)V_{in}}{1-(n+2)D_S}$ $V_{C2} = \dfrac{(n+1)D_S V_{in}}{1-(n+2)D_S}$	$I_{L1} = I_{in}$ $I_{Lm} = I_{in}$	$C_1 = \dfrac{D_S(n+1)(1-(n+2)D_S)I_{in}}{V_{in}(1-D_S)k_{C1}f_S}$ $C_2 = \dfrac{(1-(n+2)D_S)I_{in}}{V_{in}(n+1)k_{C2}f_S}$	$L_1 = \dfrac{V_{in}(n+1)D_S(1-D_S)}{I_{in}(1-(n+2)D_S)k_{L1}f_S}$ $L_m = \dfrac{V_{in}D_S(1-D_S)}{I_{in}(1-(n+2)D_S)k_{Lm}f_S}$	$\dfrac{V_{in}}{1-(n+2)D_S}$	$\dfrac{V_{in}(1-D_S)}{1-(n+2)D_S}$
f	$V_{C1} = \dfrac{(1-D_S)V_{in}}{1-(n+1)D_S}$ $V_{C2} = \dfrac{D_S n V_{in}}{1-(n+1)D_S}$	$I_{L1} = I_{in}$ $I_{Lm} = -I_{in}$	$C_1 = \dfrac{D_S n(1-(n+1)D_S)I_{in}}{V_{in}(1-D_S)k_{C1}f_S}$ $C_2 = \dfrac{(1-(n+1)D_S)I_{in}}{V_{in}nk_{C2}f_S}$	$L_1 = \dfrac{V_{in}D_S(1-D_S)}{I_{in}(1-(n+1)D_S)k_{L1}f_S}$ $L_m = \dfrac{V_{in}D_S n(1-D_S)}{I_{in}(1-(n+1)D_S)k_{L2}f_S}$	$\dfrac{V_{in}}{1-(n+1)D_S}$	$\dfrac{V_{in}(1-D_S)}{1-(n+1)D_S}$

[1] for Figure 1d, $n = N_1/N_3$; for Figure 1e, $n = N_2/N_1$.

The table shows the voltage across the capacitors, average current across inductors, nominal values of capacitances and inductances that depend on the voltage high frequency ripple factor k_C and the current high frequency ripple factor k_L, the boost factor in dc-dc application, and the gain factor in the dc-ac application. Based on the derived equations, the main parameters, such as the inductance and maximum current for inductors and capacitance and peak voltage for capacitors presented in Equations (1)–(5), can be obtained.

4. Results of Comparison

In a very general case, any boost converter can provide a very wide range of the boost factor and the input voltage regulation, respectively. The main limitations usually consist in the size of passive elements and losses in semiconductors.

The main outcome of the proposed comparative approach is in the possibility to define the maximum stored energy in the passive components as a function of the input voltage taking into account the constant ripples in the voltage across capacitors and current across inductors. For normal operation of the converter, the output RMS voltage level has to be constant despite the input voltage variation. As it was mentioned before, in the dc-dc application, the dc-link voltage is equal to the peak voltage across the IS network, while in the dc-ac application, the effective dc-link voltage is less than the peak voltage across the IS network.

As a result, at the constant dc-link voltage or constant effective (average) dc-link voltage across the IS networks, the ST duty cycle strictly depends on the input voltage. The closed loop control system sets up the ST duty cycle to maintain the necessary dc-link voltage. In other words, the ST duty cycle can be replaced in expressions presented above by the input voltage.

For example, in the case of qZSI, the ST duty cycle can be expressed as

$$D_{qZS} = \frac{V_0 - V_{in}}{2V_0 - V_{in}},\tag{6}$$

Taking into account Equation (6) and equations shown in Table 1, Figures 4 and 5 show the dependences of the passive elements values as a function of the input voltage and inductor turns ratio (n), respectively.

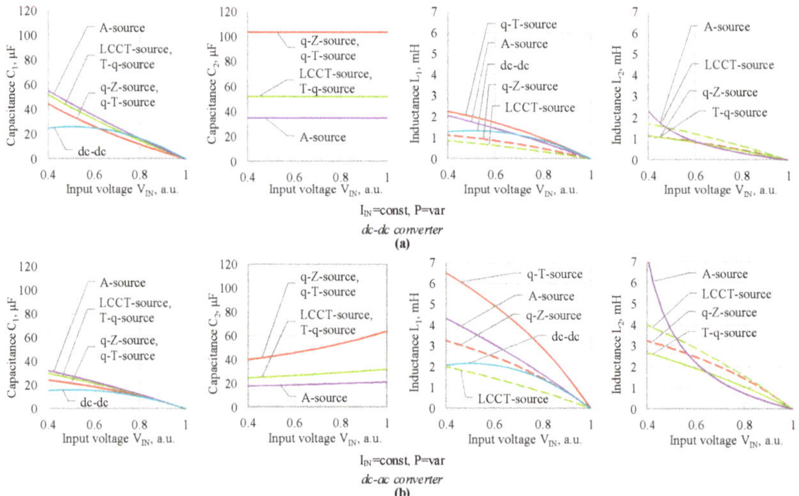

Figure 4. Value of the passive components as a function of the input voltage for (**a**) the dc-dc converter and (**b**) the dc-ac application.

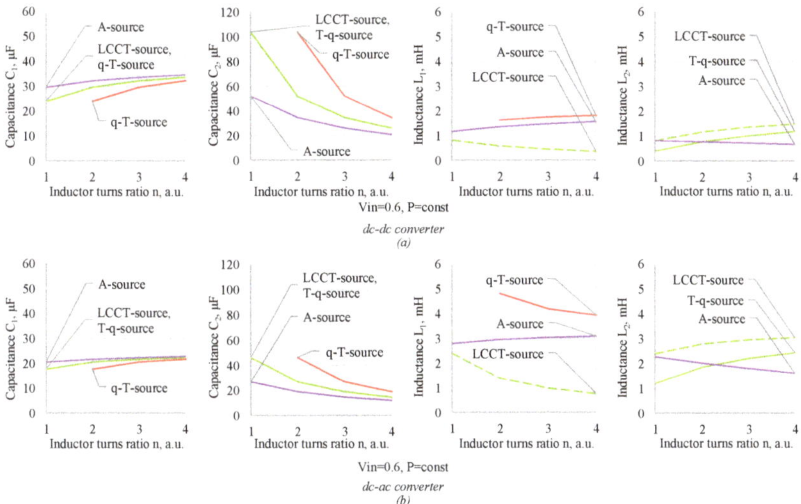

Figure 5. Value of the passive components as a function of the different value of turns ratio (n) for (**a**) the dc-dc converter and (**b**) the inverter, with $V_{in} = 0.6$.

Figure 4 shows the required values of the passive components to provide a constant current ripple in magnetic components (L_1 and L_2) and voltage ripple across the capacitors (C_1 and C_2) as a function of the input voltage. Figure 4a illustrates the dc-dc application while Figure 4b corresponds to the dc-ac application. It can be seen that in all cases, the increasing input voltage leads to the capacitance C_1 and inductances L_1, L_2 decreasing. This result is expected because of no need for the boosting voltage. If the input voltage corresponds to the nominal dc-link voltage, only simple VSI is enough to provide energy injection to the grid or ac load.

The most interesting conclusion is that the classical solution in the worst case ($V_{in} = 0.4$ a.u.) requires lower capacitance and inductance than all other IS solutions.

A separate study was conducted in order to verify the hypothesis that an increase in the turns ratio may improve the characteristics of the IS based converter. Figure 5 shows that certain improvements in the value of capacitance and inductance are not evident. In particular, it can be seen that the behavior of the topologies is different, but a common conclusion is that an increase in the turns ratio leads to an opportunity to decrease one capacitor and inductance but at the same time, increasing inductance and capacitance of another one is required.

The only evident advantages of increasing the turns ratio lie in lower peak dc-link voltage for the dc-ac application.

At the same time, the value of inductance or capacitance is not an objective criterion. The value of stored energy should be analyzed. It is assumed to be proportional to the size and cost. Figure 6 shows the maximum energy stored in the passive components during the switching period. It can be seen that in all cases, this energy is decreasing with the input voltage decreasing. If the input voltage equals the nominal dc-link voltage, the energy stored in the IS network is equal to zero. It means that no boost is required and the IS network is not required at all. A direct conclusion from this graph is that in order to provide a higher boost, the size and the volume of the passive components must be increased.

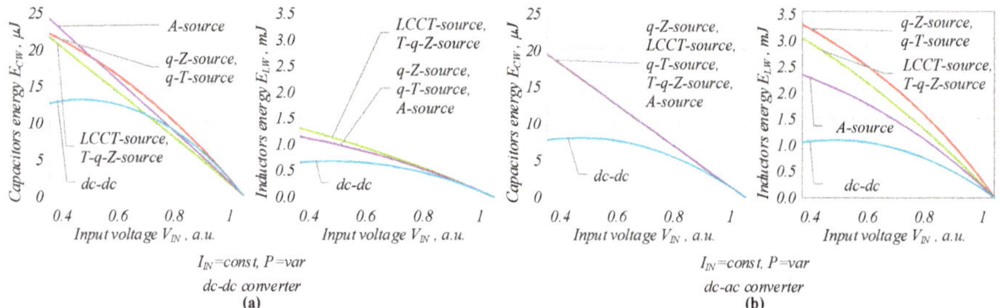

Figure 6. Energy stored in the passive components in (**a**) the dc-dc converter and (**b**) the inverter.

It can also be seen that the conventional solution based on the conventional boost circuit is better almost in all points. This conclusion is even more evident in the dc-ac application.

Another interesting conclusion is that high-gain IS based solutions such as the A-source network have a slightly smaller size of inductances but still higher than in the conventional solution. There is also an interesting feature of the total energy stored in capacitors in the dc-ac application: All IS based solutions have the same capacitance energy and they are independent of the turns ratio of coupled inductors. It means that different IS circuits have the same nature.

Figure 7 shows the dependence of conduction losses versus the input voltage. It is based on the assumption that conduction losses are proportional to the RMS current value in transistors and the average current in diodes. The same type of transistors and diodes were selected for all topologies ($V_F = 0.8$, $R_{DS} = 0.27$). It can be seen that despite a constant value of the input current, the high power losses correspond to the high boost. It means that the ST current is increasing and leads to the conduction losses increasing in semiconductors. It can also be seen that the value of the turns ratio leads to the conduction losses increasing as well. It is explained by the current ripple increasing in the semiconductors.

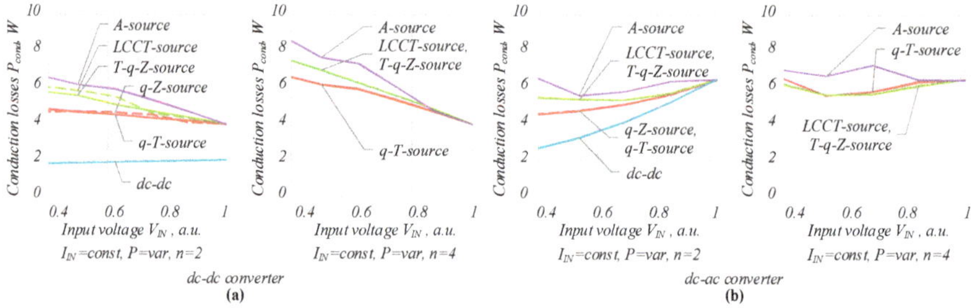

Figure 7. Conduction losses by different (n) in (**a**) the dc-dc converter and (**b**) the inverter.

Finally, in order to summarize the results of the comparison, Figure 8 shows several spider diagrams. These diagrams include the capacitors energy, inductors energy, voltage stress across semiconductors (diodes and transistors), and conduction losses in relative units. Among all the discussed topologies, the conventional solution with the boost circuit, the qZS network, and the A-source network were selected for the comparison. Similar to the previous figures, both dc-dc and dc-ac applications were considered.

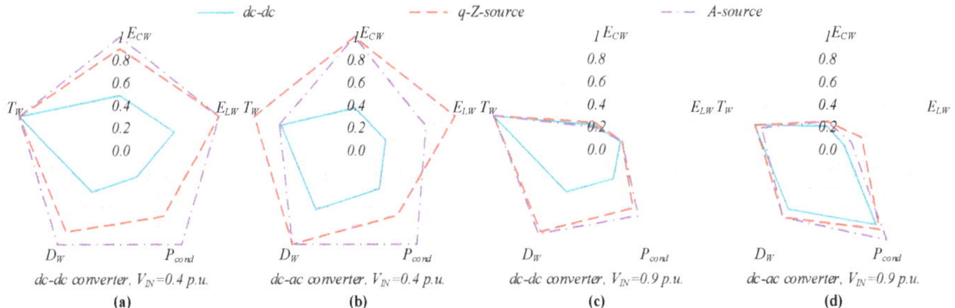

Figure 8. Comparative diagrams of the boost circuit, quasi-Z-source (qZS) networks, and A-source network: (**a**) dc-dc application with V_{in} = 0.4 p.u, (**b**) dc-ac application with V_{in} = 0.4 p.u., (**c**) dc-dc application with V_{in} = 0.9 p.u., and (**d**) dc-ac application with V_{in} = 0.9 p.u.

Figure 8a shows a diagram for the dc-ac application when the input voltage is relatively low (0.4 p.u.). In this case, the size of the passive components in all IS solutions was significantly larger.

Figure 8b shows a diagram for the same input voltage but for the dc-dc application. In this case, the size of the passive components was larger as well, but the difference was not so evident.

1 p.u. of the input voltage corresponds to the boundary input voltage between the buck and the boost mode. In the case of the dc-dc converter, 1 p.u. of the input voltage is equal to the reference output voltage. All the other parameters that are shown in Figure 8 were normalized to the maximum value.

Similar results are demonstrated in Figure 8c,d. In this case, only a minor boost is required and the difference in the capacitors and inductors energy is not significant.

As an intermediate conclusion, it can be claimed that IS networks require larger sizes of the passive components with the same current and voltage ripples. It also means that in the case of the equal components, the ripple will be higher. Among IS network topologies, the simple qZS topology can be recommended for the dc-dc and dc-ac application. The A-source has lower inductors energy where the inductance L_1 and the magnetizing inductance L_M of the coupled inductor are taken into account, but also the ideal transformer should be considered. It does not contribute in terms of accumulated energy, but it contributes in terms of size and costs.

At the same time, it should be mentioned that the IS network is a more complex solution with a larger number of passive components. It gives some freedom for size components optimization. For example, the second inductor of the qZS network can be smaller at higher ripples, which does not define the input current ripple. There is also many techniques for efficiency optimization.

5. Pros and Cons for Application in Galvanically Isolated Converters

The IS technology was applied to the galvanically isolated dc-dc converter right after the introduction of the qZS network in 2009 [2], which was the first network with a continuous input current [6]. Later on, this research area mostly focused on high step-up dc-dc converters due to the requirements of the wide input voltage range needed in emerging applications [30]. The basic competitor for these IS converters was CS counterparts, in particular, those with active clamping utilized, as shown in Figure 9.

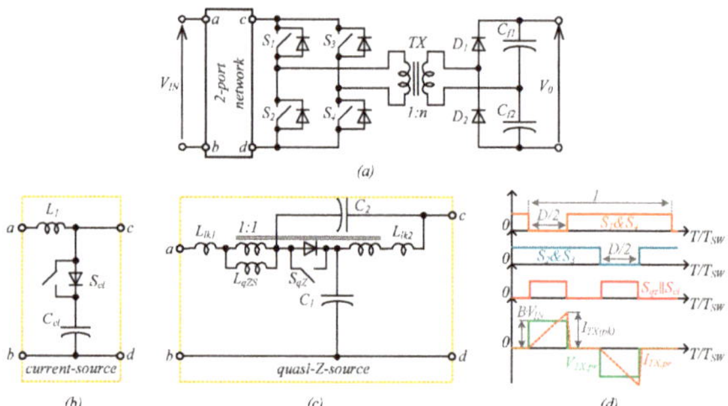

Figure 9. Generalized topology of high step-up galvanically isolated dc-dc converters (**a**) that can be implemented as either (**b**) active-clamped current-source (CS) or (**c**) qZS and feature (**d**) a triangular transformer current.

This section compares galvanically isolated qZS and CS full-bridge converters from Figure 9. Both converters feature a capacitive filter and triangular transformer current resulting from the presence of a transformer leakage inductance in real conditions. Both of them could be controlled by means of ST generation through the overlap of active states with the cumulative ST duty cycle of D. The qZS converter contains a higher number of passive components than that in the CS counterpart and a similar number of the semiconductor components. In such implementation, both converters will suffer from the same voltage stress of switches, but require a different ST duty cycle to achieve the required input voltage boost factor B that defines the transformer voltage amplitude. Obviously, the qZS converter requires lower duty cycle to step up the input voltage:

$$D_{qZS} = \frac{B-1}{2B} \tag{7}$$

At the same time, the CS counterpart requires twice higher duty cycle:

$$D_{CS} = \frac{B-1}{B} \tag{8}$$

Dependences (7) and (8) are visualized in Figure 10. The curve described by Equation (7) is asymptotic to the maximum value of $D_{CS} = 0.5$, while the one of Equation (8) is asymptotic to $D_{CS} = 1$. Hence, the qZS converter provides better transformer utilization, since the active state is never less than half of the switching period: $(1 - D_{qZS}) \geq 0.5$. Evidently, the CS converter should transfer energy through isolation within very narrow pulses at a high boost factor. As a result, the CS counterpart can suffer from a higher RMS current in the transformer as well as from a wider spectrum harmonic content of this current, resulting in higher skin and proximity losses.

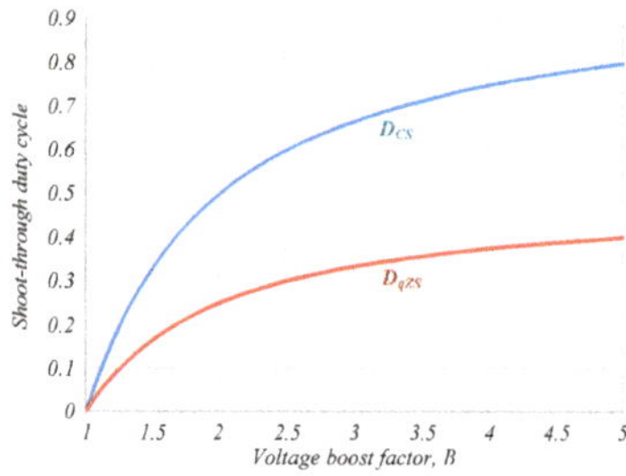

Figure 10. Dependences of the ST duty cycles on the input voltage boost factor.

Applications where the given converters compete are usually of low power at the sub-kW level. This enables implementation of the qZS network with a coupled inductor instead of the two discrete inductors. Usually, the leakage inductances L_{lk1} and L_{lk2} are considered to be equal. This assumption results in an ideal case, when the magnetizing current ripple is equally shared by the windings: $\Delta I_{L_{qZS}}/2 = \Delta I_{L_{lk1}} = \Delta I_{L_{lk2}}$. However, in practice, the current ripples could be calculated as follows [56]:

$$\Delta I_{L_{lk1}} = \frac{V_{in} \cdot L_{lk2} \cdot D_{qZS} \cdot (1 - D_{qZS})}{f_{SW} \cdot (L_{lk1} \cdot L_{lk2} + L_{qZS} \cdot (L_{lk1} + L_{lk2})) \cdot (1 - 2 \cdot D_{qZS})}, \tag{9}$$

$$\Delta I_{L_{lk2}} = \frac{V_{in} \cdot L_{lk1} \cdot D_{qZS} \cdot (1 - D_{qZS})}{f_{SW} \cdot (L_{lk1} \cdot L_{lk2} + L_{qZS} \cdot (L_{lk1} + L_{lk2})) \cdot (1 - 2 \cdot D_{qZS})}. \tag{10}$$

From Equations (9) and (10), it follows that the input current ripple of the qZS converter can be reduced greatly when $L_{lk1} \gg L_{lk2}$. In this case, the magnetizing current ripple is distributed among the windings asymmetrically. The asymmetry is inversely proportional to the ratio of the leakage inductances, as shown in Figure 11. In practice, the tight coupling with very low leakage inductance is relatively easy to achieve in the coupled inductor. Therefore, a small wiring inductance at the input can result in a tremendous reduction of the input current ripple. Therefore, in low power galvanically isolated applications, the coupled inductor in the qZS network could be designed to be comparable in size than the inductor of the competing CS converter. This possibility to divert the current ripple from the input source is not available in the CS converter and, hence, it gives additional flexibility over the CS counterpart.

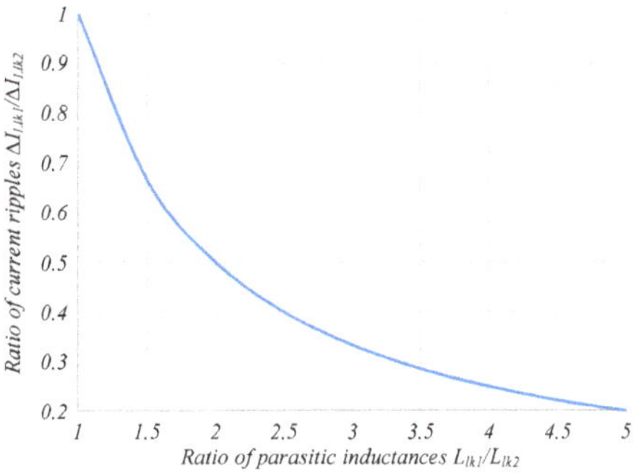

Figure 11. Dependence of the asymmetry of the current ripples in the qZS network coupled inductor.

One of the main advantages of the qZS converter is a lower RMS current of the transformer winding. The normalized value of the transformer RMS current in the qZS converter can be calculated as follows:

$$I_{qZS,TX(RMS)} = \left. \frac{I_{TX(RMS)}}{I_{in}} \right|_{D=D_{qZS}} = \frac{2 \cdot \sqrt{2 \cdot (1 - D_{qZS})}}{\sqrt{3} \cdot B \cdot (1 - D_{qZS})} = \frac{4}{\sqrt{3 \cdot B \cdot (B + 1)}} \tag{11}$$

A similar equation could be derived for the corresponding CS converter:

$$I_{CS,TX(RMS)} = \left. \frac{I_{TX(RMS)}}{I_{in}} \right|_{D=D_{CS}} = \frac{2 \cdot \sqrt{2 \cdot (1 - D_{CS})}}{\sqrt{3} \cdot B \cdot (1 - D_{CS})} = \frac{2 \cdot \sqrt{6}}{3 \cdot \sqrt{B}} \tag{12}$$

Dependences (11) and (12) are compared in Figure 12. Evidently, the RMS current stress of the transformer is up to 45% lower in the qZS converter than that in the CS converter at high boost factors.

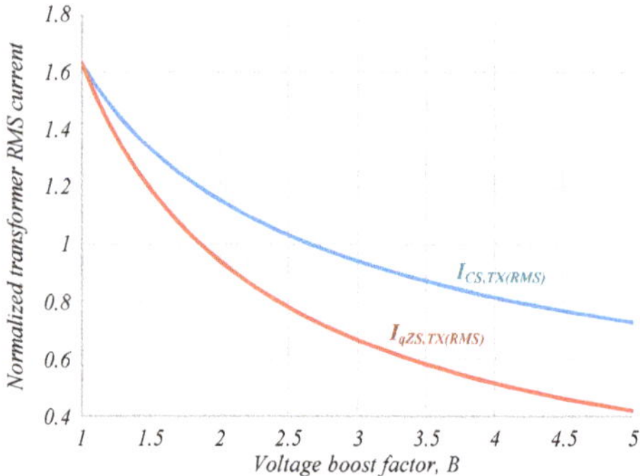

Figure 12. Normalized RMS current of the transformer input winding as a function of the input voltage boost factor B.

Another advantage of the qZS converter is the lower peak current of the transformer. The normalized value of the transformer peak current in the qZS converter can be calculated as follows:

$$I_{qZS,TX(pk)} = \left.\frac{I_{TX(pk)}}{I_{in}}\right|_{D=D_{qZS}} = \frac{2}{B \cdot (1 - D_{qZS})} = \frac{4}{(B+1)}. \tag{13}$$

Similar to that, it can be shown that the peak transformer current of the CS converter is double of the average input current:

$$I_{CS,TX(pk)} = \left.\frac{I_{TX(pk)}}{I_{in}}\right|_{D=D_{CS}} = \frac{2}{B \cdot (1 - D_{CS})} = 2. \tag{14}$$

Dependences (13) and (14) are compared in Figure 13. It can be appreciated from the figure that the qZS converter features up to a 70% lower transformer peak current, while the normalized transformer peak current of the CS converter is constant. It should be mentioned that these results derived under the assumption of the same current ripple in the input stage inductors.

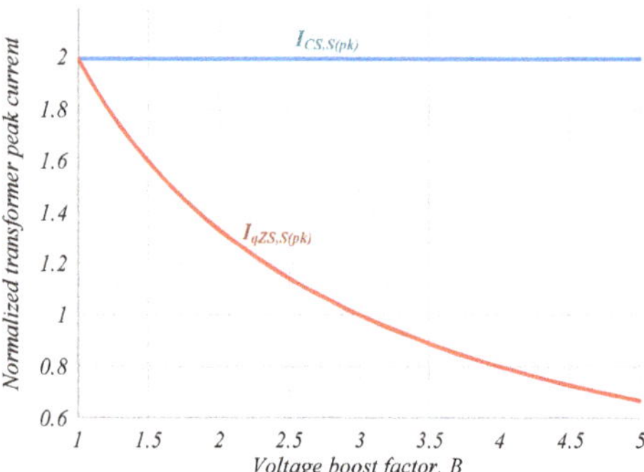

Figure 13. Normalized peak current of the transformer input winding as a function of the input voltage boost factor B.

Operation of the qZS converter with lower ST duty cycles is associated with twice higher current stress during the ST states in the inverter bridge. As a result, the normalized RMS switch current of the qZS converter is calculated as:

$$I_{qZS,S(RMS)} = \frac{I_{S(RMS)}}{I_{in}}\bigg|_{D=D_{qZS}} = \sqrt{D_{qZS} + \frac{2}{3 \cdot B^2 \cdot (1 - D_{qZS})}} = \sqrt{\frac{3 \cdot B^2 + 5}{6 \cdot B \cdot (B + 1)}}, \tag{15}$$

is higher than that of the CS converter calculated as:

$$I_{CS,S(RMS)} = \frac{I_{S(RMS)}}{I_{in}}\bigg|_{D=D_{CS}} = \sqrt{\frac{D_{CS}}{4} + \frac{2}{3 \cdot B^2 \cdot (1 - D_{CS})}} = \sqrt{\frac{3 \cdot B + 5}{12 \cdot B}}, \tag{16}$$

which results in increased RMS losses in the qZS converter, as shown in Figure 14. Moreover, this stress is the same for both topologies at the low boost factors up to 1.7, while the gap increases to up to only 15% at very high boost factors. Hence, this small increase in the current stress of the switches is not a challenge in low power high step-up systems, where low voltage Si MOSFETs with very low on-state resistance are commonly utilized.

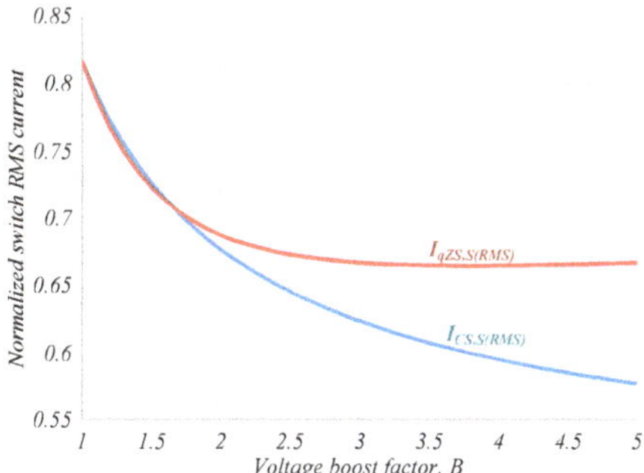

Figure 14. Normalized RMS current of the input side switches as a function of the input voltage boost factor B.

Another disadvantage of the qZS converters that becomes apparent as compared to the CS counterparts is the higher flux density swing in the transformer resulting from a longer active state of the qZS converter. The core flux swing ΔB_S is proportional to the transformer voltage and the duty cycle: $\Delta B_S \propto B \cdot V_{in} \cdot (1 - D)$. This results in up to a three times higher core flux density swing, as shown in Figure 15. This implies that the transformer should be designed differently for the qZS and CS converters.

From the considerations described above it follows that IS converters optimize the operation of the magnetic components, take advantage of wiring inductance to achieve ripple-free input current, and provide buck regulation mode, considerably enhancing the input voltage regulation range—all of these features are unavailable in the CS converters. However, design of the IS converters is more complicated due to the design constraints described above.

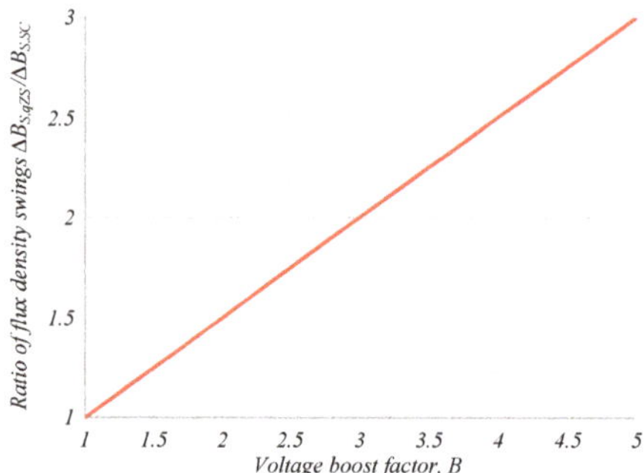

Figure 15. Ratio of the flux density swings at different input voltage boost factors.

6. Simulation and Experimental Study

In order to verify the theoretical conclusions, a simulation and an experimental study were performed. The qZS and A-source networks were selected for practical realization as the most evident representatives of the discussed network topologies. The parameters of the selected topologies are presented in Table 2. A three-phase dc-ac system was selected for simulation and experimental verification.

Table 2. Parameters of the Network Used for Simulation and Experimental Verification.

Parameters	qZS			A-Source		
	Theory	Simulation	Experiment	Theory	Simulation	Experiment
Input voltage V_{in}	183 V	183 V	183 V	183 V	183 V	183 V
Input average current I_{in}	3.14 A	3.4 A	3.14 A	2.64 A	2.6 A	2.6 A
Input current ripple ΔI_{in}	1.83 A	1.87 A	1.5 A	1.08 A	1.07 A	1.1 A
Transformer current ripple ΔI_{trans}	-	-	-	-	6.66 A	6.5 A
Output ac RMS voltage V_{OUT}	110 V	117.7 V	116.6 V	110 V	116 V	110.5 V
dc-link peak voltage V_0	467 V	479.3 V	480 V	396 V	395 V	396 V
Voltage across capacitor C_1	325 V	330 V	341.6 V	325 V	323.8 V	306.8 V
Voltage across capacitor C_2	142 V	147 V	155.8 V	142 V	140.8 V	127.7 V
ST duty cycle D_S	0.3	0.3	0.3	0.18	0.18	0.2
Switching frequency f	60 kHz					

Table 2. *Cont.*

Parameters	qZS			A-Source		
	Theory	Simulation	Experiment	Theory	Simulation	Experiment
Passive components	Value	Energy (Rated energy)	Size	Value	Energy (Rated energy)	Size
Capacitance value of the capacitor C_1	0.47 mF (400 V)	25 J (37.6 J)	38.5 sm^3	0.47 mF (400 V)	24 J (37.6 J)	38.5 sm^3
Capacitance value of the capacitor C_2	1.5 mF (200 V)	16.2 J (30 J)	43.3 sm^3	1.5 mF (200 V)	14.7 J (30 J)	43.3 sm^3
Inductance value of the inductor L_1	900 µH (9 A)	4.4 mJ (36.5 mJ)	110 sm^3	1.8 mH (9 A)	6 mJ (73 mJ)	220 sm^3
Inductance value of the inductors L_2	900 µH (9 A)	4.4 mJ (36.5 mJ)	110 sm^3	-		-
Output side inductor filter L_g	0.2 mH			0.2 mH	-	
Capacitor filter C_f	0.47 µF			0.47 µF	-	
Transformer T_r						230 sm^3

Figure 16 shows the experimental setup. It consisted of the three-phase VSI, qZS or A-source network, and inductive output filter. The control board was based on a field-programmable gate array (FPGA) which is able to provide any modulation technique. The measurement system was also involved for general monitoring. The experimental setup depicted in Figure 16a belonged to a qZS inverter, while the setup depicted in Figure 16b belonged to an A-source inverter. It can be seen, that it is the same setup and only IS networks were replaced.

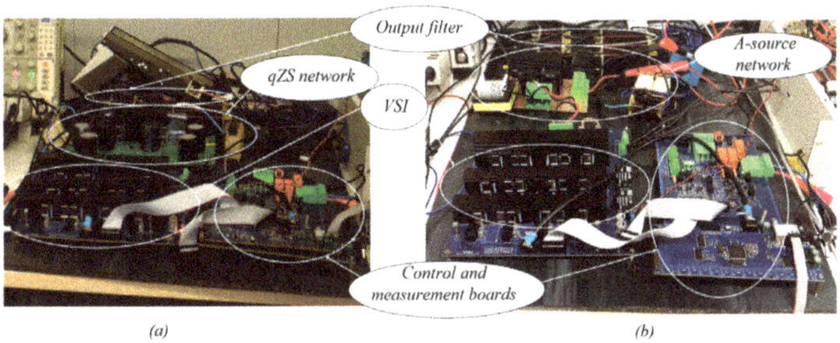

Figure 16. Experimental setup: (**a**) The qZS three-phase inverter and (**b**) the A-source three-phase inverter.

Selected topologies were compared by the same parameters of the input voltage, load, capacitors, and input inductor value. The input voltage was equal to 183 V with a RMS ac output voltage of about 110 V.

Figure 17 shows the simulation results for the qZS network, while Figure 18 shows similar results for the A-source network. It can be seen that approximately the same output voltage was achieved for different circuits. The ST duty cycle $D_{qZS} = 0.2$ of the A-source was smaller than $D_{qZS} = 0.27$ of the qZS.

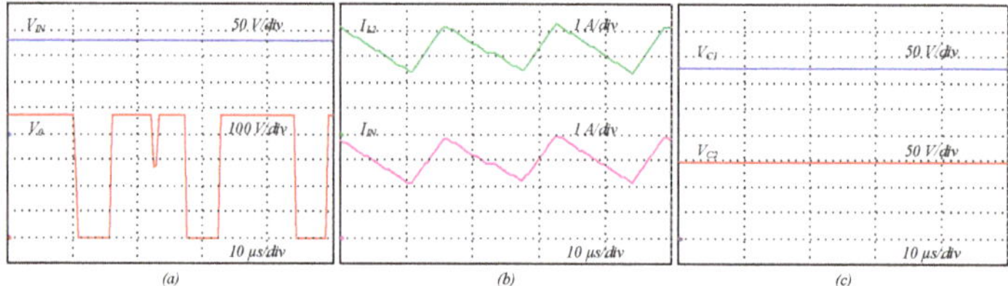

Figure 17. Simulation results of the qZS inverter: (a) Input voltage (V_{in}) along with the dc-link voltage (V_o), (b) input current (I_{in}) along with the inductor current (I_{L2}), and (c) voltage across capacitors V_{C1} and V_{C2}.

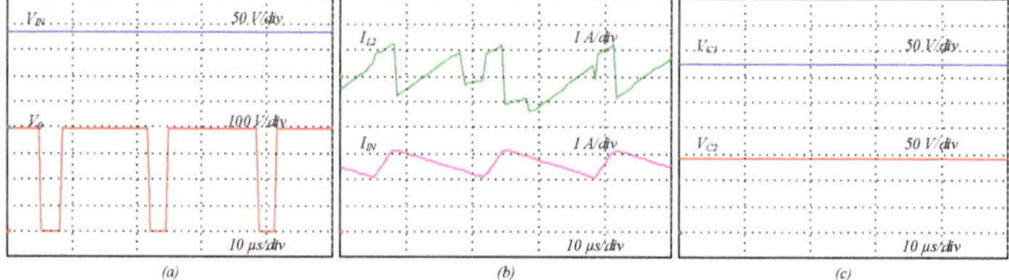

Figure 18. Simulation results of the A-source inverter: (a) Input voltage (V_{in}) along with the dc-link voltage (V_o), (b) input current (I_{in}) along with the inductor current (I_{L2}), and (c) voltage across capacitors V_{C1} and V_{C2}.

The input current quality of the A-source was slightly better. The input average current value of the A-source was lower than qZS: The average value of the input current was I_{in} = 2.6 A and I_{in} = 3.14 A, respectively.

However, it can be seen that the capacitor voltages of the A-source were almost the same as the capacitor voltages of qZS: The average value of V_{C1} = 307 V, V_{C2} = 128 V and V_{C1} = 342 V, V_{C2} = 156 V, respectively. Generally, the above confirms the theoretical results. The average value of the transformer current was approximately zero. This indicates that the transformer core was not saturated.

In order to finalize the verification, Figures 19 and 20 show the experimental results that fully corresponded to the simulation results. Figure 19 illustrates the experimental results of the qZS circuit board. At the same time, Figure 20 shows the obtained results of the A-source circuit board.

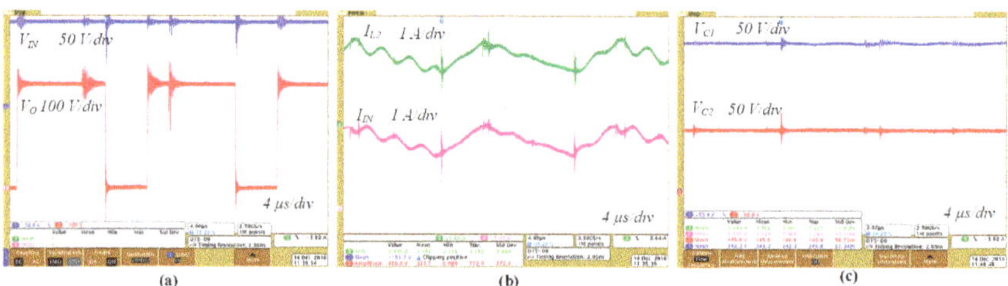

Figure 19. Experimental results of the qZS inverter: (a) Input voltage (V_{in}) along with the dc-link voltage (V_o), (b) input current (I_{in}) along with the inductor current (I_{L2}), and (c) voltage across capacitors V_{C1} and V_{C2}.

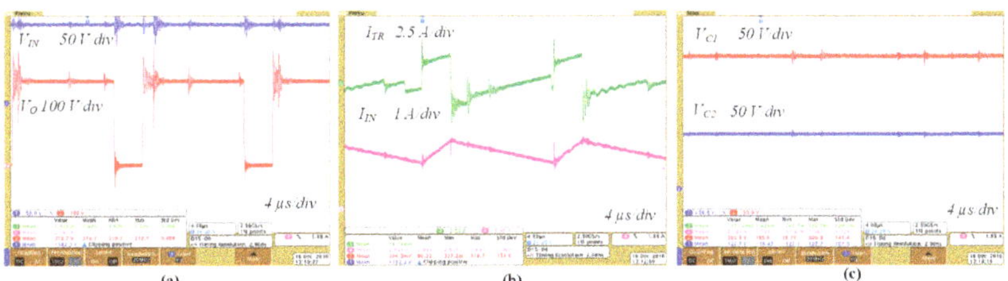

Figure 20. Experimental results of the A-source inverter: (**a**) Input voltage (V_{in}) along with the dc-link voltage (V_o), (**b**) input current (I_{in}) along with the inductor current (I_{L2}), and (**c**) voltage across capacitors V_{C1} and V_{C2}.

Finally, all the results are summarized in Table 2. Except for simulation and experimental verification, the theoretical results are represented.

It can be seen that circuits were tested under different power levels, which is due to the limited power of the transformer of the A-source network, but the main idea of the simulation and the experimental study was to confirm the feasibility of the derived equations in Table 1. It has very good coincidence with the simulation results. It means that previously shown diagrams based on the same approach are valid as well. Minor differences between the simulation and the experimental results can be explained by the tolerance of the passive components and losses in the power conversion.

Table 2 also has parameters of the passive components including maximum stored energy according to the specification and actual size of the element.

Except for typical values of the passive components, voltage, and current across passive elements, Table 2 shows the size of the passive elements and energy that could be stored in them. First of all, it can be concluded that the size of the passive elements is proportional to the energy and the assumption described above is correct. Secondly, it was also obvious that some of the passive components are overdesigned. They were chosen for the experimental setup just because of their availability in the lab. At the same time, in order to demonstrate a more objective value of energy, real voltage across the capacitors and the inductors current were taken into account.

It clearly showed that required energy was stored in the passive element in a certain operation point. The obvious conclusion that the A-source inverter may require a slightly smaller overall inductor's size as compared to the qZS solution, but the transformer has to be taken into account as well.

Despite the different power level, it could be seen that experimentally obtained values were correlating with those theoretically estimated in Figure 8. The capacitance energy was almost the same while the inductance energy was smaller in the A-source network. At the same time, the additional transformer, the size of which is significant, was used in the A-source network. The instantiations value of the energy stored in the transformer was zero, but the size and cost should be taken into account.

7. Conclusions

IS networks are popular in the research area, in particular in a single-stage buck-boost dc-dc and dc-ac applications.

This paper presented a comprehensive analytical and experimental comparison of the IS-based buck-boost solutions in terms of the passive components and semiconductors.

The main criterion for this comparison was the stored energy in the passive elements, which was considered under a constant and predefined high frequency current ripple in the inductors and the voltage ripple across the capacitors. Semiconductor stress and the range of input voltage variation were considered as well. First of all, it was clearly demonstrated that in order to provide higher boost of the input voltage with the same ripples, much larger passive components are required for any solution. Secondly, almost in any case, the classical solution requires lower capacitance and inductance than all other IS solutions.

Also, an interesting conclusion is that the overall size of the similar IS based converters designed for identical operating conditions is the same. Some differences can be obtained for high boost IS solutions. Most of them are based on the coupled inductor or transformer, which in turn adds some volume and cost.

The main drawback of any IS based inverter lies in the increased voltage stress across semiconductors. A high-voltage gain solution may mitigate this drawback, but such solutions demand additional magnetics.

It should also be mentioned that the calculation approach of the proposed passive components was very simplified. In practical experience, some of the capacitors or inductors can be smaller or with an increased ripple. Different optimization techniques can be applied. The most evident advantages of IS networks application correspond to the galvanically isolated converters. It consists in the magnetic components optimization.

Author Contributions: O.H. designed the systems and supervised the development of the experiments. T.S. implemented the mathematical basis and performed the tests. D.V. conducted the state of the art updating and supervised the results obtained. C.R.-C. carried out the simulation study and E.R.-C. assisted with the development of the idea and paper writing. A.C. was in response for description of galvanically isolated converters.

Funding: This research work was financed in part by the Estonian Centre of Excellence in Zero Energy and Resource Efficient Smart Buildings and Districts, ZEBE, grant 2014-2020.4.01.15- 0016 funded by the European Regional Development Fund and by the Estonian Research Council under Grant PUT1443.

Conflicts of Interest: The authors declare no conflict of interest.

References

1. Peng, F.Z. Z-Source Inverter. *IEEE Trans. Ind. Appl.* **2003**, *39*, 504–510. [CrossRef]
2. Anderson, J.; Peng, F.Z. Four quasi-Z-Source inverter. In Proceedings of the 2008 IEEE Power Electronics Specialists Conference (PESC'08), Rhodes, Greece, 5–19 June 2008; pp. 2743–2749. [CrossRef]
3. Peng, F.Z.; Joseph, A.; Wang, J.; Shen, M.; Chen, L.; Pan, Z.; Rivera, E.O.; Huang, Y. Z-source inverter for motor drives. *IEEE Trans. Power Electron.* **2005**, *20*, 857–863. [CrossRef]
4. Peng, F.Z.; Yuan, X.; Fang, X.; Qian, Z. Z-source inverter for adjustable speed drives. *IEEE Trans. Power Electron.* **2003**, *1*, 33–35. [CrossRef]
5. Yushan, L.; Abu-Rub, H.; Baoming, G. Z-source/quasi-Z-source inverters: Derived networks, modulations, controls, and emerging applications to photovoltaic conversion. *IEEE Ind. Electron. Mag.* **2014**, *8*, 32–44. [CrossRef]
6. Vinnikov, D.; Roasto, I. Quasi-Z-Source-Based Isolated DC/DC Converters for Distributed Power Generation. *IEEE Trans. Ind. Electron.* **2011**, *58*, 192–201. [CrossRef]
7. Liu, Y.; Ge, B.; Abu-Rub, H.; Peng, F.Z. Phase-shifted pulse-width-amplitude modulation for quasi-Z-source cascade multilevel inverter-based photovoltaic power system. *IET Power Electron.* **2014**, *7*, 1444–1456. [CrossRef]
8. Rajakaruna, S.; Jayawickrama, L. Steady-state analysis and designing impedance network of Z-source inverters. *IEEE Trans. Ind. Electron.* **2010**, *57*, 2483–2491. [CrossRef]
9. Yu, Y.; Zhang, Q.; Liu, X.; Cui, S. DC-link voltage ripple analysis and impedance network design of single-phase Z-source inverter. In Proceedings of the 2011 14th European Conference on Power Electronics and Applications (EPE 2011), Birmingham, UK, 30 August–1 September 2011; pp. 1–10.
10. Husev, O.; Liivik, L.; Blaabjerg, F.; Chub, A.; Vinnikov, D.; Roasto, I. Galvanically Isolated Quasi-Z-Source DC-DC Converter with a Novel ZVS and ZCS Technique. *IEEE Trans. Ind. Electron.* **2015**, *62*, 7547–7556. [CrossRef]
11. Sun, D.; Baoming, G.; Yan, X.; Bi, D.; Abu-Rub, H.; Peng, F.Z. Impedance design of quasi-Z source network to limit double fundamental frequency voltage and current ripples in single-phase quasi-Z source inverter. In Proceedings of the 2013 Energy Conversion Congress and Exposition (ECCE), Denver, CO, USA, 15–19 September 2013; pp. 2745–2750. [CrossRef]

12. Husev, O.; Chub, A.; Romero-Cadaval, E.; Roncero-Clemente, C.; Vinnikov, D. Hysteresis current control with distributed shoot-through states for impedance source inverters. *Int. J. Circuit Theory Appl.* **2015**, *44*, 783–797. [CrossRef]

13. Roncero-Clemente, C.; Romero-Cadaval, E.; Ruiz-Cortés, M.; Husev, O. Carrier Level-Shifted Based Control Method for the PWM 3L-T-Type qZS Inverter with Capacitor Imbalance Compensation. *IEEE Trans. Ind. Electron.* **2018**, *65*, 8297–8306. [CrossRef]

14. Shults, T.; Husev, O.; Blaabjerg, F.; Roncero, C.; Romero-Cadaval, E.; Vinnikov, D. Novel Space Vector Pulse Width Modulation Strategies for Single-Phase Three-Level NPC Impedance-Source Inverters. *IEEE Trans. Power Electron.* **2018**. [CrossRef]

15. Shen, M. Z-source Inverter Design, Analysis, and Its Application in Fuel Cell Vehicles. Ph.D. Thesis, Michigan State University, East Lansing, Michigan, 2006.

16. Husev, O.; Stepenko, S.; Roncero-Clemente, C.; Romero-Cadaval, E.; Vinnikov, D. Single phase three-level quasi-z-source inverter with a new boost modulation technique. In Proceedings of the 2007 European Conference on Power Electronics and Applications, Aalborg, Denmark, 2–5 September 2007; pp. 1–10.

17. Peng, F.Z.; Shen, M.; Qian, Z. Maximum boost control of the Z-source inverter. *IEEE Trans. Power Electron.* **2005**, 833–838. [CrossRef]

18. Husev, O.; Strzelecki, R.; Blaabjerg, F.; Chopyk, V.; Vinnikov, D. Novel family of single-phase modified impedance-source buck-boost multilevel inverters with reduced switch count. *IEEE Trans. Power Electron.* **2016**, *31*, 7580–7591. [CrossRef]

19. Loh, P.C.; Gao, F.B. Embedded EZ-Source Inverter. *IEEE Trans. Ind. Appl.* **2010**, *46*, 256–267. [CrossRef]

20. Loh, P.C.; Li, D.; Blaabjerg, F. T-Z-Source Inverters. *IEEE Trans. Power Electron.* **2013**, *28*, 4880–4884. [CrossRef]

21. Nguyen, M.K.; Lim, Y.C.; Kim, Y.G. TZ-Source Inverters. *IEEE Trans. Ind. Electron.* **2013**, *60*, 5686–5695. [CrossRef]

22. Nguyen, M.K.; Lim, Y.C.; Park, S.J.; Jung, Y.G. Cascaded TZ-source inverters. *IET Power Electron.* **2014**, *8*, 2068–2080. [CrossRef]

23. Zhu, M.; Yu, K.; Luo, F.L. Switched Inductor Z-Source Inverter. *IEEE Trans. Power Electron.* **2010**, *25*, 2150–2158. [CrossRef]

24. Loh, P.C.; Gao, F.; Blaabjerg, F.; Goh, A.L. Buck-boost impedance networks. In Proceedings of the 2007 European Conference on Power Electronics and Applications, Aalborg, Denmark, 2–5 September 2007; pp. 1–10.

25. Shen, M.; Wang, J.; Joseph, A.; Peng, F.Z.; Tolbert, L.M.; Adams, D.J. Constant boost control of the Z-source inverter to minimize current ripple and voltage stress. *IEEE Trans. Ind. Appl.* **2006**, *42*, 770–778. [CrossRef]

26. Ellabban, O.; van Mierlo, J.; Lataire, P. Comparison between different PWM control methods for different Z-source inverter topologies. In Proceedings of the 13th IEE European Conference on Power Electronics and Applications (EPE'09), Barcelona, Spain, 8–10 September 2009; pp. 1–11.

27. Ellabban, O.; Abu-Rub, H. Z-Source Inverter: Topology Improvements Review. *IEEE Ind. Electron. Mag.* **2016**, *10*, 6–24. [CrossRef]

28. Gupta, A.K.; Samuel, P.; Kumar, D. A state of art review and challenges with impedance networks topologies. In Proceedings of the 2016 IEEE 7th Power India International Conference (PIICON), Rajasthan, India, 25–27 November 2016; pp. 1–6.

29. Siwakoti, Y.P.; Peng, F.; Blaabjerg, F.; Loh, P.C.; Town, G.E. Impedance Source Networks for Electric Power Conversion Part-I: A Topological Review. *IEEE Trans. Power Electron.* **2015**, *2*, 699–716. [CrossRef]

30. Siwakoti, Y.P.; Peng, F.; Blaabjerg, F.; Loh, P.; Town, G.E. Impedance Source Networks for Electric Power Conversion Part-II: Review of Control Method and Modulation Techniques. *IEEE Trans. Power Electron.* **2015**, *30*, 1887–1906. [CrossRef]

31. Husev, O.; Blaabjerg, F.; Roncero-Clemente, C.; Romero-Cadaval, E.; Vinnikov, D.; Siwakoti, Y.P.; Strzelecki, R. Comparison of Impedance-Source Networks for Two and Multilevel Buck-Boost Inverter Applications. *IEEE Trans. Power Electron.* **2016**, *31*, 7564–7579. [CrossRef]

32. Chub, A.; Vinnikov, D.; Blaabjerg, F.; Peng, F.Z. A Review of Galvanically Isolated Impedance-Source DC-DC Converters. *IEEE Trans. Power Electron.* **2016**, *31*, 2808–2828. [CrossRef]

33. Franke, W.T.; Mohr, M.; Fuchs, F.W. Comparison of a Z-source inverter and a voltage-source inverter linked with a DC/DC-boost-converter for wind turbines concerning their efficiency and installed semiconductor power. In Proceedings of the Power Electronics Specialists Conference, Rhodes, Greece, 15–19 June 2008; pp. 1814–1820. [CrossRef]

34. Battiston, A.; Martin, J.P.; Miliani, E.D.; Nahid-Mobarakeh, B.; Pierfederici, S.; Meibody-Tabar, F. Comparison criteria for electric traction system using z-source/quasi z-source inverter and conventional architectures. *IEEE J. Emerg. Sel. Top. Power Electron.* **2014**, *3*, 467–476. [CrossRef]

35. Shen, M.; Joseph, A.; Wang, J.; Peng, F.Z. Comparison of Traditional Inverters and Z-Source Inverter Power Electronics. In Proceedings of the Specialists Conference PESC'05, Recife, Brazil, 12–16 June 2005; pp. 1692–1698. [CrossRef]

36. Burkart, R.; Kolar, J.W.; Griepentrog, G. Comprehensive comparative evaluation of single- and multi-stage three-phase power converters for photovoltaic applications. In Proceedings of the Intelec, Scottsdale, AZ, USA, 30 September–4 October 2012; pp. 1–8. [CrossRef]

37. Panfilov, D.; Husev, O.; Blaabjerg, F.; Zakis, J.; Khandakji, K. Comparison of three-phase three-level voltage source inverter with intermediate dc-dc boost converter and quasi-Z-source inverter. *IET Power Electron.* **2014**, *7*, 1444–1456. [CrossRef]

38. Gadalla, B.; Schaltz, E.; Siwakoti, Y.; Blaabjerg, F. Thermal performance and efficiency investigation of conventional boost, Z-source and Y-source converters. In Proceedings of the 16th International Conference on Environment and Electrical Engineering (EEEIC), Florence, Italy, 7–10 June 2016; pp. 1–6. [CrossRef]

39. Wolski, K.; Zdanowski, M.; Rabkowski, J. High-frequency SiC-based inverters with input stages based on quasi-Z-source and boost topologies—Experimental comparison. *IEEE Trans. Power Electron.* **2019**. [CrossRef]

40. Siwakoti, Y.P.; Loh, P.C.; Blaabjerg, F.; Town, G.E. Y-Source Impedance Network. *IEEE Trans. Power Electron.* **2014**, *29*, 3250–3254. [CrossRef]

41. Mo, W.; Loh, P.C.; Blaabjerg, F. Voltage type T-source Inverters with Continuous Input Current and Enhanced Voltage Boost Capability. In Proceedings of the 15th International Power Electronics and Motion Control Conference, EPE-PEMC 2012 ECCE Europe, Novi Sad, Serbia, 4–6 September 2012; pp. 1–8.

42. Siwakoti, Y.P.; Blaabjerg, F.; Loh, P.C. New Magnetically Coupled Impedance (Z-) Source Networks. *IEEE Trans. Power Electron.* **2016**, *31*, 7419–7435. [CrossRef]

43. Strzelecki, R.; Adamowicz, M.; Balkowski, B.; Strzelecka, N. The Buck-Boost Inverter Circuit Especially Designed for Single-Stage Power Conversion. PL Patent Application P386084, 9 September 2008.

44. Strzelecki, R.; Adamowicz, M.; Strzelecka, N.; Bury, W. New type T-source inverter. In Proceedings of the CPE 2009, Badajoz, Spain, 20–22 May 2009; pp. 191–195. [CrossRef]

45. Strzelecki, R.; Adamowicz, M.; Balkowski, B.; Strzelecka, N. Multi-Level Inverter Circuit Especially for Voltage Boost. PL Patent Application P386085, 29 March 2010.

46. Mo, W.; Loh, P.C.; Blaabjerg, F.; Wang, P. Trans-Z-source and Γ-Z-source neutral-point-clamped inverters. *IET Power Electron.* **2015**, *8*, 371–377. [CrossRef]

47. Adamowicz, M.; Guzinski, J.; Vinnikov, D.; Strzelecka, N. Trans-Z-source-like Inverter with Built-in DC Current Blocking Capacitors. In Proceedings of the CPE 2011, Tallinn, Estonia, 1–3 June 2011; pp. 137–143. [CrossRef]

48. Adamowicz, M.; Strzelecki, R.; Peng, F.Z.; Guzinski, J.; Abu-Rub, H. New type LCCT-Z-source inverters. In Proceedings of the 14th European Conference on Power Electronics, Birmingham, UK, 30 August–1 September 2011; pp. 1–10.

49. Shults, T.; Husev, O.; Blaabjerg, F. Design and Comparison of Three-Level Three-Phase T-Source Inverters. In Proceedings of the International Conference POWERENG'2015, At Riga, Latvia, 11–13 May 2015; pp. 1–6. [CrossRef]

50. Adamowicz, M.; Guzinski, J.; Strzelecki, R.; Peng, F.Z.; Abu-Rub, H. High step-up continuous input current LCCT-Z-source inverters for fuel cells. In Proceedings of the IEEE Energy Conversion Congress and Exposition, ECCE, Phoenix, AZ, USA, 17–22 September 2011; pp. 2276–2282. [CrossRef]

51. Adamowicz, M. LCCT-Z-source inverters. In Proceedings of the 10th International Conference on Environment and Electrical Engineering, EEEIC, Rome, Italy, 8–11 May 2011; pp. 1–6. [CrossRef]

52. Park, J.-K.; Shin, Y.-S.; Jung, Y.-G.; Lim, Y.-C. LCCT Z-Source DC-DC Converter with the Bipolar Output Voltages for Improving the Voltage Stress and Ripple. *Trans. Korean Inst. Power Electron.* **2013**, *18*, 91–102. [CrossRef]
53. Shults, T.; Husev, O.; Blaabjerg, F.; Zakis, J.; Khandakji, K. LCCT-Derived Three-Level Three-Phase Inverters. *IET Power Electron.* **2017**, *10*, 996–1002. [CrossRef]
54. Siwakoti, Y.P.; Blaabjerg, F.; Galigekere, V.P.; Kazimierczuk, M.K. A-source impedance network. In Proceedings of the IEEE Energy Conversion Congress and Exposition (ECCE), Milwaukee, Wisconsin, 18–22 September 2016; pp. 1–6. [CrossRef]
55. Siwakoti, Y.P.; Blaabjerg, F.; Galigekere, V.P.; Ayachit, A.; Kazimierczuk, M.K. A-source impedance network. *IEEE Trans. Power Electron.* **2016**, *31*, 8081–8087. [CrossRef]
56. Vinnikov, D.; Chub, A.; Liivik, E.; Roasto, I. High-Performance Quasi-Z-Source Series Resonant DC–DC Converter for Photovoltaic Module-Level Power Electronics Applications. *IEEE Trans. Power Electron.* **2017**, *32*, 3634–3650. [CrossRef]

Article

Control of MMC-Based STATCOM as an Effective Interface between Energy Sources and the Power Grid

Fatemeh Shahnazian [1], Ebrahim Adabi [2], Jafar Adabi [1], Edris Pouresmaeil [3,*],
Kumars Rouzbehi [4], Eduardo M. G. Rodrigues [5,*] and João P. S. Catalão [6]

[1]   Faculty of Electrical and Computer Engineering, Babol (Noshirvani) University of Technology, PO Box 484, Babol, Iran; f_shahnazian@yahoo.com (F.S.); jafar.adabi@gmail.com (J.A.)
[2]   Intelligent Electrical Power Grids at Department of Electrical Sustainable Energy, Delft University of Technology, 5031, 2600 GA Delft, The Netherlands; ebrahim.adabi@tudelft.nl
[3]   Department of Electrical Engineering and Automation, Aalto University, 02150 Espoo, Finland
[4]   Department of System Engineering and Automatic Control, University of Seville, 41004 Seville, Spain; krouzbehi@us.es
[5]   Management and Production Technologies of Northern Aveiro—ESAN, Estrada do Cercal 449, Santiago de Riba-Ul, 3720-509 Oliveira de Azeméis, Portugal
[6]   Faculty of Engineering of the University of Porto, and INESC TEC, 4200-465 Porto, Portugal; catalao@fe.up.pt
*   Correspondence: edris.pouresmaeil@aalto.fi (E.P.); emgrodrigues@ua.pt (E.M.G.R.)

Received: 2 October 2019; Accepted: 30 October 2019; Published: 1 November 2019

Abstract: This paper presents a dynamic model of modular multilevel converters (MMCs), which are considered as an effective interface between energy sources and the power grid. By improving the converter performance, appropriate reactive power compensation is guaranteed. Modulation indices are calculated based on detailed harmonic evaluations of both dynamic and steady-state operation modes, which is considered as the main contribution of this paper in comparison with other methods. As another novelty of this paper, circulating current control is accomplished by embedding an additional second harmonic component in the modulation process. The proposed control method leads to an effective reduction in capacitor voltage fluctuation and losses. Finally, converter's maximum stable operation range is modified, which provides efficiency enhancements and also stability assurance. The proficiency and functionality of the proposed controller are demonstrated through detailed theoretical analysis and simulations with MATLAB/Simulink.

Keywords: circulating current control; modular multilevel converter (MMC); static synchronous compensator (STATCOM)

---

1. Introduction

The growing global demand for energy consumption has led to fundamental changes in power grids. The need for improved reliability and efficiency, as well as reduced transmission and distribution losses and cost [1], has led to the widespread use of distributed generation (DG) units. In this regard, increasing utilization of DGs due to the aforementioned benefits, along with the development of high-voltage direct current (HVDC) transmission systems, has led to major power electronics advancements in terms of switching technology and converter design [2,3].

Early power conversion technology started with 2-level converters. Additional filter demand in order to overcome the high total harmonic distortion (THD) reflected in the output voltage plus limited voltage tolerance of switches have made these structures inefficient for high voltage applications [4]. Producing high voltage in stair-case waveform by means of separate DC sources [5], multilevel converters provide lower voltage stress, losses, and filter costs effectively [6]. Reference [7] provides a comprehensive study of various multilevel structures along with the most recent advances in their modulation techniques, control methods, and applications.

However, modular converters mainly replaced all other multilevel structures due to their advantages such as modularity, scalability, and performance compatibility with one common DC source [8]. Modular multilevel converter (MMC) was first proposed by Marquardt and Lesnicar in 2003 and it has been effectively providing a variety of applications since then [9]. The attractive features of MMC such as the capability of transformer-less operation, ease of scalability to higher voltage levels, low expenses and robustness in redundancy strategies and fault tolerant operation, high reliability, and good quality of the output waveforms have made this topology one of the most beneficial structures in various medium/high voltage applications [10,11]. In this regard, studying different aspects of MMC operation such as modelling, modulation, circulating current, and capacitor voltage fluctuation seems to be noteworthy.

These non-linear power electronics converters improve power quality at consumer's end, i.e., an affordable consistent supply with fixed voltage level and suitable power factor (PF) and thus necessitate a local reactive power regulation. In this regard, limiting the grid side current flow, voltage regulation, and efficiency enhancement can be accomplished at the same time [12]. Since early fulfilling procedures include huge expensive capacitor banks, active compensation methods are becoming more common in high power applications [13]. As one of the most practical components in a flexible AC transmission system (FACTS), static synchronous compensator (STATCOM) can efficiently improve the power quality [14]. STATCOM is not only capable of fast reactive current injection but also performs voltage compensation with lower loss and lower output harmonics [15].

The conventional structures of STATCOM are usually based on two-level converters and step-up coupling transformers [16]. The embedded transformer makes the overall system expensive, bulky, and unreliable [17]. These restrictions necessitate the use of multilevel converters in high voltage applications. The most popular modular structures employed in STATCOM applications are flying-capacitor multilevel converters (FCMC), diode-clamped multilevel converters (DCMC), and cascaded H-bridge multilevel converters (CHMC) [18]. The excessive cost of flying capacitors in FCMCs, as well as the need for an additional auxiliary voltage balancing algorithm in order to solve the inherent voltage imbalances in DCMCs, is considered as the major disadvantages of these two topologies. Furthermore, the complexity of converter topology and controller design also gets significantly increased in high voltage applications. In this regard, the CHMC configuration seems more suitable among others, however the structure is unable to operate continuously under unbalanced situations. Therefore, the necessity of adding a DC voltage supply through multi-winding transformers in such situations imposes similar imperfections as the line-frequency transformer does [19].

In this regard, various MMC layouts are extensively employed in STATCOM applications [20], providing a high voltage transformer-less structure with improved efficiency and better fault tolerance [21]. Several studies have been performed in order to discuss MMC-based STATCOM schemes from different aspects. Reference [8] proposes a basic control method in which reactive power demand is supplied based on DC link voltage stabilization using PI controllers. Therefore, detailed studies on converter dynamics are neglected. Reference [22] studies MMC capability as an efficient STATCOM interface in HVDC applications. Considering a stable AC power grid, the abc-frame power flow control is presented while MMC is islanded from the DC grid. The controller guarantees an appropriate operation of the converter; however, the minimum advantageous DC voltage across MMC terminals should be evaluated. The theory of a decoupled control method applied to the active and reactive power of the system is also pursued in [23]. Studying cascaded MMC structure, output voltages of sub-modules (SMs) are controlled separately since reactive power compensation is distributed between them. This imposes an extra computation burden on the MMC simulation scenarios. Circulating currents, DC link voltage, and output AC currents are considered as state variables of the proposed controller in [24]. Multiple control loops are employed in this method, which lead to accuracy as well as complexity of the control method. Circulating current and capacitor voltage balancing algorithms are also exclusively discussed in [20,25] and [26] respectively. The proposed methods are mostly designed based on PI controllers and imposed on MMC structures with particular applications such as delta-configuration. It should be noted that various converter structures based on different SM configurations have also been employed in MMC-based STATCOM studies in order to configure the pros and cons [18,27]. Besides all, modelling has an enormous impact while studying any of the MMC operation aspects. Providing an accurate and at the same time efficient model can be considered as one of the most important technical challenges [28,29].

This paper presents a control strategy in order to guarantee the stable performance of a three-phase MMC-based STATCOM. Simplifying converter stability studies and controller design, dq frame dynamic and steady-state modelling are chosen over conventional abc models in this paper. This accurate dynamic model provides an efficient computation of modulation indices, which is considered as the first contribution of this paper. Circulating current control is also accomplished based on second harmonic expressions of the proposed model as the second contribution. In addition to that, the maximum range of the converter stable operation is determined as another novelty of this paper. The specified stable region is then used in power factor correction controller design such that the MMC would supply all reactive power demand of the load within the feasible power range, and the excess amount of power remaining from total power capacity of the converter is utilized to provide the active power demand of the load. The main advantages of this proposed control method are the ability of simultaneously providing dq-frame modulation, as well as the elimination of circulating current and the capacitor voltage fluctuations. Moreover, the proposed controller provides an efficient MMC-based STATCOM operation where all reactive power demands of the load, as well as the most available active power demands, can be provided considering the maximum stable operation region.

The rest of the paper is organized as follows. Section 2 provides an overall view of the studied configuration along with the basic assumptions and dynamic equations. Steady-state operation is investigated in Section 3 where modulation indices are also calculated based on the proposed model. Section 3.1 includes circulating current control algorithm while the beneficial effects of this proposed method are clarified through the presented procedure for loss calculation of MMC in Section 3.2. Furthermore, maximum stable operation range of the converter is established in Section 3.3. Time-domain simulations of the proposed model are carried out in MATLAB/Simulink environment and the results are discussed in Section 4. Finally, the overall conclusions are specified in Section 5.

2. General Configuration and Dynamic Analysis of the Proposed Model

The most conventional configuration of a modern three-phase MMC is shown in Figure 1a. It should be noted that the improvement of grid imbalance studies is not considered as the purpose of the MMC utilization in this configuration and, therefore, the network structure is assumed as a strong grid. Herein, MMC is utilized as an efficient STATCOM interface between the energy sources and the power grid. Thus, it can well improve the overall network performance by mainly providing reactive power demands of the load. Moreover, considering the MMC superiorities, the proposed MMC-based STATCOM structure can be used to renovate conventional power networks.

Using voltage and current components at the point of common coupling (PCC) as well as the converter and grid parameters, the proposed controller is designed based on dq frame dynamic and steady-state modelling of the MMC-based STATCOM structure. In this regard, MMC is supposed to provide all reactive power demands of the load within the feasible power range since this power factor correction behavior of the MMC is considered as the first priority of the proposed controller. Moreover, the maximum stable operation region is evaluated. Based on that, the MMC can also provide maximum available active power demands of the load in order to reduce the load stress imposed on the power grid. Besides, a circulating current controller is also considered in order to improve converter operation using second harmonic current elimination, which also leads to a reduction in the capacitor voltage fluctuations as well.

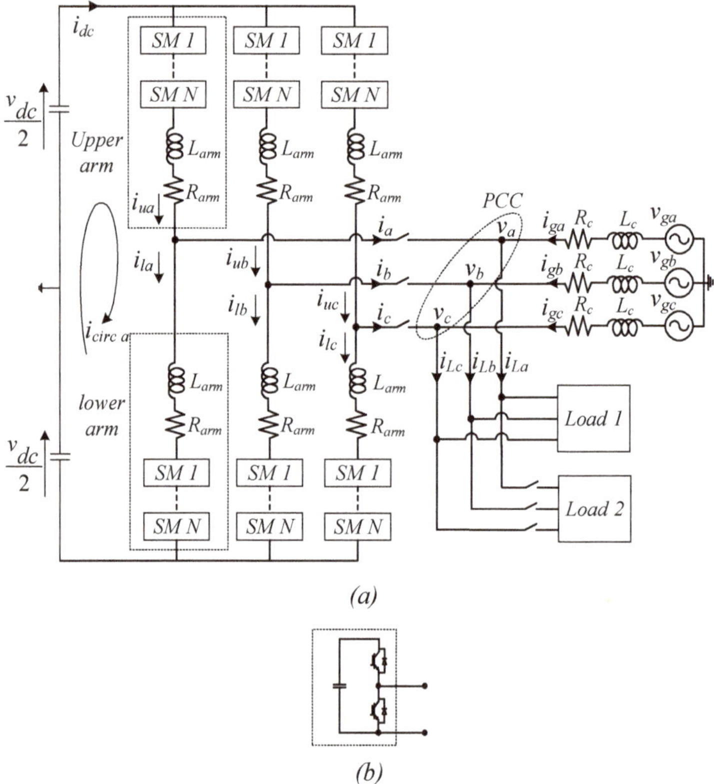

Figure 1. General schematic of a modular multilevel converter (MMC)-based static synchronous compensator (STATCOM): (**a**) Circuit structure; (**b**) Sub-module (SM) configuration.

As can be seen in Figure 1a, each phase leg is composed of two arms in which N SMs are connected in series along with one parasitic resistance R_{arm} and one arm inductor L_{arm}. The SMs are commonly based on half-bridge structures, consisting of a DC capacitor and two insulated gate bipolar transistor (IGBT)/Diode switches as shown in Figure 1b. It should be noted that arm inductors are embedded in order to limit fault based or inherent harmonic components of each arm-current. Besides, R_{arm} stands for inductor inner resistance and converter losses and thus varies with operational conditions. Also, it should be noted that the grid model here is considered to be a 10 KV feeder.

According to Figure 1a, arm currents of each phase can be described as:

$$i_{ux} = i_{circx} + \frac{i_x}{2} \tag{1}$$

$$i_{lx} = i_{circx} - \frac{i_x}{2} \tag{2}$$

where the circulating current (which is flowing between the DC link and each phase leg) is supposed to have DC and second harmonic components [30,31], while i_x is considered as the sinusoidal current of phase x at the fundamental frequency.

On the other hand, the instant number of inserted SMs in each arm determines the arm voltage and thus directly affects the output voltage as well. In other words, each arm voltage is controlled through modulation index (n) which is defined as the ratio between sum capacitor voltages (U_c^Σ) and desired voltages of each arm (U_c). Given that n varies from zero to one, we have:

$$
\begin{cases}
U_{cux} = n_{ux} U_{cux}^\Sigma \\
U_{clx} = n_{lx} U_{clx}^\Sigma
\end{cases}
\tag{3}
$$

The control modulation indices and sum capacitor voltages are considered to include DC, fundamental, and second harmonic components. This is due to the presence of different harmonic components in both upper and lower arm currents which inevitably affects the SM capacitor voltages and thus should be considered in the control modulation indices as well. Therefore, these signals can be represented in dq0 frame as follows:

$$
n_u = \left(\frac{1}{2}\right) + \left(\frac{-m_d}{2}\right)\cos(\omega t) + \left(\frac{-m_q}{2}\right)\sin(\omega t) + \left(\frac{-m_{d2}}{2}\right)\cos(2\omega t) + \left(\frac{-m_{q2}}{2}\right)\sin(2\omega t)
\tag{4}
$$

$$
n_l = \left(\frac{1}{2}\right) + \left(\frac{m_d}{2}\right)\cos(\omega t) + \left(\frac{m_q}{2}\right)\sin(\omega t) + \left(\frac{-m_{d2}}{2}\right)\cos(2\omega t) + \left(\frac{-m_{q2}}{2}\right)\sin(2\omega t)
\tag{5}
$$

$$
U_{cu}^\Sigma(t) = U_{cum0}^\Sigma + U_{cum1}^\Sigma \cos(\omega t - \theta_u) + U_{cum2}^\Sigma \cos(2\omega t - \theta_{u2}) \equiv U_{cu0}^\Sigma + U_{cud}^\Sigma + U_{cuq}^\Sigma + U_{cud2}^\Sigma + U_{cuq2}^\Sigma
\tag{6}
$$

$$
U_{cl}^\Sigma(t) = U_{clm0}^\Sigma + U_{clm1}^\Sigma \cos(\omega t - \theta_l) + U_{clm2}^\Sigma \cos(2\omega t - \theta_{l2}) \equiv U_{cl0}^\Sigma + U_{cld}^\Sigma + U_{clq}^\Sigma + U_{cld2}^\Sigma + U_{clq2}^\Sigma
\tag{7}
$$

In addition, the dynamic equation describing energy storage capability of SM capacitors in each arm can be written as:

$$
W_{cu,l}^\Sigma = \frac{1}{2}\frac{C}{N}\left(U_{cu,l}^\Sigma\right)^2
\tag{8}
$$

Following that, the energy derivative is equal to the power injected into each arm. Therefore:

$$
\frac{dW_{cu,l}^\Sigma}{dt} = \frac{C}{N}\left(U_{cu,l}^\Sigma\right)\left(\frac{dU_{cu,l}^\Sigma}{dt}\right) = \left(U_{cu,l}\right)\left(i_{u,l}\right)
\tag{9}
$$

Substituting $U_{cu,l} = n_{u,l}U_{cu,l}^\Sigma$, Equation (10) can be obtained as:

$$
\frac{C}{N}\frac{dU_{cu,l}^\Sigma}{dt} = n_{u,l}i_{u,l}
\tag{10}
$$

Then, harmonic components of the upper arm current, modulation index, and sum capacitor voltage are considered based on Equations (1), (4), and (6) respectively. Thus, for the upper arm, Equation (10) can be rewritten in dq0 frame as follows:

$$
\frac{C}{N}\frac{d}{dt}\left(U_{cu0}^\Sigma + U_{cud}^\Sigma\cos\omega t + U_{cuq}^\Sigma\sin\omega t + U_{cud2}^\Sigma\cos 2\omega t + U_{cuq2}^\Sigma\sin 2\omega t\right) =
$$
$$
\left(\tfrac{1}{2} + \tfrac{-m_d}{2}\cos\omega t + \tfrac{-m_q}{2}\sin\omega t + \tfrac{-m_{d2}}{2}\cos 2\omega t + \tfrac{-m_{q2}}{2}\sin 2\omega t\right)\left(i_{u0} + i_{ud}\cos\omega t + i_{uq}\sin\omega t + i_{ud2}\cos 2\omega t + i_{uq2}\sin 2\omega t\right)
\tag{11}
$$

The next step would be applying the dq0 frame derivation term in the left-hand side as well as multiplying the two expressions in right-hand side of Equation (11). Representing different harmonic components of the product signal in different matrix rows leads to:

$$
\frac{C}{N}\frac{d}{dt}
\begin{bmatrix}
U_{cud}^\Sigma \\
U_{cuq}^\Sigma \\
U_{cud2}^\Sigma \\
U_{cuq2}^\Sigma \\
U_{cu0}^\Sigma
\end{bmatrix}
=
\frac{C}{N}
\begin{bmatrix}
0 & \omega & 0 & 0 & 0 \\
-\omega & 0 & 0 & 0 & 0 \\
0 & 0 & 0 & 2\omega & 0 \\
0 & 0 & -2\omega & 0 & 0 \\
0 & 0 & 0 & 0 & 0
\end{bmatrix}
\begin{bmatrix}
U_{cud}^\Sigma \\
U_{cuq}^\Sigma \\
U_{cud2}^\Sigma \\
U_{cuq2}^\Sigma \\
U_{cu0}^\Sigma
\end{bmatrix}
+
\begin{bmatrix}
-\frac{m_d}{2}i_{u0} + \frac{1}{2}i_{ud} - \frac{m_{d2}}{4}i_{ud} - \frac{m_{q2}}{4}i_{uq} - \frac{m_d}{4}i_{ud2} - \frac{m_q}{4}i_{uq2} \\
-\frac{m_q}{2}i_{u0} + \frac{1}{2}i_{uq} + \frac{m_{d2}}{4}i_{uq} - \frac{m_{q2}}{4}i_{ud} + \frac{m_q}{4}i_{ud2} - \frac{m_d}{4}i_{uq2} \\
-\frac{m_d}{4}i_{ud} + \frac{m_q}{4}i_{uq} - \frac{m_{d2}}{2}i_{u0} + \frac{1}{2}i_{ud2} \\
-\frac{m_q}{4}i_{ud} - \frac{m_d}{4}i_{uq} - \frac{m_{q2}}{2}i_{u0} + \frac{1}{2}i_{uq2} \\
\frac{1}{2}i_{u0} - \frac{m_d}{4}i_{ud} - \frac{m_q}{4}i_{uq} - \frac{m_{d2}}{4}i_{ud2} - \frac{m_{q2}}{4}i_{uq2}
\end{bmatrix}
\tag{12}
$$

On the other hand, v_x is considered as the MMC output voltage at PCC. In this regard, applying KVL to the circuit structure of the proposed MMC-based STATCOM represented in Figure 1a yields to:

$$\frac{v_{dc}}{2} - n_{ux}U_{cux}^{\Sigma} - L_{arm}\frac{di_{ux}}{dt} - R_{arm}i_{ux} = v_x \tag{13}$$

According to Equations (4)–(7), dynamic state-space equations of the proposed model are derived in dq0 frame as:

$$L_{arm}\frac{d}{dt}\begin{bmatrix} i_{ud} \\ i_{uq} \\ i_{u0} \end{bmatrix} = \begin{bmatrix} -R_{arm} & L_{arm}\omega & 0 \\ -L_{arm}\omega & -R_{arm} & 0 \\ 0 & 0 & -R_{arm} \end{bmatrix}\begin{bmatrix} i_{ud} \\ i_{uq} \\ i_{u0} \end{bmatrix} - \begin{bmatrix} v_d \\ v_q \\ 0 \end{bmatrix} + \begin{bmatrix} 0 \\ 0 \\ \frac{v_{dc}}{2} \end{bmatrix} - \begin{bmatrix} -\frac{m_d}{2}U_{cu0}^{\Sigma} + \frac{1}{2}U_{cud}^{\Sigma} - \frac{m_{d2}}{4}U_{cud}^{\Sigma} - \frac{m_{q2}}{4}U_{cuq}^{\Sigma} - \frac{m_d}{4}U_{cud2}^{\Sigma} - \frac{m_q}{4}U_{cuq2}^{\Sigma} \\ -\frac{m_q}{2}U_{cu0}^{\Sigma} + \frac{1}{2}U_{cuq}^{\Sigma} + \frac{m_{d2}}{4}U_{cuq}^{\Sigma} - \frac{m_{q2}}{4}U_{cud}^{\Sigma} + \frac{m_q}{4}U_{cud2}^{\Sigma} - \frac{m_d}{4}U_{cuq2}^{\Sigma} \\ \frac{1}{2}U_{cu0}^{\Sigma} - \frac{m_d}{4}U_{cud}^{\Sigma} - \frac{m_q}{4}U_{cuq}^{\Sigma} - \frac{m_{d2}}{4}U_{cud2}^{\Sigma} - \frac{m_{q2}}{4}U_{cuq2}^{\Sigma} \end{bmatrix} \tag{14}$$

3. Steady-State Operation Analysis and Controller Design

In order to study the stable performance of the converter, dynamic equations of sum capacitor voltages are evaluated in the proposed model. Rewriting (12) in steady-state, all voltage derivatives are equated to zero while zero component of sum capacitor voltages (U_{cu0}^{Σ}) is replaced with the reference value of DC-link voltage (V_r). Also, second harmonic components of the currents, as well as the control modulation indices (m_{d2} and m_{q2}), can be neglected due to the embedded circulating current control algorithm. Therefore, based on the last row of (12), i_{u0} can be expressed as:

$$i_{u0} = \frac{m_{ds}}{2}i_{ud} + \frac{m_{qs}}{2}i_{uq} \tag{15}$$

which then yields to:

$$U_{cud}^{\Sigma} = \frac{-N\left(m_{qs}^2 - 2\right)}{4C\omega}i_{uq} \tag{16}$$

$$U_{cuq}^{\Sigma} = \frac{N\left(m_{ds}^2 - 2\right)}{4C\omega}i_{ud} \tag{17}$$

$$U_{cud2}^{\Sigma} = -\frac{Nm_{qs}}{8C\omega}i_{ud} - \frac{Nm_{ds}}{8C\omega}i_{uq} \tag{18}$$

$$U_{cuq2}^{\Sigma} = \frac{Nm_{ds}}{8C\omega}i_{ud} - \frac{Nm_{qs}}{8C\omega}i_{uq} \tag{19}$$

On the other hand, Equation (14) should also be studied under stable operating mode in order to evaluate the steady-state modulation indices. Concerning this, the instantaneous arm currents are substituted with their reference values and the derivations are also considered in order to improve controller dynamics. Consequently:

$$\frac{di_{ud}}{dt} = I_{avud}^* = \frac{I_{avd}^*}{2}, \frac{di_{uq}}{dt} = I_{avuq}^* = \frac{I_{avq}^*}{2} \tag{20}$$

$$i_{ud} = I_{ud}^* = \frac{I_d^*}{2}, i_{uq} = I_{uq}^* = \frac{I_q^*}{2}, i_{u0} = I_0^* \tag{21}$$

Finally, applying the abovementioned assumptions to Equation (14) yields to the steady-state expressions of the proposed model as follows:

$$\begin{bmatrix} \frac{L_{arm}}{2} & 0 & 0 \\ 0 & \frac{L_{arm}}{2} & 0 \\ \frac{m_{ds}}{4}L_{arm} & \frac{m_{qs}}{4}L_{arm} & 0 \end{bmatrix}\begin{bmatrix} I_{avd}^* \\ I_{avq}^* \\ V_r \end{bmatrix} = -\begin{bmatrix} V_d^* \\ V_q^* \\ 0 \end{bmatrix} + \begin{bmatrix} -\frac{R_{arm}}{2} & \frac{L_{arm}}{2}\omega + \frac{N\left(3m_{qs}^2 - m_{ds}^2 - 8\right)}{64C\omega} & \frac{m_{ds}}{2} \\ -\frac{L_{arm}}{2}\omega - \frac{N\left(3m_{ds}^2 - m_{qs}^2 - 8\right)}{64C\omega} & -\frac{R_{arm}}{2} & \frac{m_{qs}}{2} \\ -\frac{8m_{ds}C\omega R_{arm} + 2Nm_{qs}}{32C\omega} & -\frac{8m_{qs}C\omega R_{arm} - 2Nm_{ds}}{32C\omega} & 0 \end{bmatrix}\begin{bmatrix} I_d^* \\ I_q^* \\ V_r \end{bmatrix} \tag{22}$$

Last row of Equation (22) can be rewritten as:

$$m_{ds}\underbrace{\left(\frac{L_{arm}}{4}I^*_{avd}+\frac{R_{arm}}{4}I^*_d-\frac{N}{16C\omega}I^*_q\right)}_{A}=m_{qs}\underbrace{\left(\frac{L_{arm}}{4}I^*_{avq}+\frac{R_{arm}}{4}I^*_q-\frac{N}{16C\omega}I^*_d\right)}_{B} \tag{23}$$

As can be seen, constant values of A and B leads to a linear relation between modulation indices. Applying this to the first row of Equation (22), a quadratic equation in terms of m_{ds} is obtained as:

$$\frac{\left(3NA^2-B^2\right)I^*_q}{64C\omega B^2}m^2_{ds}+\frac{V_r}{2}m_{ds}=V^*_d+\frac{L_{arm}}{2}I^*_{avd}+\frac{R_{arm}}{2}I^*_d-\frac{L_{arm}}{2}\omega I^*_q+\frac{1}{8C\omega}I^*_q \tag{24}$$

Finally, m_{ds} and m_{qs} can be calculated by solving the above equation as follows:

$$\begin{cases} m_{ds}=\dfrac{-16C\omega B^2 V_r}{(3NA^2-B^2)I^*_q}\pm\dfrac{\sqrt{\frac{V_r^2}{4}-\left(\frac{(3NA^2-B^2)I^*_q}{16C\omega B^2}\right)\left(V^*_d+\frac{L_{arm}}{2}I^*_{avd}+\frac{R_{arm}}{2}I^*_d-\frac{L_{arm}}{2}\omega I^*_q+\frac{1}{8C\omega}I^*_q\right)}}{\frac{(3NA^2-B^2)I^*_q}{32C\omega B^2}}\\[3ex] m_{qs}=\left(\dfrac{\frac{L_{arm}}{4}I^*_{avd}+\frac{R_{arm}}{4}I^*_d-\frac{N}{16C\omega}I^*_q}{-\frac{L_{arm}}{4}I^*_{avq}-\frac{R_{arm}}{4}I^*_q-\frac{N}{16C\omega}I^*_d}\right)m_{ds} \end{cases} \tag{25}$$

In order to improve the proposed model performance, a circulating current control algorithm will be described in advance while the maximum available range of power transfer is also determined for the studied MMC-based STATCOM. The former eliminates the adverse effects of circulating currents on the model operation while the latter is used to achieve the best active and reactive power compensation in the MMC-based STATCOM structure.

3.1. Circulating Current Control

Although circulating currents have no effect on output quantities of the interfaced converter, different elimination algorithms are suggested in this field. These internal currents are originated from instantaneous voltage differences between each phase and the DC-link. The most prominent components of this current are DC and second harmonic. The DC part is responsible for the DC to AC power transfer while the second harmonic component increases arm currents and losses as well. In order to design an efficient controller based on second harmonic equations of the converter, KVL should be considered for each phase loop based on Figure 1a. Given that the phase sequence of transformation to the rotational frame at $\theta=2\omega t$ is a-c-b, the aforementioned KVL can be expressed as:

$$L_{arm}\frac{d}{dt}\begin{bmatrix} i_{ud2}\\ i_{uq2} \end{bmatrix}=\begin{bmatrix} -R_{arm} & 2L_{arm}\omega\\ -2L_{arm}\omega & -R_{arm} \end{bmatrix}\begin{bmatrix} i_{ud2}\\ i_{uq2} \end{bmatrix}-\begin{bmatrix} -\frac{m_d}{4}U^\Sigma_{cud}+\frac{m_q}{4}U^\Sigma_{cuq}-\frac{m_{d2}}{2}U^\Sigma_{cu0}+\frac{1}{2}U^\Sigma_{cud2}\\ -\frac{m_q}{4}U^\Sigma_{cud}-\frac{m_d}{4}U^\Sigma_{cuq}-\frac{m_{q2}}{2}U^\Sigma_{cu0}+\frac{1}{2}U^\Sigma_{cuq2} \end{bmatrix} \tag{26}$$

Second harmonic current elimination can be accomplished using corresponding second harmonic components of modulation indices. Therefore, m_{d2} and m_{q2} are directly evaluated based on Equation (26) as follows:

$$m_{d2}=\frac{2}{U^\Sigma_{cu0}}\left[L_{arm}\frac{d}{dt}i_{ud2}+R_{arm}i_{ud2}-2L_{arm}\omega i_{uq2}-\frac{m_d}{4}U^\Sigma_{cud}+\frac{m_q}{4}U^\Sigma_{cuq}+\frac{1}{2}U^\Sigma_{cud2}\right] \tag{27}$$

$$m_{q2}=\frac{2}{U^\Sigma_{cu0}}\left[L_{arm}\frac{d}{dt}i_{uq2}+R_{arm}i_{uq2}+2L_{arm}\omega i_{ud2}-\frac{m_q}{4}U^\Sigma_{cud}-\frac{m_d}{4}U^\Sigma_{cuq}+\frac{1}{2}U^\Sigma_{cuq2}\right] \tag{28}$$

The abovementioned equations can be utilized in order to design the appropriate controller. As demonstrated in Figure 2, PI controllers are included in order to eliminate the evaluated second harmonic components of modulation indices and thus minimize the second harmonic components of the currents.

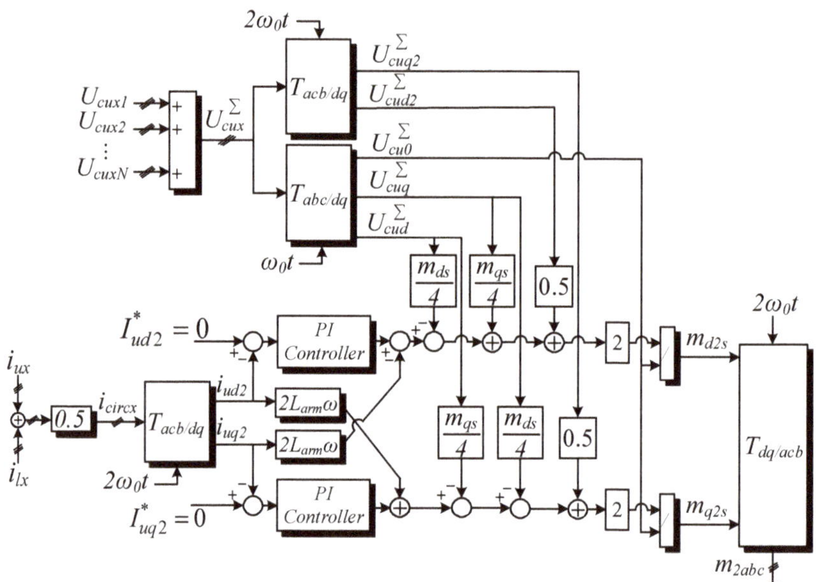

Figure 2. Circulating current control algorithm.

3.2. MMC Loss Evaluations

In this section the procedure for loss calculation of MMC is presented in order to evaluate the effect of the proposed circulating current control on semiconductor losses. In this regard, first the proper switch has been selected for this application. Then, converter losses have been calculated for the proposed MMC structure with and without circulating current control.

Figure 3 shows the block diagram for loss calculation in each IGBT/Diode switch structure. It should be noted that semiconductor losses can be divided into three parts: conduction losses, switching losses, and blocking losses. Since blocking losses are normally not considered, it can be concluded that total semiconductor losses are composed of conduction and switching losses. In this paper, the loss calculation is based on information extracted from manufacturer datasheet; more details are given in [32].

Conduction loss calculations for both IGBTs and diodes are achieved through multiplying the voltage across the semiconductor by the current passing through it as follows:

$$\Delta P_{conIGBT/D} = V_{con\,IGBT/D} \times i_{S/D}(t) \tag{29}$$

However, switching losses of IGBTs and diodes are calculated through different procedures. Considering IGBTs, switching losses are divided into turn-on and turn-off losses. These losses can be achieved based on turn-on and turn-off energy losses during turn-on and turn-off commutation times respectively. In this regard, first energy losses can be calculated as follows:

$$E_{onIGBT} = \int_{T}^{T+t_{on}} V_{CE}(t) \times i_S(t)dt \tag{30}$$

$$E_{offIGBT} = \int_{T}^{T+t_{off}} V_{CE}(t) \times i_S(t)dt \tag{31}$$

Then, the turn-on and turn-off switching losses can be achieved through dividing turn-on and turn-off energy losses by simulation time step as follows:

$$\Delta P_{onIGBT} = \frac{E_{onIGBT}}{T_s} \tag{32}$$

$$\Delta P_{offIGBT} = \frac{E_{offIGBT}}{T_s} \tag{33}$$

On the other hand, reverse recovery loss of diode is also achieved based on the reverse recovery energy losses during reverse recovery time of diode as follows:

$$E_{recD} = \int_{T}^{T+t_{rec}} V_D(t) \times i_D(t)dt \tag{34}$$

$$\Delta P_{recD} = \frac{E_{recD}}{T_s} \tag{35}$$

Finally, the sum of all these calculated loss quantities gives the total loss amount in each IGBT/Diode switch. Considering all switches together, this method can be used in order to evaluate the beneficial effects of circulating current control on MMC loss reductions.

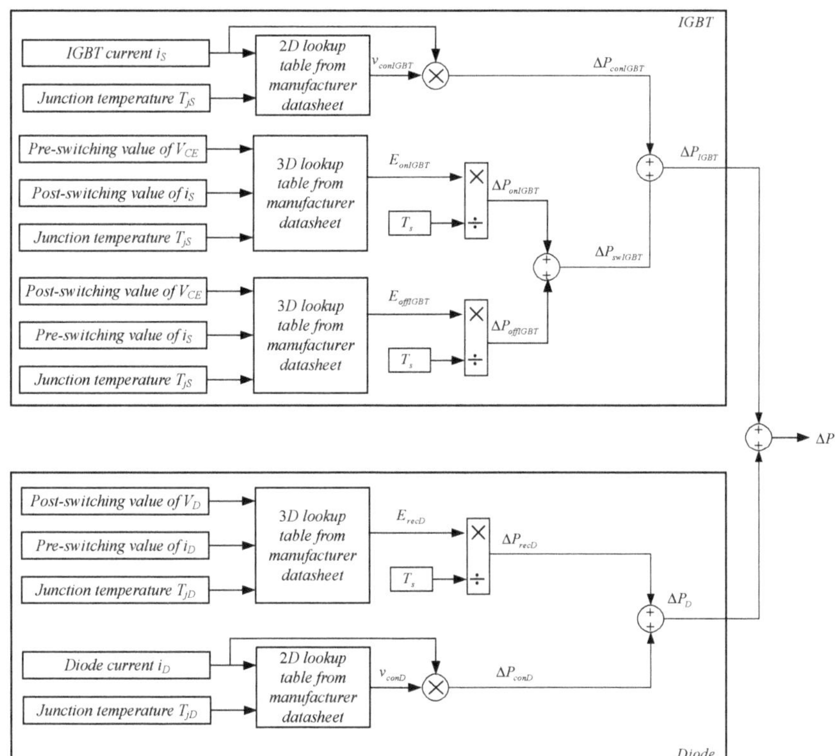

Figure 3. Loss calculation procedure for each insulated gate bipolar transistor (IGBT)/Diode switch.

3.3. Maximum Stable Operation Range

Acting as STATCOM, MMC should provide the maximum available reactive power demand in order to improve the grid-side power factor. In consequence, the maximum stable power supply range should be determined based on different converter and energy source parameters.

Input DC power inserted to MMC would be equal to output AC power if the losses are neglected. Therefore,

$$v_{dc}i_{dc} = -\left(v_d i_d + v_q i_q\right) \tag{36}$$

On the other hand, applying KVL to Figure 1a results in the following dynamics:

$$v_d = v_{gd} + R_c i_{gd} + L_c\frac{di_{gd}}{dt} - \omega L_c i_{gq} \tag{37}$$

$$v_q = v_{gq} + R_c i_{gq} + L_c\frac{di_{gq}}{dt} + \omega L_c i_{gd} \tag{38}$$

Given that $i_{gd} = i_{Ld} - i_d$ and $i_{gq} = i_{Lq} - i_q$, we can substitute Equations (37) and (38) into (36), which leads to:

$$\left(i_d + \frac{L_c(i_{avd}-i_{avLd})-R_c i_{Ld}-v_{gd}}{2R_c}\right)^2 + \left(i_q + \frac{L_c(i_{avq}-i_{avLq})-R_c i_{Lq}-v_{gq}}{2R_c}\right)^2 = \\ \left(\left(L_c(i_{avd} - i_{avLd}) - R_c i_{Ld} - v_{gd}\right)^2 + \left(L_c(i_{avq} - i_{avLq}) - R_c i_{Lq} - v_{gq}\right)^2 + 4R_c v_{dc}i_{dc}\right)/\left(4R_c^2\right) \tag{39}$$

Equation (39) has been developed based on converter power capacity equations represented in (36). Therefore, the equation represents current (and thus power) margins of the MMC in a stable operation mode based on converter and grid parameters. Note that these calculations study converter power capacity from the fundamental harmonic point of view.

On the other hand, considering the first row of Equation (26) in steady-state operating condition leads to:

$$\frac{m_d}{4}U_{cud}^{\Sigma} - \frac{m_q}{4}U_{cuq}^{\Sigma} - \frac{1}{2}U_{cud2}^{\Sigma} = 0 \tag{40}$$

which then can be rewritten using Equations (16)–(18) as follows:

$$\frac{m_d}{m_q} = -\frac{I_d^*}{I_q^*} \tag{41}$$

Replacing (23) in (41) gives:

$$\frac{I_d^*}{I_q^*} = \frac{\frac{L_{arm}}{4}I_{avq}^* + \frac{R_{arm}}{4}I_q^* + \frac{N}{16C\omega}I_d^*}{\frac{L_{arm}}{4}I_{avd}^* + \frac{R_{arm}}{4}I_d^* - \frac{N}{16C\omega}I_q^*} \tag{42}$$

which can also be written as follows:

$$\left(I_q^* + \frac{L_{arm}}{2R_{arm}}I_{avq}^*\right)^2 - \left(I_d^* + \frac{L_{arm}}{2R_{arm}}I_{avd}^*\right)^2 = \frac{L_{arm}^2}{4R_{arm}^2}\left(I_{avq}^{*2} - I_{avd}^{*2}\right) \tag{43}$$

Equation (43) describes a hyperbolic while (39) is the equation of a circle with the centre of

$$\left(-\frac{L_c(i_{avd} - i_{avLd}) - R_c i_{Ld} - v_{gd}}{2R_c}, -\frac{L_c(i_{avq} - i_{avLq}) - R_c i_{Lq} - v_{gq}}{2R_c}\right)$$

and the radius measure of

$$\sqrt{\left(\left(L_c(i_{avd} - i_{avLd}) - R_c i_{Ld} - v_{gd}\right)^2 + \left(L_c(i_{avq} - i_{avLq}) - R_c i_{Lq} - v_{gq}\right)^2 + 4R_c v_{dc}i_{dc}\right)/\left(4R_c^2\right)}$$

Drawing them together, the current and thus power flow can be estimated in order to guarantee MMC stability as STATCOM in the proposed model. Therefore, the stable operation range of the converter is demonstrated as the dashed area in Figure 4. It is clear that the stable operation can be obtained for current amounts which are inside the specified criteria since the margins are obtained based on converter power capacity.

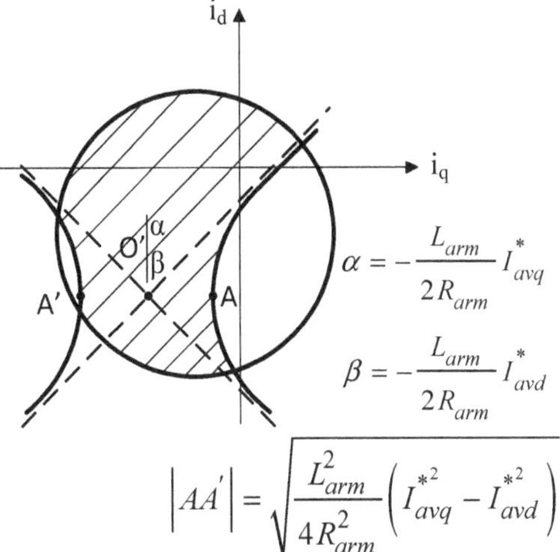

Figure 4. Stable operation region (hatched) for MMCs in STATCOM operating mode.

4. Results and Discussion

The detailed configuration of the MMC-based STATCOM and proposed control method shown in Figure 5 has been simulated in MATLAB/Simulink. The converter structure, as well as the proposed control method, have been simulated using various blocks of the Simulink library. The main blocks used in the simulation of the grid-connected MMC structure and the measurements have been derived from the following section of the MATLAB/Simulink library:

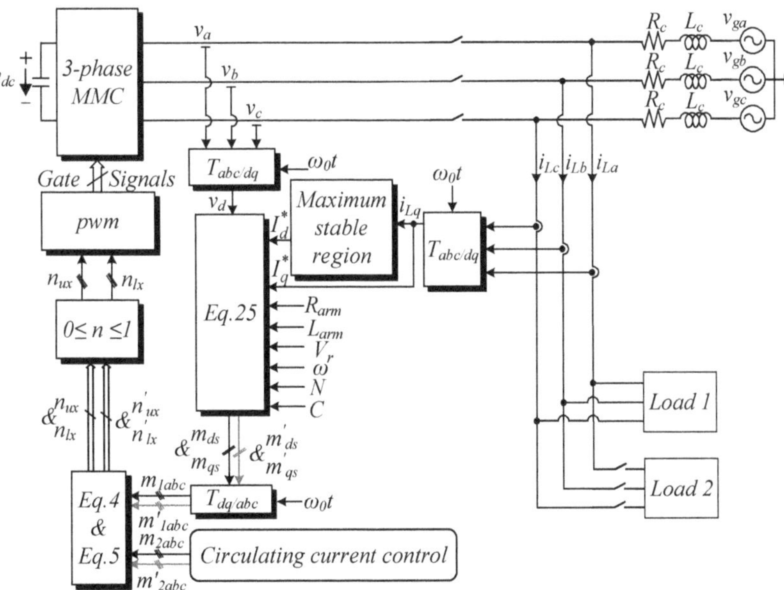

Figure 5. General configuration of the proposed MMC-based STATCOM controller.

"Simscape/ Power Systems/ Specialized Technology/ Fundamental Blocks".

Besides, the values of different circuit components and operational conditions applied to simulations are listed in Table 1.

Table 1. MMC parameters and operational conditions applied to simulations.

Items	Values
AC system voltage V_g^* (L-L,rms)	10 KV
AC system inductance L_c	0.5 mH
AC system resistance R_c	1 mΩ
DC bus voltage V_r	20 KV
number of SMs per arm N	20
SM capacitance C	10,000 μF
arm inductance L_{arm}	40 mH
arm equivalent resistance R_{arm}	1.5 Ω
SM capacitor voltage V_{csm}	1 KV
carrier frequency f_{sw}	5 KHz
simulation time step T_s	83.3×10^{-6} s
Load1 active power P_{ref1}	0.5 MW
Load1 reactive power Q_{ref1}	4 MVar
Load2 active power P_{ref2}	3.5 MW
Load2 reactive power Q_{ref2}	0.8 MVar

In this paper, the MMC has been considered as a STATCOM interface between DC-link and the grid. Considering the maximum stable operation range of the converter, MMC is supposed to provide all reactive power demands as well as maximum available active power demands of the load within the feasible power range. Therefore, MMC will be able to efficiently provide the power factor correction behavior of a STATCOM interface. Moreover, a circulating current control algorithm is also considered in the proposed control method. The proposed controller is expected to improve MMC performance by eliminating the second harmonic components of currents, which also leads to a reduction in the capacitor voltage fluctuations.

In order to validate the efficient performance of the proposed control and switching method, four different simulation time intervals have been discussed separately. At the first time interval of simulation, the MMC has been disconnected, leaving load demands to the network. As can be seen in Figure 6a–c, the load current is initially provided by the grid, while the disconnected converter structure delivers maximum voltage during this no-load period (demonstrated in Figure 6d). At the second time interval, the energy source is connected to the grid through PCC at 0.2 s. Figure 6c shows that the grid current reaches to the minimum value as the load current is supplied by energy source through the efficient MMC connection (see Figure 6a,b respectively). This feature has a noticeable impact on voltage stress reduction of the grid, especially in peak demand times, since MMC can help the grid in providing load demands.

During the third time interval, active and reactive power demands of the load are increased due to the addition of the second load at 0.4 s. As can be seen in Figure 6, MMC supplies maximum available load based on the stability margin and the rest is accomplished by the grid. The most important studied capability of the proposed model at this time period is the priority of reactive power compensation, which will be discussed further in detail.

It should be noted that the proposed controller can provide stable operation during all these dynamic changes. Thus, the output voltages and currents are stable as shown in Figure 6d,e respectively.

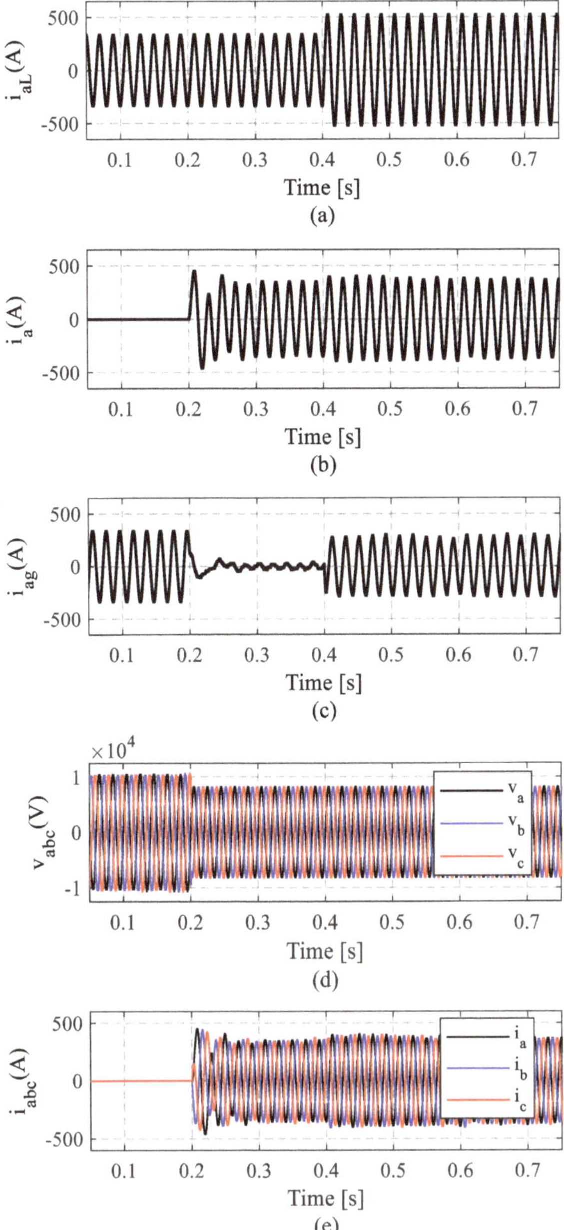

Figure 6. Simulated waveforms of the MMC-based STATCOM structure: (**a**) Phase-a waveform of load current; (**b**) Phase-a waveform of converter output current; (**c**) Phase-a waveform of grid current; (**d**) Output voltages of MMC; and (**e**) Output currents of MMC.

In order to verify MMC capabilities as STATCOM, reactive power compensation can be accurately discussed based on a more detailed analysis of current components. As demonstrated in Figure 5, the maximum available power capacity of MMC based on converter and grid parameters have been considered in the simulation process as follows:

Considering the maximum available power capacity of the converter, the MMC supplies all reactive power demand of the load within the feasible power range. This power factor correction behavior of the converter is considered as a priority in the proposed control method of this MMC-based STATCOM structure. Therefore, with the assumption of converter power capacity being way more than load reactive power demand, it is considered that $I_q^* = i_{lq}$ (which means that load reactive power demand is completely supplied by the MMC-based STATCOM structure). On the other hand, the excess amount of power remaining from total power capacity of the converter is utilized to provide the active power demand of the load. This way the grid provides less active power for the load and the load stress imposed on power grid can be reduced. In this regard, active power reference of the converter is modified based on maximum stable operation range of the converter. The desired range is specified by (39) and (43) in which the converter and the grid parameters, as well as the reactive current components, have been modified.

Figure 7 illustrates reactive current components of the simulated system in dq0 frame. As mentioned before, load reactive power demands (even after load increment at 0.4 s) are considered to be less than the converter's maximum power transfer capability. At the same time, providing reactive power demands of the load is considered as the first priority of the proposed MMC-based STATCOM model. Therefore, it can be seen that after MMC connection at 0.2 s all reactive current demands of the load are compensated through the energy source (see Figure 7a,b respectively). This yields to an almost zero reactive current flow in the grid side after the MMC connection, which is clearly shown in Figure 7.

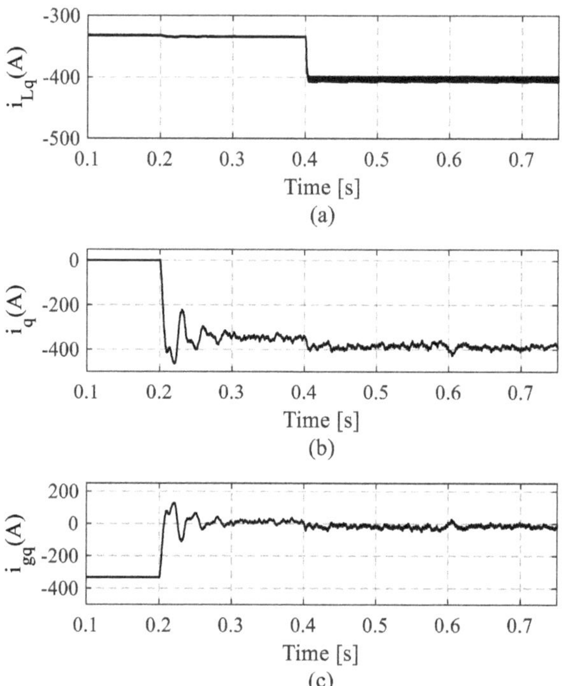

Figure 7. Reactive current components of the STATCOM model: (**a**) q component of load current; (**b**) q component of MMC current; and (**c**) q component of grid current.

On the other hand, active current components of the simulated system in dq0 frame are demonstrated in Figure 8. As it can be seen, grid current supplies active power demands of the load before 0.2 s as the energy source is not connected yet. During the time interval of 0.2 < t < 0.4 s the energy source provides maximum established active power, keeping the grid current at the lowest level. As the load demands are increased after 0.4 s, converter current waveform remains almost constant, which can be seen in Figure 8a,b respectively. This is due to the active power margin specification of the MMC, which depends on the maximum available power left after providing all load reactive power demands. Therefore, the grid current increment after 0.4 s is justifiable due to the abovementioned explanations and it is less worrying since the grid power factor is not decayed.

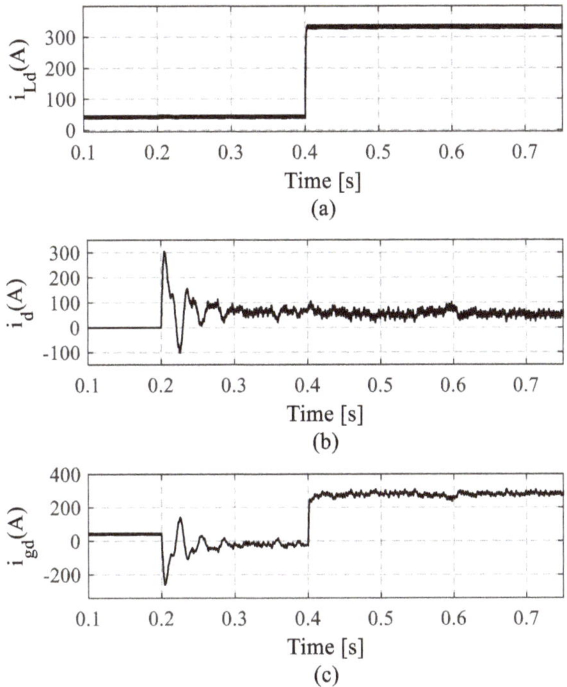

Figure 8. Active current components of the STATCOM model: (**a**) d component of load current; (**b**) d component of MMC current; and (**c**) d component of grid current.

The power factor correction capability of the proposed model can be seen in Figure 9. An obvious phase difference between the grid voltages and currents can be detected before 0.2 s when the MMC-based STATCOM structure is not connected yet and the load active and reactive power demands are completely provided by the grid. During the time interval of 0.2 < t < 0.4 s, MMC provides all active and reactive power demands of the load as it can be completely covered within the converter stable operation range. Consequently, the grid current is kept at its lowest amount during this time interval. Following that, with an increase in load active and reactive power demands after 0.4 s, the MMC provides all reactive as well as the most available active power demands of the load within the stable operation range. Therefore, providing the rest of load active power demands will be the only responsibility of the grid at this time. It is obvious that the reactive power compensating method is working properly since the phase differences between the grid voltages and currents are reduced effectively.

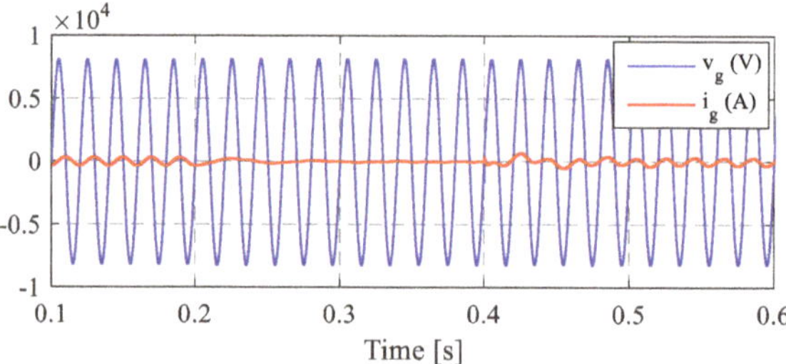

Figure 9. Grid voltage and current of phase-a before and after reactive power compensation.

As another contribution of this paper, circulating current control was applied during the 4th interval of the simulation time. As can be seen in Figure 10a, applying the proposed circulating current control after 0.6 sec can lead to a 90% reduction in the second harmonic component of arm current. This is while most recent studies of the same field have been unable to provide such mitigations in harmonic components of arm currents. For example, reference [33] proposes a unified on-line calculation scheme by using the instantaneous information of MMC in order to achieve different circulating current injection targets under different conditions. In this regard, four circulating current control targets have been considered. The comparisons between the obtained circulating current waves and the uncontrolled one show that the largest decrease in the circulating current value can be achieved under the first scenario and with the amount of 50%. As another example, reference [34] employs a feedback linearization-based current control strategy in order to regulate the output and inner differential currents of the MMC. Furthermore, the conventional cascaded controller is also considered where the differential current is only regulated by PI controller. A comparison of the obtained results shows that the harmonics in the differential current are reduced from 25% to 3.22% with the aid of the feedback linearization method. This is while the proposed circulating current control in this paper leads to a final 0.5% circulating current remaining in the total arm current, and therefore, shows an effective improvement in circulating current mitigation over the previous methods represented in various recent studies. It should also be noted that the additional second harmonic component of modulation indices generated by the controller can improve arm current adjustments as well as reduce the capacitor voltage fluctuations. The abovementioned results are demonstrated in Figure 10b,c respectively. It should be noted that circulating currents are considered as interior quantities. Therefore, the applied control algorithm has no direct effect on converter outer dynamic performances which is clearly reflected in Figure 6b,d,e.

In order to evaluate the beneficial effects of the proposed circulating current control on MMC loss reductions, several studies have been accomplished. In this regard, MMC loss calculations have been performed considering a fixed load active power demand of 0.5 MW as well as various load reactive power demands. Figure 11 demonstrates the MMC losses, with and without circulating current control for five different values of reactive power. As it can be observed from the figure, the losses are decreased by more than 10% (and up to 16% in the first two reactive power amounts) due to mitigation of circulating current. This is while most recent studies in this field are suggesting less than 10% reductions in converter losses, using various second harmonic current control approaches [35,36]. Therefore, MMC loss evaluations provided in Figure 11 confirm that the proposed method can effectively improve converter operation in power loss terms, as well.

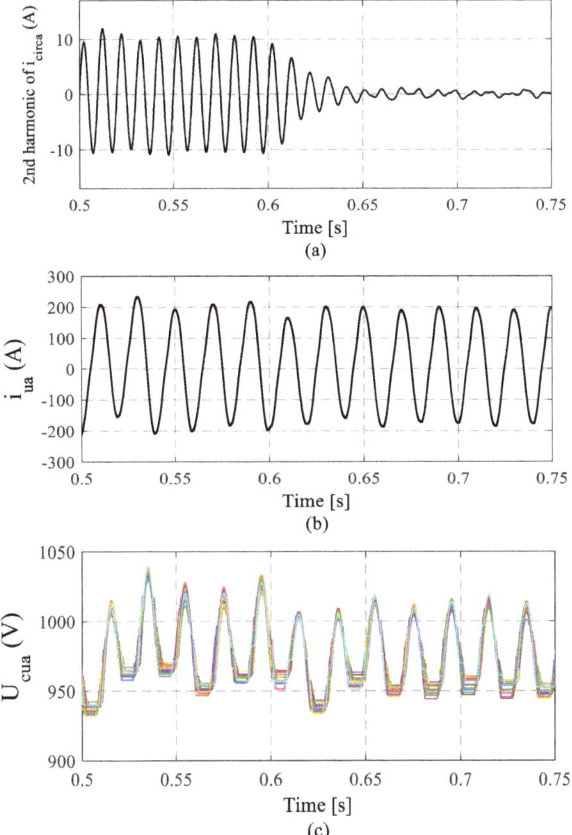

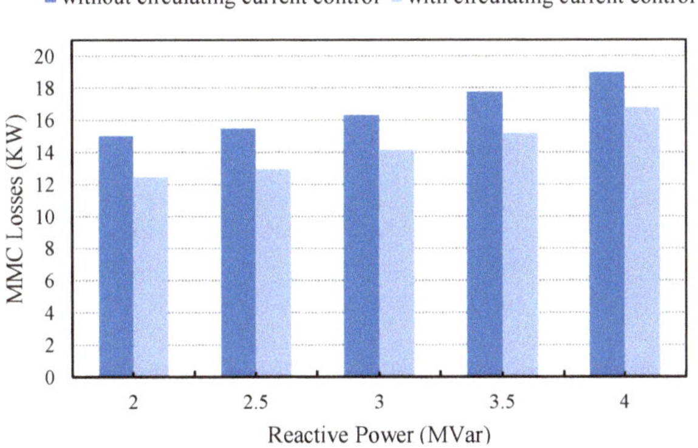

Figure 10. Simulated waveforms of the MMC using circulating current control: (**a**) Second harmonic component of the circulating current; (**b**) Upper arm current of phase-a; and (**c**) Upper arm SM capacitor voltages of phase-a.

Figure 11. MMC losses with and without circulating current control.

5. Conclusions

This paper presented a dynamic model of MMC in STATCOM operating mode. Detailed harmonic evaluations were performed in dq0 frame in order to provide an efficient control method. Based on steady-state equations, modulation indices have been obtained so that the stable connection between the energy source and the power grid was securely established. Circulating current control was provided through second harmonic component additions to the overall modulation process. As a novelty of this paper, the abovementioned method helped improve capacitor voltage regulations, as well as arm current harmonic characteristics and losses. Also, the maximum stability margin of the converter was developed based on dynamic studies, which is considered as another contribution of this paper. Reactive power compensation is considered as a priority in the proposed MMC-based STATCOM performance. Thus, controller design was focused on the maximum reactive power supply, leaving an upper limit for the active power coverage ability of the converter based on the modified maximum stable operation range. It has been shown that MMC can efficiently provide reactive power demands of the load, resulting in grid power factor correction. It should be noted that since the improvement of the grid imbalance studies is not considered as the purpose of MMC utilization in this configuration, the proposed control method becomes vulnerable if grid voltage imbalance situations occur. The main reason of this, which can be addressed in future work, is that the appeared negative sequence components in grid imbalance studies are not considered in the proposed control method. Despite this limitation, advantageous MMC superiorities lead to various applications where the proposed MMC-based STATCOM structure can be used to renovate conventional power networks.

Author Contributions: Formal analysis, E.A.; Investigation, K.R.; Methodology, J.A.; Supervision, E.P.; Visualization, J.P.S.C.; Writing—original draft, F.S.; Writing—review & editing, E.A. and E.M.G.R.

Funding: This research received no external funding.

Conflicts of Interest: The authors declare no conflict of interest.

Nomenclature

Indices

x	a, b, c
u, l	upper and lower arm

Variables

i_{ux}, i_{lx}	upper and lower arm currents in abc frame
i_{circx}	circulating current in abc frame
i_x	converter output current in abc frame
i_{Lx}	load current in abc frame
U_{cux}, U_{clx}	upper and lower arm desired voltages in abc frame
n_{ux}, n_{lx}	upper and lower arm modulation indices in abc frame
$U^{\Sigma}_{cux}, U^{\Sigma}_{clx}$	upper and lower arm sum capacitor voltages in abc frame
m_{1x}	fundamental frequency component of modulation indices in abc frame
m_{2x}	second harmonic component of modulation indices in dq0 frame
m_d, m_q	fundamental frequency component of modulation indices in dq0 frame
m_{d2}, m_{q2}	second harmonic component of modulation indices in dq0 frame
$U^{\Sigma}_{cum0}, U^{\Sigma}_{clm0}$	DC components of upper and lower arm sum capacitor voltages in abc frame
$U^{\Sigma}_{cum1}, U^{\Sigma}_{clm1}$	fundamental frequency components of upper and lower arm sum capacitor voltages in abc frame
$U^{\Sigma}_{cum2}, U^{\Sigma}_{clm2}$	second harmonic components of upper and lower arm sum capacitor voltages in abc frame
ω	angular frequency
ω_0	reference angular frequency
t	time
θ_u, θ_l	phase angle of the fundamental frequency components of upper and lower arm sum capacitor voltages in abc frame

Variables

θ_{u2},θ_{l2}	phase angle of the second harmonic frequency components of upper and lower arm sum capacitor voltages in abc frame
$U^{\Sigma}_{cu0},U^{\Sigma}_{cl0}$	zero component of upper and lower arm sum capacitor voltages in dq0 frame
$U^{\Sigma}_{cud},U^{\Sigma}_{cuq}$	fundamental frequency components of upper arm sum capacitor voltage in dq0 frame
$U^{\Sigma}_{cld},U^{\Sigma}_{clq}$	fundamental frequency components of lower arm sum capacitor voltage in dq0 frame
$U^{\Sigma}_{cud2},U^{\Sigma}_{cuq2}$	second harmonic frequency components of upper arm sum capacitor voltage in dq0 frame
$U^{\Sigma}_{cld2},U^{\Sigma}_{clq2}$	second harmonic frequency components of lower arm sum capacitor voltage in dq0 frame
$W^{\Sigma}_{cu,l}$	total energy stored in SM capacitors of each upper and lower arm
i_{u0}	zero component of upper arm current in dq0 frame
i_{ud},i_{uq}	fundamental frequency components of upper arm current in dq0 frame
i_{ud2},i_{uq2}	second harmonic components of upper arm current in dq0 frame
v_{dc}	DC-link voltage
V_r	reference value of DC-link voltage
i_{dc}	DC-link current
v_x	MMC output voltage at PCC in abc frame
v_{gx}	phase voltage of the grid in abc frame
v_d,v_q	fundamental frequency components of MMC output voltage at PCC in dq0 frame
v_{gd},v_{qg}	fundamental frequency components of grid voltage at PCC in dq0 frame
i_d,i_q	fundamental frequency components of MMC output current in dq0 frame
i_{gd},i_{gq}	fundamental frequency components of grid current in dq0 frame
i_{Ld},i_{Lq}	fundamental frequency components of load current in dq0 frame
i_{avd},i_{avq}	fundamental frequency components of average MMC output current in dq0 frame
i_{avLd},i_{avLq}	fundamental frequency components of average load current in dq0 frame
m_{ds},m_{qs}	steady-state fundamental frequency components of modulation indices in dq0 frame
m_{d2s},m_{q2s}	steady-state second harmonic frequency components of modulation indices in dq0 frame
$I^{*}_{avud},I^{*}_{avuq}$	reference values of average upper arm currents in dq0 frame
I^{*}_{avd},I^{*}_{avq}	reference values of average MMC output currents in dq0 frame
I^{*}_{ud},I^{*}_{uq}	reference values of instantaneous upper arm currents in dq0 frame
I^{*}_{d},I^{*}_{q}	reference values of MMC output currents in dq0 frame
I^{*}_{0}	reference value of upper arm zero component current in dq0 frame
V^{*}_{d},V^{*}_{q}	reference value of MMC output voltages in dq0 frame
ΔP	total losses in each IGBT/diode switch
$\Delta P_{conIGBT/D}$	conduction loss for IGBT/diode
$V_{con\,IGBT/D}$	voltage across IGBT/diode during conduction
$i_{S/D}$	current passing through IGBT/diode during conduction
ΔP_{IGBT}	total losses of IGBT
ΔP_{swIGBT}	switching losses in IGBTs
ΔP_{onIGBT}	turn-on switching losses of IGBTs
$\Delta P_{offIGBT}$	turn-off switching losses of IGBTs
E_{onIGBT}	turn-on energy losses of IGBTs
$E_{offIGBT}$	turn-off energy losses of IGBTs
t_{on}	turn-on commutation time
t_{off}	turn-off commutation time
V_{CE}	voltage across IGBT
V_D	voltage across diode
$T_{jS/D}$	junction temperature in IGBT and diode
T_s	simulation time step
ΔP_D	total diode losses
ΔP_{recD}	reverse recovery loss of diode
E_{recD}	reverse recovery energy loss of diode
t_{rec}	reverse recovery time of diode

Parameters

C	SM capacitance
N	total number of SMs per arm
R_{arm}	arm equivalent resistance
L_{arm}	arm inductance
R_c	ac system resistance
L_c	ac system inductance

Abbreviation

MMC	modular multilevel converter
STATCOM	static synchronous compensator
DG	distributed generation
HVDC	high voltage direct current
THD	total harmonic distortion
PF	power factor
FACTS	flexible AC transmission system
PI	proportional integral
SM	sub-modules
PCC	point of common coupling
KVL	Kirchhoff's voltage law
IGBT	insulated gate bipolar transistor

References

1. Ochoa, L.F.; Harrison, G.P. Minimizing energy losses: Optimal accommodation and smart operation of renewable distributed generation. *IEEE Trans. Power Syst.* **2010**, *26*, 198–205. [CrossRef]
2. Shaaban, M.F.; Atwa, Y.M.; El-Saadany, E.F. DG allocation for benefit maximization in distribution networks. *IEEE Trans. Power Syst.* **2012**, *28*, 639–649. [CrossRef]
3. Debnath, S.; Qin, J.; Bahrani, B.; Saeedifard, M.; Barbosa, P. Operation, control, and applications of the modular multilevel converter: A review. *IEEE Trans. Power Electron.* **2014**, *30*, 37–53. [CrossRef]
4. Sahoo, A.K.; Otero-De-Leon, R.; Mohan, N. Review of modular multilevel converters for teaching a graduate-level course of power electronics in power systems. In Proceedings of the 2013 North American Power Symposium (NAPS), Manhattan, KS, USA, 22–24 September 2013; pp. 1–6.
5. Nami, A.; Liang, J.; Dijkhuizen, F.; Demetriades, G.D. Modular multilevel converters for HVDC applications: Review on converter cells and functionalities. *IEEE Trans. Power Electron.* **2014**, *30*, 18–36. [CrossRef]
6. Franquelo, L.G.; Rodriguez, J.; Leon, J.I.; Kouro, S.; Portillo, R.; Prats, M.A. The age of multilevel converters arrives. *IEEE Ind. Electron. Mag.* **2008**, *2*, 28–39. [CrossRef]
7. Kouro, S.; Malinowski, M.; Gopakumar, K.; Pou, J.; Franquelo, L.G.; Wu, B.; Rodriguez, J.; Pérez, M.A.; Leon, J.I. Recent advances and industrial applications of multilevel converters. *IEEE Trans. Ind. Electron.* **2010**, *57*, 2553–2580. [CrossRef]
8. Lawan, A.U.; Abbas, H.; Khor, J.G.; Karim, A.A. Dynamic performance improvement of MMC inverter with STATCOM capability interfacing PMSG wind turbines with grid. In Proceedings of the 2015 IEEE Conference on Energy Conversion (CENCON), Johor Bahru, Malaysia, 19–20 October 2015; pp. 492–497.
9. Lesnicar, A.; Marquardt, R. An innovative modular multilevel converter topology suitable for a wide power range. In Proceedings of the 2003 IEEE Bologna Power Tech Conference Proceedings, Bologna, Italy, 23–26 June 2003.
10. Liu, Z.; Hu, X.; Liao, Y. Vehicle-Grid System Stability Analysis Based on Norm Criterion and Suppression of Low-Frequency Oscillation With MMC-STATCOM. *IEEE Trans. Transp. Electr.* **2018**, *4*, 757–766. [CrossRef]
11. Farias, J.V.M.; Cupertino, A.F.; Pereira, H.A.; Junior, S.I.S.; Teodorescu, R. On the redundancy strategies of modular multilevel converters. *IEEE Trans. Power Deliv.* **2017**, *33*, 851–860. [CrossRef]
12. Yue, Y.; Ma, F.; Luo, A.; Xu, Q.; Xie, L. A circulating current suppressing method of MMC based STATCOM for negative-sequence compensation. In Proceedings of the 2016 IEEE 8th International Power Electronics and Motion Control Conference (IPEMC-ECCE Asia), Hefei, China, 22–26 May 2016.
13. Muñoz, J.A.; Espinoza, J.R.; Baier, C.R.; Morán, L.A.; Guzmán, J.I.; Cárdenas, V.M. Decoupled and modular harmonic compensation for multilevel STATCOMs. *IEEE Trans. Ind. Electr.* **2013**, *61*, 2743–2753. [CrossRef]

14. Zhu, J.; Li, L.; Pan, M. Study of a novel STATCOM based on modular multilevel inverter. In Proceedings of the IECON 2012-38th Annual Conference on IEEE Industrial Electronics Society, Montreal, QC, Canada, 25–28 October 2012; pp. 1428–1432.

15. Zhang, W.; Gao, Q.; Su, B.; Jin, M.; Xu, D.; Liu, J. Research on the control strategy of STATCOM based on modular multilevel converter. In Proceedings of the 2014 International Power Electronics Conference (IPEC-Hiroshima 2014-ECCE ASIA), Hiroshima, Japan, 18–21 May 2014.

16. Vivas, J.H.; Bergna, G.; Boyra, M. Comparison of multilevel converter-based STATCOMs. In Proceedings of the 2011 14th European Conference on Power Electronics and Applications, Birmingham, UK, 30 August–1 September 2011.

17. Nwobu, C.J.; Efika, I.B.; Oghorada, O.J.K.; Zhang, L. A modular multilevel flying capacitor converter-based STATCOM for reactive power control in distribution systems. In Proceedings of the 2015 17th European Conference on Power Electronics and Applications (EPE'15 ECCE-Europe), Geneva, Switzerland, 8–10 September 2015.

18. Liu, X.; Lv, J.; Gao, C.; Chen, Z.; Chen, S. A novel STATCOM based on diode-clamped modular multilevel converters. *IEEE Trans. Power Electr.* **2016**, *32*, 5964–5977. [CrossRef]

19. Yang, X.; Li, J.; Fan, W.; Wang, X.; Zheng, T.Q. Research on modular multilevel converter based STATCOM. In Proceedings of the 2011 6th IEEE Conference on Industrial Electronics and Applications, Beijing, China, 21–23 June 2011; pp. 2569–2574.

20. Pereira, H.A.; Haddioui, M.R.; De Oliveira, L.O.; Mathe, L.; Bongiorno, M.; Teodorescu, R. Circulating current suppression strategies for D-STATCOM based on modular Multilevel Converters. In Proceedings of the 2015 IEEE 13th Brazilian Power Electronics Conference and 1st Southern Power Electronics Conference (COBEP/SPEC), Fortaleza, Brazil, 29 November–2 December 2015; pp. 1–6.

21. Nieves, M.; Maza, J.; Mauricio, J.; Teodorescu, R.; Bongiorno, M.; Rodriguez, P. Enhanced control strategy for MMC-based STATCOM for unbalanced load compensation. In Proceedings of the 2014 16th European Conference on Power Electronics and Applications, Lappeenranta, Finland, 26–28 August 2014; pp. 1–10.

22. António-Ferreira, A.; Gomis-Bellmunt, O.; Teixidó, M. HVDC-based modular multilevel converter in the statcom operation mode. In Proceedings of the 2016 18th European Conference on Power Electronics and Applications (EPE'16 ECCE Europe), Karlsruhe, Germany, 5–9 September 2016; pp. 1–10.

23. Liu, L.; Li, H.; Xue, Y.; Liu, W. Decoupled active and reactive power control for large-scale grid-connected photovoltaic systems using cascaded modular multilevel converters. *IEEE Trans. Power Electr.* **2014**, *30*, 176–187.

24. Mehrasa, M.; Pouresmaeil, E.; Zabihi, S.; Catalao, J.P.S. Dynamic model, control and stability analysis of MMC in HVDC transmission systems. *IEEE Trans. Power Deliv.* **2016**, *32*, 1471–1482. [CrossRef]

25. Mehrasa, M.; Pouresmaeil, E.; Zabihi, S.; Vechiu, I.; Catalão, J.P.S. A multi-loop control technique for the stable operation of modular multilevel converters in HVDC transmission systems. *Int. J. Electr. Power Energy Syst.* **2018**, *96*, 194–207. [CrossRef]

26. Xu, C.; Dai, K.; Kang, Y.; Liu, C. Characteristic analysis and experimental verification of a novel capacitor voltage control strategy for three-phase MMC-DSTATCOM. In Proceedings of the 2015 IEEE Applied Power Electronics Conference and Exposition (APEC), Charlotte, NC, USA, 15–19 March 2015; pp. 1528–1533.

27. Cupertino, A.F.; Farias, J.V.M.; Pereira, H.A.; Seleme, S.I.; Teodorescu, R. Comparison of dscc and sdbc modular multilevel converters for statcom application during negative sequence compensation. *IEEE Trans. Ind. Electr.* **2019**, *66*, 2302–2312. [CrossRef]

28. Luo, F.; Wang, J.; Li, Z.; Duan, Q.; Lv, Z.; Ji, Z.; Gu, W.; Wu, Z. A circuit-oriented average-value model of modular multilevel converter based on sub-module modelling method. In Proceedings of the 2017 12th IEEE Conference on Industrial Electronics and Applications (ICIEA), Siem Reap, Cambodia, 18–20 June 2017; pp. 7–12.

29. Mehrasa, M.; Pouresmaeil, E.; Taheri, S.; Vechiu, I.; Catalão, J.P.S. Novel control strategy for modular multilevel converters based on differential flatness theory. *IEEE J. Emerg. Sel. Top. Power Electr.* **2018**, *6*, 888–897. [CrossRef]

30. Sun, Y.; Teixeira, C.A.; Holmes, D.G.; McGrath, B.P.; Zhao, J. Low-order circulating current suppression of PWM-based modular multilevel converters using DC-link voltage compensation. *IEEE Trans. Power Electr.* **2017**, *33*, 210–225. [CrossRef]

31. Yang, S.; Wang, P.; Tang, Y.; Zagrodnik, M.; Hu, X.; Tseng, K.J. Circulating current suppression in modular multilevel converters with even-harmonic repetitive control. *IEEE Trans. Ind. Appl.* **2017**, *54*, 298–309. [CrossRef]

32. Adabi, M.E.; Martinez-Velasco, J.A. MMC-based solid-state transformer model including semiconductor losses. *Electr. Eng.* **2018**, *100*, 1613–1630. [CrossRef]

33. Wang, J.; Han, X.; Ma, H.; Bai, Z. Analysis and injection control of circulating current for modular multilevel converters. *IEEE Trans. Ind. Electr.* **2019**, *66*, 2280–2290. [CrossRef]

34. Yang, S.; Wang, P.; Tang, Y. Feedback linearization-based current control strategy for modular multilevel converters. *IEEE Trans. Power Electr.* **2018**, *33*, 161–174. [CrossRef]

35. Xiang, M.; Hu, J.; He, Z. Loss reduction analysis and control of AC voltage-boosted FBSM MMC by injecting second harmonic circulating current. *J. Eng.* **2018**, *2019*, 2328–2331. [CrossRef]

36. Yang, L.; Li, Y.; Li, Z.; Wang, P.; Xu, S.; Gou, R. Loss optimization of MMC by second-order harmonic circulating current injection. *IEEE Trans. Power Electr.* **2018**, *33*, 5739–5753. [CrossRef]

 *electronics*

Article

An Efficient H7 Single-Phase Photovoltaic Grid Connected Inverter for CMC Conceptualization and Mitigation Method

Mehrdad Mahmoudian [1,*], **Eduardo M. G. Rodrigues** [2,*] **and Edris Pouresmaeil** [3]

1   Firouzabad Institute of Higher Education, Firouzabad 74717, Iran
2   Management and Production Technologies of Northern Aveiro—ESAN, Estrada do Cercal 449, Santiago de Riba-Ul, 3720-509 Oliveira de Azeméis, Portugal
3   Department of Electrical Engineering and Automation, Aalto University, 02150 Espoo, Finland; edris.pouresmaeil@aalto.fi
*   Correspondence: mahmoodian.cc@fabad-ihe.ac.ir (M.M.); emgrodrigues@ua.pt (E.M.G.R.)

Received: 31 July 2020; Accepted: 31 August 2020; Published: 3 September 2020

Abstract: Transformerless inverters are the economic choice as power interfaces between photovoltaic (PV) renewable sources and the power grid. Without galvanic isolation and adequate power convert design, single-phase grid connected inverters may have limited performance due to the presence of a significant common mode ground current by creating safety issues and enhancing the negative impact of harmonics in the grid current. This paper proposes an extended H6 transformerless inverter that uses an additional power switch (H7) to improve common mode leakage current mitigation in a single-phase utility grid. The switch with a diode in series connection aims to make an effective clamp of common mode voltage at the DC link midpoint. The principles of operation of the proposed structure with bipolar sinusoidal pulse width modulation (SPWM) is presented and formulated. Laboratory tests' performance is detailed and evaluated in comparison with well-known single-phase transformer-less topologies in terms of power conversion efficiency, total harmonic distortion (THD) level, and circuit components number. The studied topology performance evaluation is completed with the inclusion of reactive power compensation functionality verified by a low-power laboratory implementation with 98.02% efficiency and 30.3 mA for the leakage current.

Keywords: common mode inverters; photovoltaic; leakage current elimination; pulse width modulation

1. Introduction

The energy demand for industrial, commercial, and residential consumers is a fact in the 21st century. One way to accomplish this demand without harming the environment has been seen through the incorporation of distributed generation (DG) systems in distribution networks. Large and as well as small-scale photovoltaic (PV) arrays for home applications have become popular rather than other DG sources due to the improvement in manufacturing techniques and significant advances in power electronic interfaces. Electrical energy produced by a PV installation is basically a DC system. Therefore, it must be converted into AC power to make it available in power utility through a grid-tie inverter [1–3]. Transformerless voltage source inverters (VSIs) are the standard choice in detriment of current source inverters (CSIs), because they show better conversion efficiency, smaller size, and lower manufacturing costs for the same power rating. Transformerless inverters generally have two classes of connection such as galvanic or non-galvanic categories. When a high-frequency low-size transformer on the DC side or low-frequency large-size transformer on the AC side are used, the electric connection is removed, which means electrical isolation between the two electrical systems [4]. This electrical separation improves operation security and reliability. However, it deteriorates power conversation efficiency. If the transformer is removed (following protection guidelines [5,6]), renewable power generation investment cost is considerably lower. Many studies have been conducted so far on this issue to explain how common mode current (CMC) could be surpassed [7]. CMC circulation is only possible, since there is no galvanic isolation which, combined with the stray capacitance that appears between PV installation and the DC side ground, allows a path for ground current to be injected into the neutral point at the AC side. The stray capacitance level may depend on:

- Class of PV cells;
- Climate type and residential localization;
- The height between PVs and ground on the DC side;
- Voltage level; and
- Electromagnetic Interference (EMI).

To guarantee grid reliability, modern grid-connected PV standards propose leakage current protection device actuation according to CMC amplitude. This means that CMC values below 300 mA are permanently acceptable, while above this point, the protection must react and trigger a break by isolating the transformerless inverter from the utility grid. Furthermore, this current must not continue over 0.3 s during the operation scheduling period. If the CMC increases, the maximum permissible time to trip will be decreased, correspondingly. Table 1 depicts the recommended tripping time as a function of CMC amplitude.

Table 1. Maximum common mode current (CMC) [7].

RMS Value	Automatic Disconnection Time
CMI > 300 mA	0.3 s
CMI > 450 mA	0.15 s
CMI > 800 mA	0.04 s

To eliminate the leakage current presence, several topologies have been studied and analyzed in the literature, which are shown in Figure 1. The Karschny structure depicted in Figure 1a [8] is based on an asymmetrical output inductor. The topology works at low voltage on the DC side being configured to be operated in two different modes. In one half cycle, its operation is very similar to a buck converter. On the other hand, when S_1 and S_5 are switching in a complementarily way, the topology operates as a buck-boost converter. This allows generating a semi-pure sinusoidal current. In practical terms, it is not easy to implement because it requires a complex control technique. In addition, the three switches that contribute to the current path lead to an increase in the active power loss. Other forms of proposed topologies are based on dual-buck topologies [9,10], which can be seen in Figure 1b,c. These structures are basically two buck converters that can be connected in series or in parallel being operated complementarily to perform DC decoupling of the current path. Switches S_{L1} and S_{L2} are commuted at a power grid frequency in order to prevent the reversal of the inductor current [10]. The half-bridge-based transformerless inverters as shown in Figure 1d may limit the CMC presence, but at the cost of operating restrictions such as the requirement of the DC input voltage being at least twice that of the utility grid voltage. In addition, from a design point of view, power switch components are subjected to higher voltage stress and require a larger output filter size [11]. In Figure 1e, the neutral point clamped (NPC) VSIs have been recently applied in grid-connected PV systems. Due to its structure, extended switching frequency is possible, allowing in turn smaller filter size calculations and reduced voltage stress on power switches. The NPC inverters can balance the voltage applied to off-biased switches. Despite their advantages, the active power loss is not negligible, taking into account the number of components used [12].

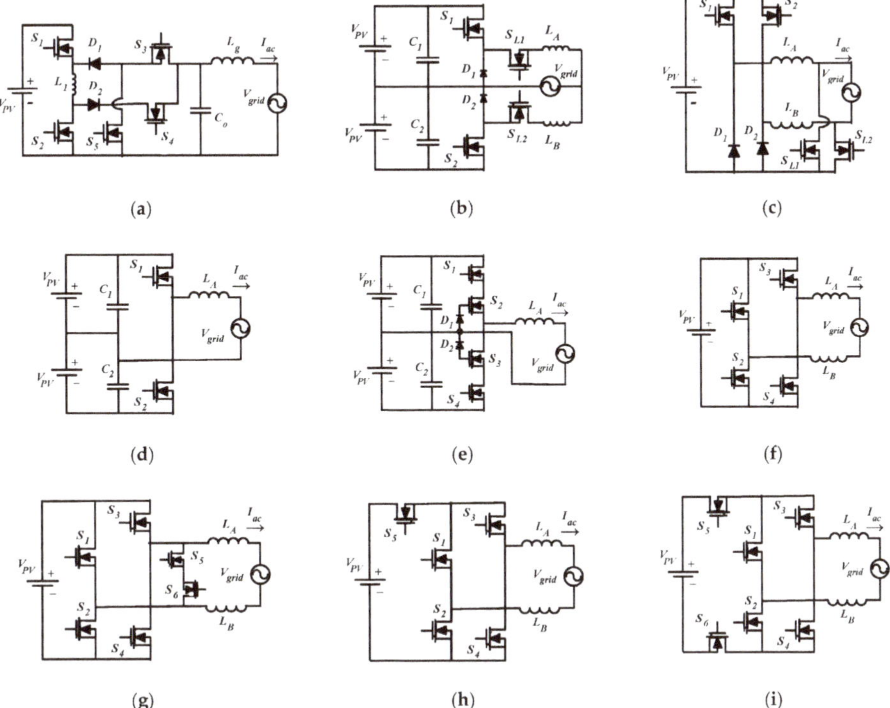

Figure 1. The main configurations of transformerless photovoltaic (PV) inverters in the literature. (**a**) Karschny [8], (**b**) series buck [9], (**c**) parallel beck [10], (**d**) half bridge [11], (**e**) neutral point clamped (NPC) [12], (**f**) full bridge [13], (**g**) highly efficient and reliable inverter (HERIC) [14], (**h**) H5 [15], (**i**) H6 [16].

The full bridge topology represented in Figure 1f that uses symmetrical inductor filters cannot effectively decouple the CMC on the AC side, since high-amplitude CMC is generated with the sinusoidal pulse width modulation (SPWM) technique [13].

A concept known as highly efficient and reliable inverter (HERIC) can be seen in Figure 1g. It comprises a full bridge configuration where two additional high-frequency switches are arranged in opposite directions on the AC side. The freewheel path is also produced by those high-frequency switches in high-efficiency operation. The main disadvantage of HERIC VSI is that the exact regulation of them results in output voltage and current chopping. The H5 topology shown in Figure 1h is obtained from a full bridge converter by incorporating a switch on the DC side to decouple it from the AC side, and therefore suppressing the CMC interference [14,15]. The three upper-side switches are controlled by the power grid frequency, while the rest are under high-frequency signal switching control. In the freewheel condition, only two semi-conductor switches are conducted, but in an active state, three of them are tuned on, which results in large active power loss consequently. Adding an extra switch to H5 topology will make those DC side switches work in pairs and will obtain the H6 topology represented in Figure 1i. However, the modulation performance varies regarding H5 topology and the most important drawback of this structure is the active power loss achieved in conduction operation mode with three switches [16].

This paper is organized as follows: Section 2 presents the CMC conceptualization for a single-phase PV inverter. The mitigation method and CMC evaluations are discussed in Sections 3 and 4, respectively. The design considerations and maximum power point tracking (MPPT) algorithm are presented in

Section 5. Section 6 analyzes the active power loss into two categories: switching and conduction. The simulation results and experiments are conducted in Section 7, and finally, the reactive power compensation is described in Section 8. Briefly, in this paper, a high-efficiency grid-tie inverter structure for non-isolated PV systems is proposed. The main contributions of the presented VSI are briefly mentioned below:

- An extended H6 topology by comprising seven switches, of which five are controlled in high frequency. The remaining switches operate in line frequency modulation to create the freewheel path.
- The 7th switch is used to clamp the common mode voltage; if the voltages ($V_{AN} \approx V_{BN}$) are higher than half of the DC link voltage, freewheeling current flows through S7 and D1 to the midpoint of the dc link, which results in V_{AN} and V_{BN} being clamped at $V_{pv}/2$.
- The maximum power point tracking (MPPT) algorithm based on perturb and observe (P&O) is considered to increase the total system efficiency.
- The AC decoupling strategy has been chosen.
- The unipolar SPWM pattern is applied to gate inputs.
- A reduced filter size is hired to decrease the unsolicited harmonic generated by output current and restrict the total harmonic distortion (THD).

2. CMC Origin in Transformerless Inverter

In the first categories of PV inverter topologies without galvanic connections mentioned before, which may be known as switching inverters or transformerless inverters, many applications are verified. The main drawback of these configurations is the CMC generation, which has appeared between the AC mains and the DC side of the inverter. A general structure of a transformerless PV inverter highlighting the path of $i_{leakage}$ (i_{cm}) is shown in Figure 2. As it can be seen clearly, the common mode current is flowed in a loop that consists of a DC input voltage source or PV arrays voltage, inverter switches, output filter inductances, AC grid, grounding impedance, and parasitic stray capacitances of PV cells. The common mode voltage is the mean of voltages of node A and node B. Besides, V_{dm} is the differential of the aforementioned nodes. Figure 3 has been depicted to represent the simple single-line circuit of the discussing topology in order to calculate the total common mode voltage. Consequently, according to these explanations, it could be expressed with the following equations.

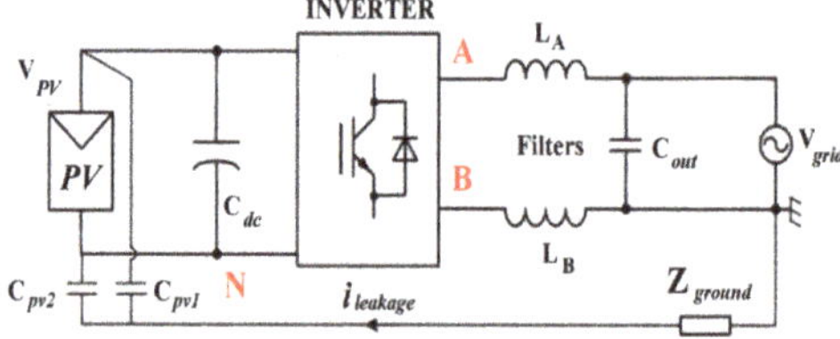

Figure 2. Equivalent circuit of a PV transformerless inverter.

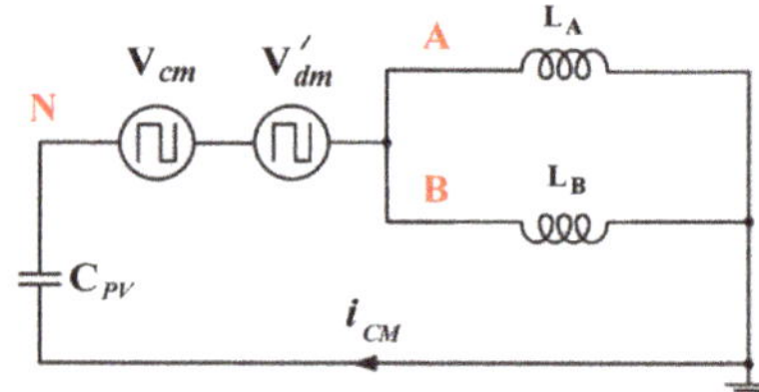

Figure 3. Common mode equivalent circuit.

$$V_{cm} = \frac{V_{AN} + V_{BN}}{2} \tag{1}$$

$$V'_{dm} = \frac{V_{dm}}{2} \frac{(L_B - L_A)}{(L_B + L_A)} \tag{2}$$

$$V_{dm} = V_{AN} - V_{BN} \tag{3}$$

Since i_{cm} appeared due to $V_{t,cm}$ flows in the abovementioned loop, it could be proved that the total common mode voltage is obtained in (4):

$$\begin{aligned} V_{t,cm} &= V_{cm} + \frac{V_{dm}}{2} \frac{(L_B - L_A)}{(L_B + L_A)} \\ &= \frac{V_{AN} + V_{BN}}{2} + \frac{V_{AN} - V_{BN}}{2} \frac{(L_B - L_A)}{(L_B + L_A)} \end{aligned} \tag{4}$$

Clearly, it is observed that the criteria for the elimination of i_{cm} is to make $V_{t,cm}$ remain constant. Thus, we should have:

$$V_{t,cm} = \frac{V_{AN} + V_{BN}}{2} + \frac{V_{AN} - V_{BN}}{2} \frac{(L_B - L_A)}{(L_B + L_A)} = cte \tag{5}$$

In some configurations such as half-bridge family or Karschny inverters and some of those indicated in Figure 1a, only one inductor is used in the output to filter the harmonics, so that the other inductor is equal to zero or does not need to be considered. Thus, by assuming that $L_A = 0$, we find the $V_{t,cm}$ as (6).

$$V_{t,cm} = \frac{V_{AN} + V_{BN}}{2} + \frac{V_{AN} - V_{BN}}{2} \frac{(L_B)}{(L_B)} = V_{AN} = cte \tag{6}$$

If $L_B = 0$, $V_{t,cm}$ will be extracted in (7).

$$V_{t,cm} = \frac{V_{AN} + V_{BN}}{2} + \frac{V_{AN} - V_{BN}}{2} \frac{(-L_A)}{(+L_A)} = V_{BN} = cte \tag{7}$$

In other PV transformerless inverter topologies, as shown in Figure 1, both inductors L_A and L_B are existing and equal to each other. This equality results in $V_{dm} = 0$. Therefore, to eliminate the leakage current flow, Equation (8) is being obtained:

$$L_A = L_B \rightarrow V_{t,cm} = \frac{V_{AN} + V_{BN}}{2} = cte \tag{8}$$

It is concluded that transformerless VSIs could be categorized into two groups. The first one is called "asymmetrical inductor based inverters", which trust in (6) and (7). Similarly, the second group is named "symmetrical inductor based inverter"; Equation (8) is applied to them.

3. H7 Topology and Formal Analysis

The H7 topology combines the main H6 family characteristic with a symmetrical inductor-based inverters group, as shown in Figure 4a. It consists of a full-bridge converter with two inductors, which can generate an appropriate sinusoidal voltage with very low THD. The switches S_1, S_2, S_3, and S_4 operate in high frequency with two remaining switches being turned on/off in line with grid frequency. The switching pattern is depicted in Figure 4b. The proposed suppression process comprises four modes of operation implemented by the control strategy in each period of the power grid frequency. As it can be observed in Figure 4b, switch S_7 is conducting when high-frequency switches S_1, S_2, S_3, and S_4 are turned off. Therefore, S_7 is conducting only in freewheel modes. The operation modes are detailed in Figure 5.

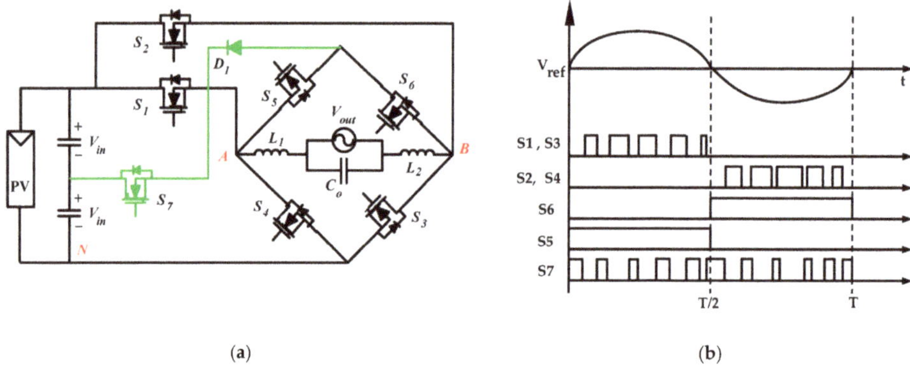

(a)   (b)

Figure 4. (**a**) The proposed topology, (**b**) Switching pattern.

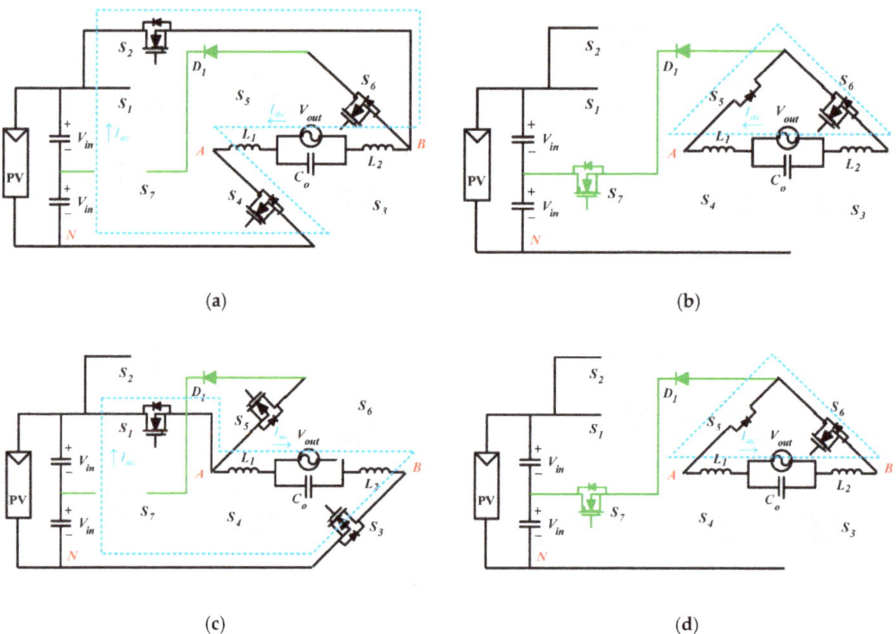

(a)   (b)

(c)   (d)

Figure 5. Currents paths and operation modes of the proposed topology. (**a**) Mode 1, (**b**) Mode 2, (**c**) Mode 3, and (**d**) Mode 4.

3.1. Mode 1, Negative AC Current

In this mode, switches S_1, S_3, and S_5 are turned off, and switch S_6 operates at grid frequency. The SPWM technique is employed to generate the gate signals of the S_2 and S_4 switches. It is supposed that switches S_2 and S_4 are conducted initially. Thus, the AC current flows from the DC part to the AC side through S_2 and S_4. Therefore, voltages V_{AN} is null and V_{BN} sets at V_{pv} input voltage. This mode is detailed in Figure 5a. If the voltages ($V_{AN} \approx V_{BN}$) are higher than half of the DC link voltage, freewheeling current flows through S_7 and D1 to the midpoint of the DC link; as a result, V_{AN} and V_{BN} are clamped at $V_{pv}/2$.

3.2. Mode 2, Negative Freewheel Current

The switches S_1, S_3, and S_5 are in an off state, while the switches S_2 and S_4 remain OFF for short periods of time, as shown in Figure 4b. The common mode leakage current as observed in Figure 5b is not allowed to circulate. Thus, it is temporarily eliminated. As a result, the AC grid is decoupled from the DC PV system. Now, voltages V_{AN} and V_{BN} are brought to the same values around $V_{pv}/2$.

3.3. Mode 3, Positive AC Current

In the next mode, switches S_2, S_4, and S_6 are turned off. Switch S_5 is working under line frequency modulation. Moreover, switches S_1 and S_3 are operated in switching frequency. The switches S_1 and S_3 are conducting, and then, switch S_5 at this moment does not have a closed loop to flow the current. Thus, the current flows from the DC side to the AC side of VSI through S_1 and S_3; therefore, we could figure out $V_{AN} = V_{pv}$ and $V_{BN} = 0$. This mode is depicted in Figure 5c. Similar to Mode 1, if the voltages ($V_{AN} \approx V_{BN}$) are higher than half of the DC link voltage, freewheeling current flows through S7 and D1 to the midpoint of the dc link; as a result, V_{AN} and V_{BN} are clamped at $V_{pv}/2$.

3.4. Mode 4

Mode 4 is characterized by switches S_1 and S_3 being in a turn-off state. This means that the AC current closes through the freewheel loop that comprises the S_5 body diode, switch S_6 in conduction mode, and AC grid (Figure 5d). Due to the closed freewheel path, there is no way for the i_{cm} current path. This operating state leads to $V_{AN} = V_{BN} \approx V_{pv}/2$.

4. H7 Topology Evaluation

In order to evaluate the common mode leakage current impact on the proposed topology, we derived the equivalent circuit of the loop from Figure 6 that imports the novel structure presented in Figure 6. Since the input capacitances C_{PV} are very small, they are not considered for calculation purposes. In addition, C_{out} has no influence in i_{cm} derivation, being discarded, too. Manipulating Equations (1) and (3) can be reorganized as:

$$V_{AN} = V_{cm} + \frac{V_{dm}}{2} \tag{9}$$

$$V_{BN} = V_{cm} - \frac{V_{dm}}{2} \tag{10}$$

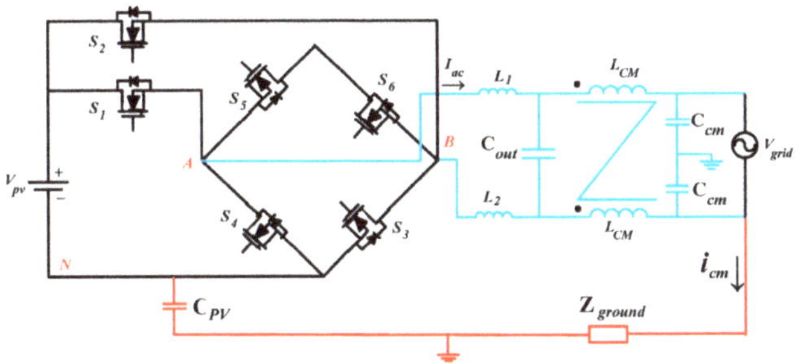

Figure 6. Proposed inverter structure with a common mode path.

Based on that, the CM equivalent circuit of Figure 6 is converted in Figure 7a. Further simplification can be obtained in the form of a single loop arrangement (Figure 7b).

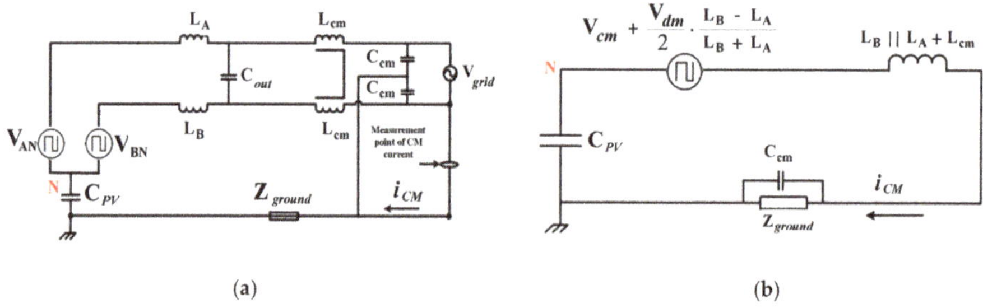

Figure 7. (**a**) CM equivalent circuit of Figure 6, (**b**) CMC single loop.

Using the single loop circuit simplification, CM voltage and current values are computed as follows:

(a) In mode 1:

$$V_{t,cm} = \frac{V_{AN}+V_{BN}}{2} + \frac{V_{AN}-V_{BN}}{2}\frac{(L_g-L_g)}{(L_B+L_A)} = \\ = \frac{0+V_{pv}}{2} = \frac{V_{pv}}{2} = cte \tag{11}$$

(b) In mode 2:

$$V_{t,cm} = \frac{V_{AN}+V_{BN}}{2} + \frac{V_{AN}-V_{BN}}{2}\frac{(L_g-L_g)}{(L_B+L_A)} = \\ = \frac{\frac{V_{pv}}{2}+\frac{V_{pv}}{2}}{2} = \frac{V_{pv}}{2} = cte \tag{12}$$

(c) In mode 3:

$$V_{t,cm} = \frac{V_{AN}+V_{BN}}{2} + \frac{V_{AN}-V_{BN}}{2}\frac{(L_g-L_g)}{(L_B+L_A)} = \\ = \frac{V_{pv}+0}{2} = \frac{V_{pv}}{2} = cte \tag{13}$$

(d) In mode 4:

$$V_{t,cm} = \frac{V_{AN}+V_{BN}}{2} + \frac{V_{AN}-V_{BN}}{2}\frac{(L_g-L_g)}{(L_B+L_A)} = \\ = \frac{\frac{V_{pv}}{2}+\frac{V_{pv}}{2}}{2} = \frac{V_{pv}}{2} = cte \tag{14}$$

5. Prototype Design Consideration

This part of the paper is organized in three sections: filter inductance design, filter capacitance selection, and semi-conductor components sketching. In these topologies, typically the maximum ripple of the current is considered between 10% and 25% of the root meat square (RMS) current [17]. Therefore, for a 36 W prototype fed a load with 30 V at 8 kHz switching frequency, the filter inductance is obtained as (15).

$$L_f = T_{on}\frac{V_{pv} - V_{ac}}{2 \times \Delta i_{ripple}^{MAX}} \sim 2\ mH \tag{15}$$

where the maximum current ripple is assumed to be 20% of the nominal current. Consequently, the capacitance of the output filter is generally designed based on the cut-off frequency. This frequency is normally between 10% and 20% of the switching frequency [18]. Equation (16) is given to help the selection of filter capacitance.

$$f_{cut-off} \leq (10\% \ \sim \ 20\%) \times f_s \\ f_{cut-off} = \frac{1}{2\pi \sqrt{L_g C_{filter}}} \tag{16}$$

Therefore, filter capacitance will achieved as approximately 220 µF. In order to select the semi-conductor components, it is noted to say that in high-frequency conditions, the switches

S_1 to S_4 should work properly. On the other hand, the switches S_5 and S_6 are operated at line frequency. According to peak inverse diode (PIV), total standing voltage (TSV), and nominal parameters of load, the MOSFET IRFP150 was selected as a power switch that could withstand up to 100 V and 40 Amperes as the nominal current. For the diodes configuration, the BY3099 reference was chosen, while the MOSFET driver role is ensured by the ICL 7667. Finally, to guarantee galvanic isolation between the gate signals and power circuit, we used a 6N137 optocoupler [19].

MPPT

The P&O approach is a well-known algorithm to perform the MPPT function. It works by applying a small disturbance in the system; then, the set-point of PV arrays are going to be changed due to tracking the maximum power point. The equation that describes the P&O algorithm operation is shown in (17) [20].

$$MPP_{k+1} = MPP_k \pm \Delta P_{PV} = MPP_k$$
$$+(MPP_k - MPP_{k-1}) \times sign(\Delta P_{PV}) \tag{17}$$

where $\Delta P_{PV} = P_k - P_{k-1}$. The MPPT flowchart of the P&O strategy is shown in Figure 8.

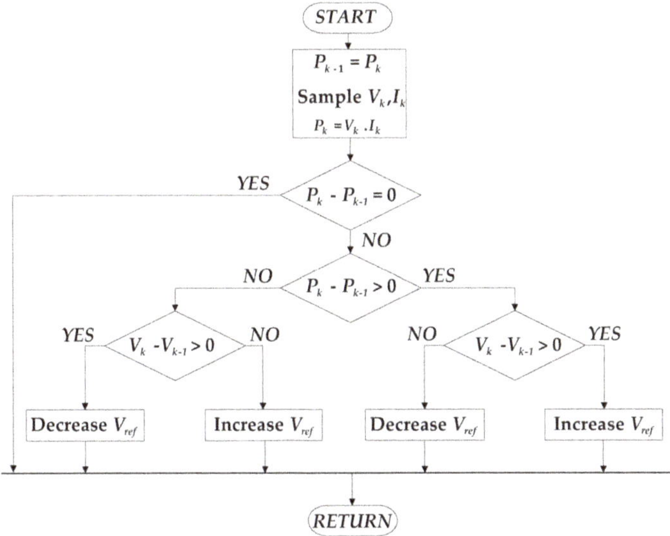

Figure 8. Maximum power point tracking (MPPT) flowchart by perturb and observe (P&O) approach [20].

6. Power Loss Analysis

Since the transformerless VSIs should be operated with maximum power conversion efficiency, the characterization of internal power losses is evaluated in the proposed topology [21]. For a modeling viewpoint, the AC current is given by Equation (18).

$$i(t) = I_m \sin(\omega t) \tag{18}$$

where I_m is the current injected in the power grid and ω is the angular frequency related to the utility frequency.

The voltage drop of these semi-conductors can be divided into two categories (19).

$$V_{ds}(t) = i(t) \times R_{ds} \rightarrow for\ MOSFETs$$
$$V_{ak}(t) = V_f + i(t) \times R_{ak} \rightarrow for\ Diodes \tag{19}$$

where R_{ds} and R_{ak} are the MOSFET and diode conducting resistance and V_f is the forward voltage of diode. Active and zero conditions regarding conduction time T_{active} and non-conduction time T_{zero} of semi-conductor components are given by (20) and (21), respectively.

$$T_{active}(t) = M\sin(\omega t) \tag{20}$$

$$T_{zero}(t) = 1 - M\sin(\omega t) \tag{21}$$

where M is the modulation index.

6.1. Steady-State Conduction Losses

The average conduction loss of MOSFETs and diodes are expressed as (22) and (23), respectively. Since the switches S_5 and S_6 are working at a line frequency, the conduction losses should not be ignored. Average power loss is calculated according to Equation (24).

$$P_{c,MOSFET} = \frac{1}{2\pi} \int_0^\pi i(t) \times V_{ds}(t) \times T_{active}(t) \times d(\omega t) = \frac{2M}{3\pi} I_m^2 R_{ds} \tag{22}$$

$$P_{c,Diode} = \frac{1}{2\pi} \int_0^\pi i(t) \times V_{ak}(t) \times T_{zero}(t) \times d(\omega t) = I_m V_f \left(\frac{1}{\pi} - \frac{M}{4}\right) + I_m^2 R_{ak}\left(\frac{1}{4} - \frac{2M}{3\pi}\right) \tag{23}$$

$$P_{c,MOSFET,line} = \frac{1}{2\pi} \int_0^\pi i(t) \times V_{ds}(t) \times T_{zero}(t) \times d(\omega t) = I_m^2 R_{ds}\left(\frac{1}{4} - \frac{2M}{3\pi}\right) \tag{24}$$

6.2. Switching Losses

Ordinarily, the switching losses are calculated by multiplying instantaneous voltage and the current of the commutation state, where they meet each other. This results in (25) and (26) for the evaluation of ON and OFF conditions, respectively.

$$P_{sw,on} = \frac{1}{2\sqrt{\pi}} f_s h I_m^k K_{gon} \frac{V_{ds}}{V_{test}} \frac{\Gamma\left(\frac{k+1}{2}\right)}{\Gamma\left(\frac{k}{2} + 1\right)} \tag{25}$$

$$P_{sw,off} = \frac{1}{2\sqrt{\pi}} f_s m I_m^k K_{goff} \frac{V_{ds}}{V_{test}} \frac{\Gamma\left(\frac{n+1}{2}\right)}{\Gamma\left(\frac{n}{2} + 1\right)} \tag{26}$$

It is proved $\Gamma\left(\frac{n+1}{2}\right) \div \Gamma\left(\frac{n}{2} + 1\right) = \frac{1}{\sqrt{\pi}} \int_0^\pi \sin(\omega t)^n d(\omega t)$. Here, coefficients h and k are turn-on energy factors; m and n are turn-off energy factors, K_g is the correction factor to take account of the gate drive impedance, and V_{test} is the test voltage for the model parameters [22]. If the paralleled capacitors of each switch are considered, the charge and discharge losses of them will be given by (27).

$$P_{cap,sw} = \frac{1}{2} C_{MOSFET} V_{ds}^2 f_s \tag{27}$$

The power loss during the reverse recovery period of diode performance due to the switching is estimated by (28). However, the diode reverse recovery current not only contributes to its loss as mentioned above, but it also generates the loss in the main switches as shown in (29) [23].

$$P_{sw,Diode} = \frac{1}{2\pi} \int_0^{\pi} \left(\frac{1}{2}V_{dc}\right) \times \left(\frac{1}{2}I_{rr}\right) \times f_s \times t_b \times d(\omega t) = \frac{V_{dc}I_{rr}f_s t_b}{8} \tag{28}$$

$$\begin{aligned} P_{D_{rr}} &= \frac{1}{2\pi} \int_0^{\pi} \left(\begin{array}{c} I_m.sin(\omega t).t_a \\ +\frac{I_{rr}}{4}(2t_a + t_b) \end{array} \right) \times V_{dc} \times f_s \times d(\omega t) \\ &= \left(\frac{I_m t_a}{\pi} + \frac{I_{rr}(2t_a + t_b)}{8} \right) V_{dc} \times f_s \end{aligned} \tag{29}$$

7. Simulation and Experimental Results

In order to evaluate the H7 topology an inverter prototype was built. It can be seen in Figure 9 the main elements of the inverter system used in bench tests. Table 2 gather the main electrical parameters that characterize the electrical tests. The switching frequency is chosen as 8 kHz and the line frequency is 50 Hz. Figure 10a,b represent the V_{AN} and V_{BN} signals obtained by simulation. Being signals with pulsating periodic waveform, this behavior makes $V_{t,cm}$ constant, as can be verified in Figure 10c. Consequently, the common mode current elimination target is accomplished. For emulating as close as possible a real scenario, it was chosen a stray capacitance of 75 nF. The experimental waveforms are shown in Figure 11, demonstrating the effectiveness of the novel photovoltaic transformerless inverter. The output voltage and load current amplitudes can be checked in Figure 11a. Voltage THD measurement at inverter output is 1.51%, in accordance with the value estimated in simulation whose result is 1.07%. Measured waveforms of V_{AN} and V_{BN} have approximately square waveforms (Figure 11b), confirming that $V_{t,cm}$ is constant (Figure 11c). Finally, in Figure 11d the CMC waveform is presented. Due to experimental device restrictions, DC input voltage of 30 volts conditioned V_{AN} and V_{BN} peak voltage around 30 V. The peak value of the output voltage and current are near 30 V and 1.2 A, respectively. From Figure 11d data the current measured in common mode is approximately 1.14 mA. Some undesirable peaks shown in these pictures appear due to non-ideality in components.

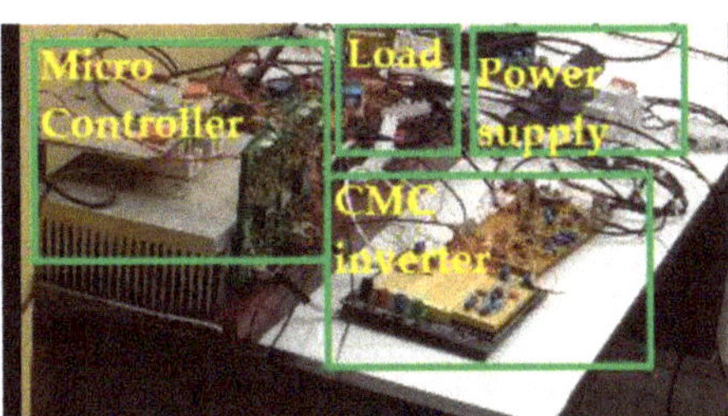

Figure 9. Laboratory setup of proposed H7-type inverter.

Table 3 shows a comparative study concerning the proposed topology with published performance data regarding some well-known single phase inverter structures. The comparison highlights that the H7-type inverter with fewer components can not only reduce significantly the CMC circulation, as well as the power conversion efficiency is one of highest. On the other hand, these results denote that the CMC of the H6 topology is the lowest of all under comparison and the two others' CMC is practically the same. However, the efficiency analysis shown in Figure 12 relies on the best performance of the proposed topology. To illustrate this, it can obviously realized by comparing our topology with those evaluated in active and freewheel conditions. In the active state, the AC current flows through only two switches in the proposed topology, similarly to HERIC-type inverter, while H6 and

H5 topologies require three power switches translating into higher active power loss. In the freewheel state, the H7-type design makes use of only one switch and one diode in line with the H6 structure. The same is not true for the H5 and HERIC converters whose operation depends on the two switches, generating consequently additional power losses.

The California efficiency that is used to calculate the diagrams shown in Figure 12 is supposed to be what is written in (30) [12–15]. Figure 13 shows the total loss of active power through the energy transmission process. As it is understood, the suggested topology has the least loss in equal condition. When the same nominal output power is selected for all topologies, the power loss in terms of nominal load percentages will be changed due to the current drawn. According to Figure 13, the proposed topology has the least power loss among all during load variations.

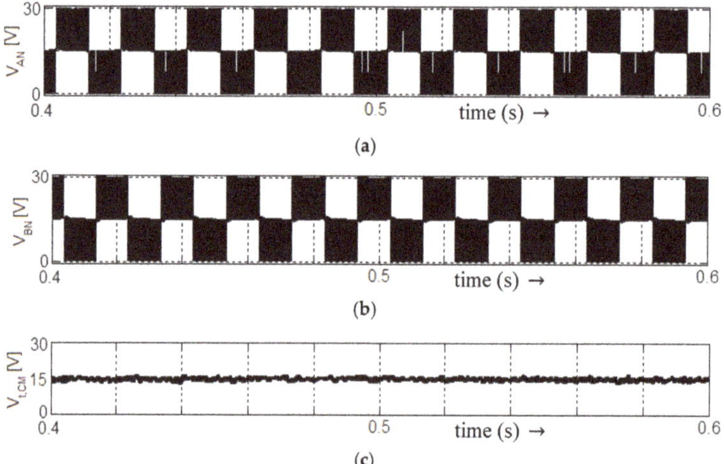

Figure 10. Simulated results of proposed inverter; (a) V_{AN}, (b) V_{BN}, and (c) $V_{t,CM}$.

Table 2. System parameters.

Parameter	Value
V_{dc}	30 V
f_{line}	50 Hz
f_{sw}	8 kHz
P_{out}	36 W
L_f	2 mH
C_f	220 µF
C_{PV}	70 nF

$$\eta = 0.04\eta_{10\%} + 0.05\eta_{20\%} + 0.12\eta_{30\%} + 0.21\eta_{50\%} + 0.53\eta_{75\%} + 0.05\eta_{100\%} \tag{30}$$

Figure 14 shows the percentage of each power loss shared during operation. As discussed before, the proposed topology has a conduction loss that is the same as the HERIC and the freewheel loss is the same as that of the H6 VSIs. The only difference between these converters appears in switching loss calculations. Whereas the rated power is loaded, the share of switching loss in the proposed structure obtains the biggest part among all quotas. This issue does not mean that lots of power transmitted through the VSI is wasted during switching transitions, but this justifies that the power loss could be decreased in terms of switching frequency. Consequently, the total loss will be mitigated in the proposed topology, and its exclusive capability does not appear in other VSIs.

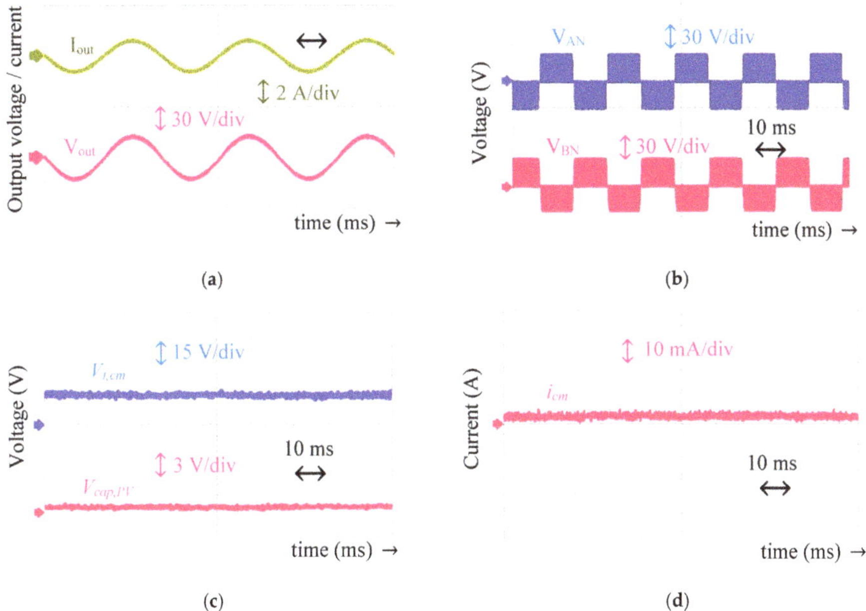

Figure 11. Experimental results; (**a**) output voltage and current; (**b**) V_{AN} and V_{BN}; (**c**) $V_{t,cm}$ (**d**) CMC.

Figure 15 compares the total loss obtained between IGBT and MOSFET utilizations. Since the switching frequency increases, the power loss growths, correspondingly, as expected. However, MOSFTE utilization can decrease the power loss about 40% rather than using IGBT in all switching scenarios. As a result, since the power loss increases as the switching frequency is enhanced, the efficiency will be decreased, subsequently. Figure 16 represents the total efficiency of proposed topology in terms of load power variations, considering switching frequency effects. If the switching frequency is chosen as 30 kHz, the maximum efficiency is obtained as 97.42% in 2.75 kW of output power. However, the maximum efficiency is being calculated as 98.02% for the other switching frequency in $P_{out} = 1.75$ kW. The average efficiency is assessed nearly as 97.1% and 97.85% for $f_s = 30$ kHz and $f_s = 8$ kHz, respectively, in all loading conditions. As the output voltage and current waveforms of experimental results are achieved at 8 kHz switching frequency, due to our laboratory equipment, the output power has been varied between 20 W and 150 W step by step. This verification implies that the proposed prototype is working successfully with high efficiency and reliable performance, as shown in Figure 17. The reactive power and its effect on efficiency is one of the most important concerns in single-phase photovoltaic inverters.

The H7-type VSI is equipped with a capability to control the active and reactive power instantly by the proposed switching pattern. The loading condition is considered as $S_{load} = 44\,W + j\,44\,VAr = 62\,VA$, and the load current is calculated as $|S_{load}|/\,V_{load} = 2.28\,A$. Then, the average and the *rms* amount of current flowing through the diodes and switches are computed, and the loss calculation is obtained as 0.886 W at 8 kHz switching frequency. Then, the effective efficiency of the proposed topology will be figured out as shown in (31).

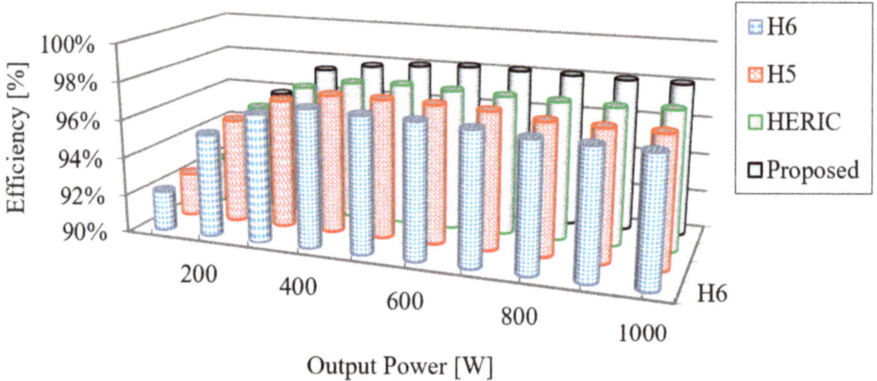

Figure 12. Efficiency evaluation in terms of output power.

Table 3. Comparison as function of components number, total harmonic distortion (THD), i_{cm}, and other properties. HERIC: highly efficient and reliable inverter.

Operation Mode	Component Numbers and Properties Evaluation								
	HERIC [14]	**H5** [15]	**H6** [16]	**oH5 **** [24]	**Ref.** [25]	**Ref.** [26]	**HBZVR *** [27]	**Ref** [28]	**H7**
Active	2 switches	3 switches	3 switches	2 switches	4 switches	2 switches	3 switches	3 switches	3 switches
Freewheel	2 switches	2 switches	1 switch and 1 diode	2 switch and 1 body diode	2 switches	1 switch and 2 diode	1 switch and 1 diode	2 switches	2 switches
THD	1.59%	1.67%	1.84%	1.61%	1.86%	1.69%	1.94%	1.96%	1.75%
Experimental i_{cm}	84.3 mA	89.4 mA	45.8 mA	30.3 mA	51.7 mA	48.9 mA	55.3 mA	67.1 mA	32.5 mA
Capacitor No.	0	0	0	0	0	2	0	0	2
Inductor No.	2	2	2	2	2	2	2	2	2
Maximum efficiency	98.00%	97.87%	97.64%	98.02%	96.36%	97.44%	95.48%	96.15%	98.01%

* HBZVR: H-bridge Zero Voltage Rectifier, ** oH5: optimized H5 topology.

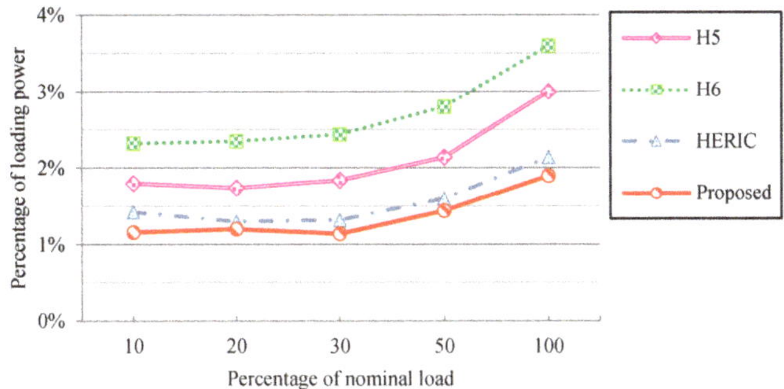

Figure 13. Total power loss through energy transmission.

It should be noted that with a rise in load power, the current ripple and the sizing of passive components are minimized due to the increment in current amplitude. Furthermore, the proposed topology has the ability to accomplish the power factor correction (PFC) operation mode as an active filter application.

$$\eta = \frac{P_{out}}{P_{out} + P_{loss}} \times 100 = \frac{44}{44 + 0.831} \times 100 = 98.02\% \tag{31}$$

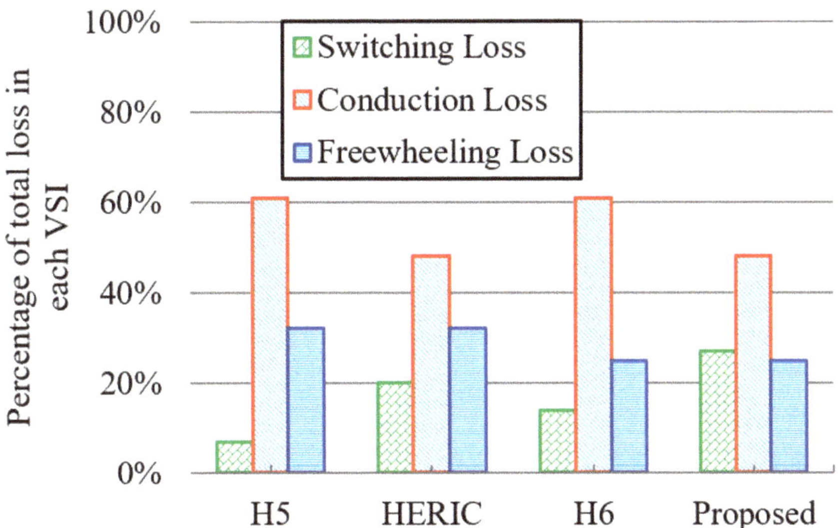

Figure 14. Percentage of self-VSI power loss sharing during operation. VSI: voltage source inverters.

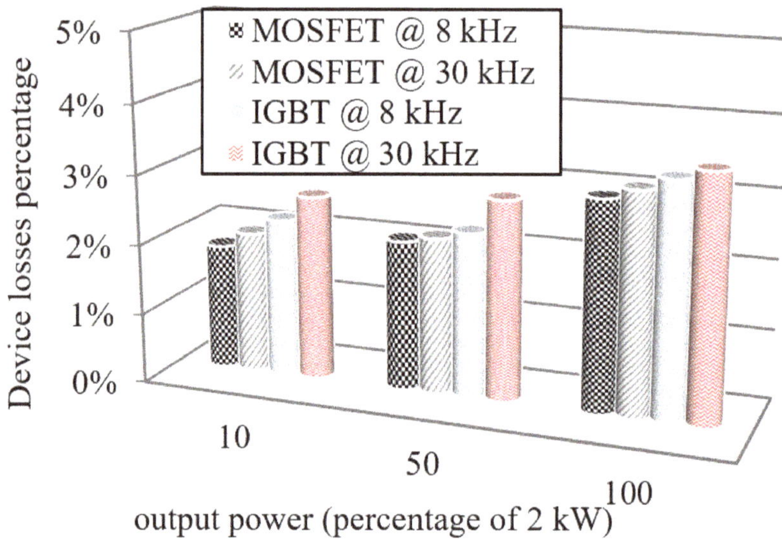

Figure 15. Percentage of device power loss in terms of output power.

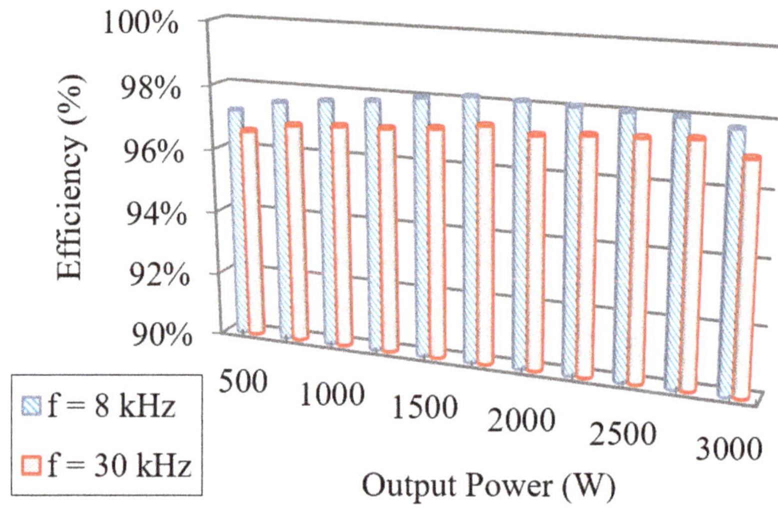

Figure 16. Efficiency analysis based on output power variations and frequency effects.

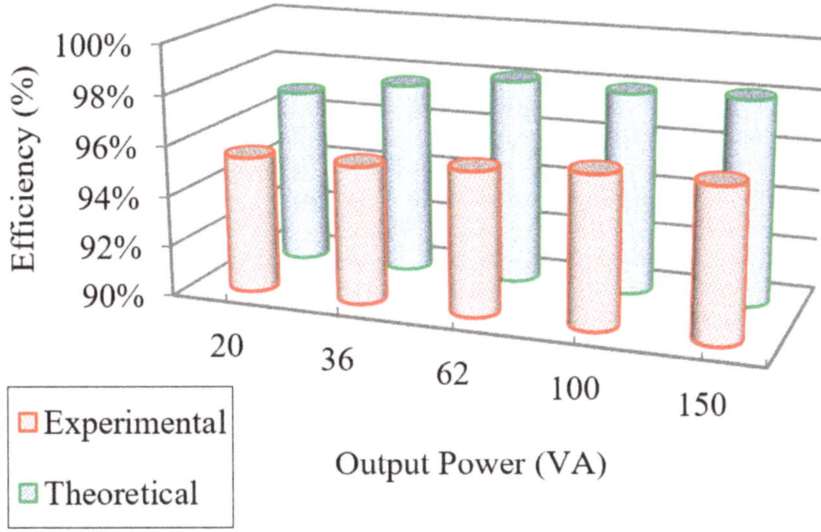

Figure 17. Experimental and theoretical efficiency analysis at 8 kHz switching frequency.

8. Reactive Power Compensation

To evaluate the active/reactive power control, an active load is connected to the system and then disconnected for a while in a time interval of 0.4 s to 1 s. At first, the system is loaded with a 50 W pure resistive load without any reactive portion. Then, at t = 0.4 s a 250 W load is paralleled to the previous load, which causes more current drawn from the DC side. While the transients resulting from connecting/disconnecting the load are inevitable, it is noticed some reactive power sharp spikes during the transient response. However, the spikes are rapidly eliminated by following the Q_{ref} reference signal as depicted in Figure 18.

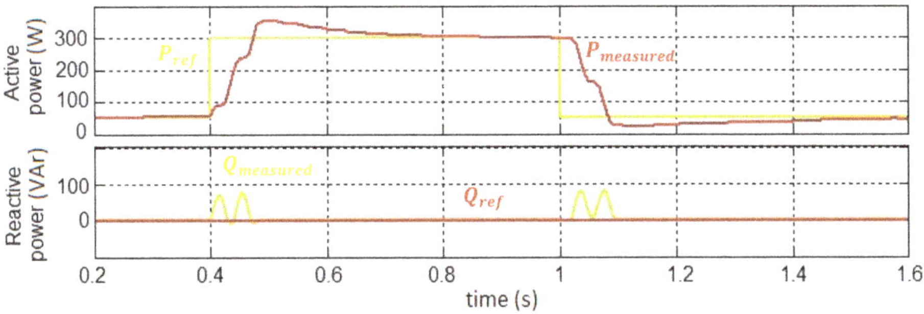

Figure 18. Active and reactive powers diagram.

9. Conclusions

This paper has presented a novel H7 architecture that was derived from the H6-type single-phase transformerless inverter to address the suppression of common mode current. In this topology, the number of components used on the current flow path is minimized, resulting in increased power conversion efficiency. In addition, the body diodes conduction in some not solicited time intervals is solved with appropriate placement of components. The prototype built for testing the concept was characterized at low power rating due to technical limitations in the laboratory infrastructure.

The results in laboratory have shown high-efficiency conversion of 98.02% efficiency. Comparing to other common mode current mitigation structures, low CMC value was achieved below 40 mA without compromising high power quality output. A satisfactory THD measurement in relation to voltage output of 1.52% proves that is in line with other power conversion structures at disposable in single-phase grid connected inverters The proposed design is satisfied with low-size LC filter, allowing high power density with low weight. Due to elimination of the CMC, this paper recommends an appropriate switching pattern with a unipolar PWM modulation technique that decouples the i_{cm} flowing path. This results in the common mode voltage being constant.

Author Contributions: Conceptualization and methodology, M.M.; investigation, review and editing, E.M.G.R.; analysis and editing, E.P. All authors contributed equally to editing and revising of this review. All authors have read and agreed to the published version of the manuscript.

Funding: This research received no external funding.

Conflicts of Interest: The authors declare no conflict of interest.

References

1. Manfredini, G.; Catania, A.; Benvenuti, L.; Cicalini, M.; Piotto, M.; Bruschi, P. Ultra-Low-Voltage Inverter-Based Amplifier with Novel Common-Mode Stabilization Loop. *Electronics* **2020**, *9*, 1019. [CrossRef]
2. Baik, J.; Yun, S.; Kim, D.; Kwon, C.; Yoo, J. Remote-State PWM with Minimum RMS Torque Ripple and Reduced Common-Mode Voltage for Three-Phase VSI-Fed BLAC Motor Drives. *Electronics* **2020**, *9*, 586. [CrossRef]
3. Jafrodi, S.T.; Ghanbari, M.; Mahmoudian, M.; Najafi, A.; Rodrigues, E.M.; Pouresmaeil, E. A Novel Control Strategy to Active Power Filter with Load Voltage Support Considering Current Harmonic Compensation. *Appl. Sci.* **2020**, *10*, 1664. [CrossRef]
4. Li, X.; Wang, N.; San, G.; Guo, X. Current source AC-Side clamped inverter for leakage current reduction in grid-connected PV system. *Electronics* **2019**, *8*, 1296. [CrossRef]
5. IEC Standard. *Photovoltaic (PV) Systems-Characteristics of the Utility Interface*; IEC: Geneva, Switzerland, 2004; IEC 61727.
6. VDE Standard. *Power Generation Systems Connected to the Low-Voltage Distribution Network*; VDE: Berlin, Germany, 2011; VDE-AR-N 4105.
7. Li, W.; Gu, Y.; Luo, H.; Cui, W.; He, X.; Xia, C. Topology review and derivation methodology of single-phase transformerless photovoltaic inverters for leakage current suppression. *IEEE Trans. Ind. Electron.* **2015**, *62*, 4537–4551. [CrossRef]
8. Dietrich, K. Wechselrichter. DE Patent DE19642522 (C1), 23 April 1998.
9. Xiao, H.; Xie, S. Transformerless split-inductor neutral point clamped three-level PV grid-connected inverter. *IEEE Trans. Power Electron.* **2011**, *27*, 1799–1808. [CrossRef]
10. Araújo, S.V.; Zacharias, P.; Mallwitz, R. Highly efficient single-phase transformerless inverters for grid-connected photovoltaic systems. *IEEE Trans. Ind. Electron.* **2009**, *57*, 3118–3128. [CrossRef]
11. González, R.; Gubía, E.; López, J.; Marroyo, L. Transformerless single-phase multilevel-based photovoltaic inverter. *IEEE Trans. Ind. Electron.* **2008**, *55*, 2694–2702. [CrossRef]
12. Luo, H.; Dong, Y.; Li, W.; He, X. Module Multilevel-Clamped Composited Multilevel Converter (M-MC^2) with Dual T-Type Modules and One Diode Module. *J. Power Electron.* **2014**, *14*, 1189–1196. [CrossRef]
13. Yu, W.; Lai, J.S.J.; Qian, H.; Hutchens, C. High-efficiency MOSFET inverter with H6-type configuration for photovoltaic non-isolated AC-module applications. *IEEE Trans. Power Electron.* **2010**, *26*, 1253–1260. [CrossRef]
14. Ji, B.; Wang, J.; Zhao, J. High-efficiency single-phase transformerless PV H6 inverter with hybrid modulation method. *IEEE Trans. Ind. Electron.* **2012**, *60*, 2104–2115. [CrossRef]
15. Victor, M.; Greizer, F.; Bremicker, S.; Hübler, U. SMA Solar Technology AG, Method of Converting a Direct Current Voltage from a Source of Direct Current Voltage, More Specifically from a Photovoltaic Source of Direct Current Voltage, into an Alternating Current Voltage. U.S. Patent 7411.802, 12 August 2008.
16. Yu, W.; Lai, J.S.; Qian, H.; Hutchens, C.; Zhang, J.; Lisi, G.; Djabbari, A.; Smith, G.; Hegarty, T. February. High-efficiency inverter with H6-type configuration for photovoltaic non-isolated ac module applications.

In Proceedings of the 2010 Twenty-Fifth Annual IEEE Applied Power Electronics Conference and Exposition (APEC), Palm Springs, CA, USA, 21–25 February 2010; pp. 1056–1061.

17. Abdelhakim, A.; Mattavelli, P.; Yang, D.; Blaabjerg, F. Coupled-inductor-based DC current measurement technique for transformerless grid-tied inverters. *IEEE Trans. Power Electron.* **2017**, *33*, 18–23. [CrossRef]

18. Liu, H.; Ran, Y.; Liu, K.; Wang, W.; Xu, D. A modified single-phase transformerless Y-source PV grid-connected inverter. *IEEE Access* **2018**, *6*, 18561–18569. [CrossRef]

19. Melkebeek, J.A. *Power Electronic Components*, 1st ed.; Springer: Berlin/Heidelberg, Germany, 2018; pp. 211–232.

20. Litrán, S.P.; Durán, E.; Semião, J.; Barroso, R.S. Single-Switch Bipolar Output DC-DC Converter for Photovoltaic Application. *Electronics* **2020**, *9*, 1171. [CrossRef]

21. Quach, T.-H.; Do, D.-T.; Nguyen, M.-K. A PWM Scheme for Five-Level H-Bridge T-Type Inverter with Switching Loss Reduction. *Electronics* **2019**, *8*, 702. [CrossRef]

22. Chung, H.S.H.; Wang, H.; Blaabjerg, F.; Pecht, M. *Reliability of Power Electronic Converter Systems*, 1st ed.; IET Press: London, UK, 2015; pp. 8885–8993.

23. Erickson, R.W.; Maksimovic, D. *Fundamentals of Power Electronics*, 1st ed.; Springer: Berlin/Heidelberg, Germany, 2007.

24. Xiao, H.; Xie, S.; Chen, Y.; Huang, R. An optimized transformerless photovoltaic grid-connected inverter. *IEEE Trans. Ind. Electron.* **2010**, *58*, 1887–1895. [CrossRef]

25. Zhang, L.; Sun, K.; Xing, Y.; Xing, M. H6 transformerless full-bridge PV grid-tied inverters. *IEEE Trans. Power Electron.* **2013**, *29*, 1229–1238. [CrossRef]

26. Cui, W.; Yang, B.; Zhao, Y.; Li, W.; He, X. A novel single-phase transformerless grid-connected inverter. In Proceedings of the IECON 2011-37th Annual Conference of the IEEE Industrial Electronics Society, Melbourne, Australia, 7–10 November 2011; pp. 1126–1130.

27. Kerekes, T.; Teodorescu, R.; Rodríguez, P.; Vázquez, G.; Aldabas, E. A new high-efficiency single-phase transformerless PV inverter topology. *IEEE Trans. Ind. Electron.* **2009**, *58*, 184–191. [CrossRef]

28. Yang, B.; Li, W.; Gu, Y.; Cui, W.; He, X. Improved transformerless inverter with common-mode leakage current elimination for a photovoltaic grid-connected power system. *IEEE Trans. Power Electron.* **2011**, *27*, 752–762. [CrossRef]

Article

A Data-Driven Based Voltage Control Strategy for DC-DC Converters: Application to DC Microgrid

Kumars Rouzbehi [1], Arash Miranian [2], Juan Manuel Escaño, Elyas Rakhshani, Negin Shariati and Edris Pouresmaeil

[1] Departamento de Ingeniería de Sistemas y Automática, Universidad de Sevilla, 41092 Sevilla, Spain; krouzbehi@us.es (K.R.); jescano@us.es (J.M.E.)
[2] Department of Electrical Engineering, University of Tehran, Tehran 1417466191, Iran; ar.miranian@gmail.com
[3] Department of Electrical Sustainable Energy, Delft University of Technology, Mekelweg 4, 2628 CD Delft, The Netherlands; E.Rakhshani@tudelft.nl
[4] Faculty of Engineering and IT, University of Technology Sydney, Sydney, NSW 2007, Australia; negin.shariati@uts.edu.au
[5] Department of Electrical Engineering and Automation, Aalto University, 02150 Espoo, Finland
* Correspondence: edris.pouresmaeil@aalto.fi; Tel.:+358-50-59-844-79

Received: 27 March 2019; Accepted: 22 April 2019; Published: 30 April 2019

Abstract: This paper develops a data-driven strategy for identification and voltage control for DC-DC power converters. The proposed strategy does not require a pre-defined standard model of the power converters and only relies on power converter measurement data, including sampled output voltage and the duty ratio to identify a valid dynamic model for them over their operating regime. To derive the power converter model from the measurements, a local model network (LMN) is used, which is able to describe converter dynamics through some locally active linear sub-models, individually responsible for representing a particular operating regime of the power converters. Later, a local linear controller is established considering the identified LMN to generate the control signal (i.e., duty ratio) for the power converters. Simulation results for a stand-alone boost converter as well as a bidirectional converter in a test DC microgrid demonstrate merit and satisfactory performance of the proposed data-driven identification and control strategy. Moreover, comparisons to a conventional proportional-integral (PI) controllers demonstrate the merits of the proposed approach.

Keywords: DC-DC power converter; Takagi–Sugeno fuzzy system; hierarchical binary tree

---

1. Introduction

DC-DC power converters have been extensively used in the infrastructure such as PV power converters, DC motor drives, and wind farm power converters [1–3]. The DC-DC converters are increasingly being used to integrate sustainable resources such as photovoltaic (PV) with high variability in their outputs to DC microgrids [4–7]. Due to the wide-ranging applications of these converters, proper modeling and control techniques are required for their voltage regulation.

Control of the power converters poses challenges to the researchers due to their nonlinear characteristics. Specifically, such difficulties stem from the following phenomena and requirements:

- DC-DC power converters are characterized by three different modes of operation, namely rising inductor current, falling inductor current and zero inductor current (which happens in discontinuous conduction operation), where each mode features linear continuous-time dynamics. Such complexities may even lead to chaotic behavior of the DC-DC converters [8].
- Converter topology requires the control input of the converter (i.e., duty cycle) to be bounded between zero and one.

- The inductor current of power converter must be non-negative when it is operating in the discontinuous conduction mode (DCM) [9].

Apart from the above-mentioned issues, for an efficient controller design of a DC-DC converter, expert designer should address wide input and output changes to guarantee the stability of the converter in any working state with a reasonable transient response [10]. In addition, general stability analysis is introduced in [11]. It is worth noting that the input voltage and output load variations change the operating point of the power converter. These issues further contribute to the complication of the DC-DC converter control problem.

Various techniques have been presented for DC-DC converters voltage control, from conventional PI and PID controllers [11,12] to fuzzy and Artificial Neural Networks (ANNs) controllers [13–15] and sliding model control [16]. Conventional PI and PID controllers present several valuable properties, for example, easily designed and being inexpensive. However, they are designed based on a locally linearized model of the system and hence their performance reduces while the operating point of the converter changes.

On the other hand, fuzzy logic controllers seem to provide satisfactory control actions, but, when there are large numbers of fuzzy rules and for high switching frequencies (small switching cycles), evaluation of all rules may turn into a problem. Moreover, proper tuning of the positions of the membership functions is a time-consuming procedure, which requires great expert knowledge.

Artificial Neural Network-based controllers are data-driven approaches that use input–output measurement of the converter to develop a valid converter model and then employ the developed model to design controller for the converters. There are also difficulties associated with the ANN-based controllers such as initialization of the ANN's parameters.

To eliminate the above-mentioned limitations, this paper proposes a data-driven strategy (DDS) that does not experience the problem associated with the conventional ANN-based control strategies, for identification and voltage control of DC-DC converters. The proposed DDS includes two stages as stated below:

- Stage 1: Identification of the power converter dynamics directly from measured data.
- Stage 2: Design of the voltage regulator based on the identified converter model.

In the first stage, the proposed DDS employs a local model network (LMN) for identification of a DC-DC converter. The LMN is made up of local linear models (LLM) and each LLM is responsible for modeling the dynamics of the converter along a specific operating regime, decided by its validity function. The LMN structure is determined directly from the input–output measurement using learning algorithm of hierarchical binary tree (HBT). Next, in the second stage, the identified LMN is used for the local linear control (LLC) considering the inverse error dynamics controller design.

In this paper, the identification and control stages are first explained for a standalone DC-DC boost converter. Then, the approach is applied on a bidirectional converter, connected to a DC microgrid with photovoltaic and storage. In both cases, the results of the proposed LLC strategy are compared to PI controllers.

Section 2 describes the framework for power converter identification. The structure of the LMN is presented in Section 3. The LLC design is described in Section 4. Simulation results and comparisons for the standalone converter are discussed in Section 5. Then, the proposed method is applied to a bidirectional converter in a DC microgrid in Section 6. Finally, the paper is concluded in Section 7.

2. Identification Methodology

As shown in Figure 1, DC-DC converter can be found in applications such as delivering the harvested energy from the PV panels to the main DC microgrid. Voltage regulation in DC microgrid is one of the critical issues for effective power delivery.

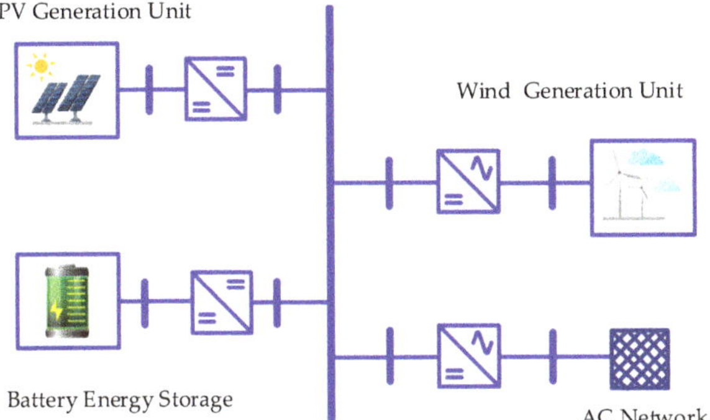

Figure 1. A typical DC microgrid integrating renewable sources.

In this paper, initially, a simple standalone DC-DC boost converter, as shown in Figure 2, is identified and controlled using the proposed DDS. Then, application of the DDS to a DC microgrid is investigated.

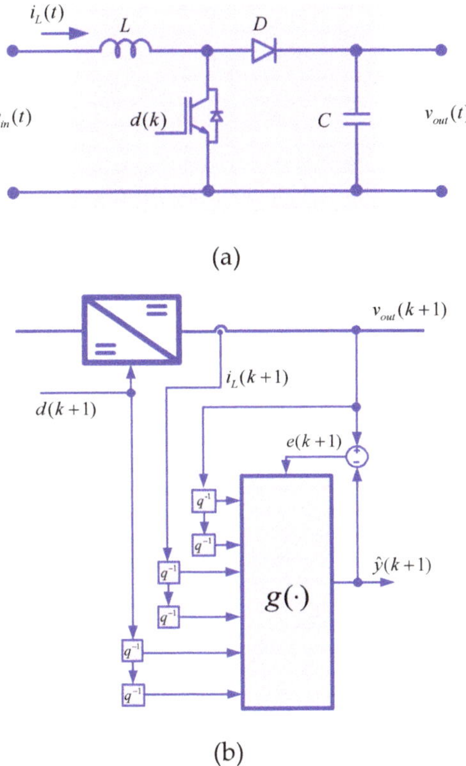

Figure 2. (**a**) Standalone DC-DC boost converter circuit; and (**b**) the identification framework using the LMN.

For the boost converter in Figure 2a, the input–output relationship is as follows:

$$v_{out}(k+1) = g(v_{out}(k-1), v_{out}(k), i_L(k-1), i_L(k), d(k-1), d(k)) \tag{1}$$

where g is a nonlinear function mapping the inputs to the output of the DC-DC boost power converter, $v_{out}(k)$ is the converter's output voltage at time kT_s (with T_s as sampling time), and i_L and d are inductor's current and duty ratio of the power converter, respectively.

The identification framework of the boost converter, based on Equation (1), is depicted in Figure 2b. In this series-parallel identification framework, symbol q^{-1} realizes unit delay.

To estimate a model that is dynamically valid from measured data, a dataset that contains rich enough information about the converter behavior should be generated. This is fulfilled by designing an appropriate excitation signal, which is the duty cycle of the system (d) in this case. To generate precise datasets, an amplitude-modulated pseudo-random binary signal (APRBS) with 59 different duty cycle levels is planned to form the excitation signal. The APRBS signals are proper choices for identification of nonlinear plants as they can extract information about the different operating modes of the nonlinear system [16,17]. The minimum hold time for this excitation signal, illustrated in Figure 3a, is 0.7 ms, which was obtained based on the step response of the boost converter.

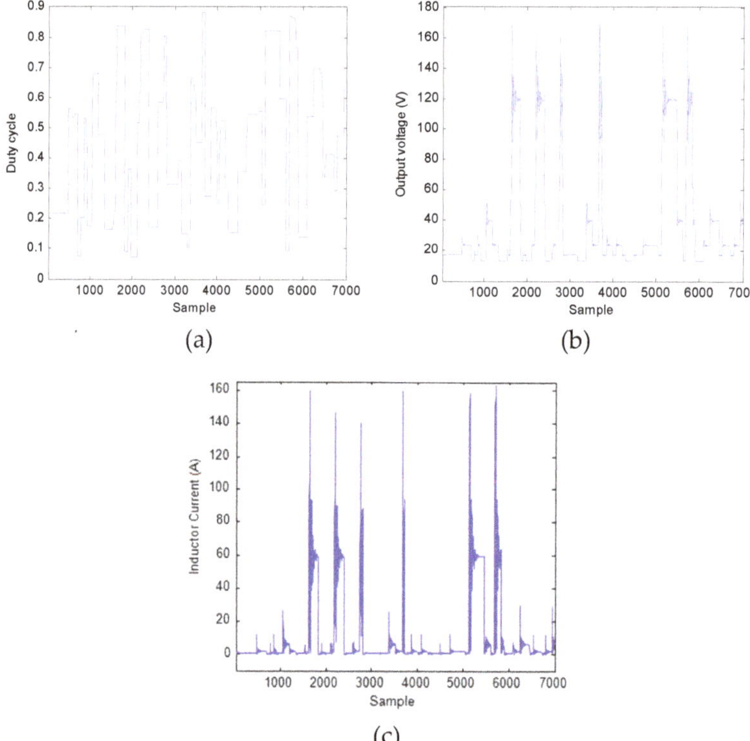

Figure 3. (**a**) The APRBS signal (duty cycle) for identification of boost converter; (**b**) waveform of converter output voltage obtained by applying the APRBS signal; and (**c**) waveform of inductor current obtained by applying the APRBS signal.

Then, the designed excitation signal is employed as the input of the boost converter to generate data for the inductor's current and the converter's output voltage, while the simulation sampling time is $T_s = 1$ µs. In this period, 7000 input–output data samples are produced, the first 4000 samples are used to build the system model by the LMN and the remaining 3000 data are considered to validate the identified model. Figure 3b, c shows the output voltage and inductor's current waveforms, respectively, achieved by applying the excitation signal.

Local model network (LMN) is utilized for modeling, identification and prediction of numerous nonlinear systems [18–20]. They are appropriate for identification and modeling of complex systems that feature several operating regimes, such as DC-DC converters, as each operating regime can be described by an LLM.

3. Framework

3.1. Network Structure

The LMN output of $\underline{u} = [u_1\ u_2\ \ldots\ u_p]^\mathrm{T}$ and M local linear models can be expressed as,

$$\hat{y} = \sum_{i=1}^{M} h_i(\underline{u}).\Phi_i(\underline{u}) \tag{2}$$

$$h_i(\underline{u}) = \theta_{i,0} + \theta_{i,1}u_1 + \ldots + \theta_{i,p}u_p \tag{3}$$

where $h_i\ (\cdot)$ describes ith local linear model (LLMi), $\theta i = [\theta_{i,0}\ \theta_{i,1}\ \ldots\ \theta_{i,p}]$ is the vector parameter of the LLMi, Φi denotes the corresponding validity function of the LLMi and $\hat{y}$ presents the LMN's output. The LMN in Equation (2) can be interpreted as a Takagi–Sugeno (TS) [21] fuzzy inference system with M rules, and $\Phi_i\ (u)$ and $h_i\ (u)$ represent the rule premises and associated rule consequents, respectively [22].

The validity functions (premises) should form a partition of unity to have a rational interpretation of local models,

$$\sum_{i=1}^{M} \Phi_i(\underline{u}) = 1 \tag{4}$$

In the structure of the LMN, two sets of parameters must be estimated from the measured data. The parameters of the consequent functions, θi, are estimated using a weighted least squares (WLS) algorithm [23]. On the other hand, the structure of the validity functions is determined by HBT.

3.2. Estimation of LLMs' Parameters

Application of the WLS algorithm for estimation of the parameters of the LLMs from N measured samples leads to the following solution,

$$\underline{\theta}_i = \left(\underline{R}_i^T.\underline{D}_i.\underline{R}_i\right)^{-1}.\underline{R}_i^T.\underline{D}_i.\underline{y} \tag{5}$$

where $\underline{y} = [y(1), \ldots, y(N)]^\mathrm{T}$ is a vector that contains N target outputs and $\underline{R}_i$ and $\underline{D}_i$ are regression and diagonal weighting matrices associated with the ith LLM, respectively, which are expressed as follows:

$$\underline{R}_i = \begin{bmatrix} 1 & u_1(1) & u_2(1) & \cdots & u_p(1) \\ \vdots & \vdots & \vdots & \cdots & \vdots \\ 1 & u_1(N) & u_2(N) & \cdots & u_p(N) \end{bmatrix}_{N\times p} \tag{6}$$

$$\underline{D}_i = \begin{bmatrix} \Phi_i(\underline{u}(2)) & 0 & \cdots & 0 \\ 0 & \Phi_i(\underline{u}(2)) & \cdots & 0 \\ \vdots & \vdots & & \vdots \\ 0 & 0 & \cdots & \Phi_i(\underline{u}(N)) \end{bmatrix}_{N\times N} \tag{7}$$

The solution in Equation (5) is obtained by local error minimization of each LLM over the N training samples [20].

3.3. Learning Algorithm

The hierarchical binary tree-learning algorithm is used to identify the parameters of the validity functions. The HBT algorithm starts by an LMN with a single LLM and then adds more LLMs and their validity functions in every iteration to refine LMN and improve its performance. This refinement is realized by partitioning of the input space into hyper-rectangles through axis-orthogonal splits [21,22]. In each iteration of HBT heuristic search, the validity region of the worst LLM is divided into two new regions. The division is tried in all dimensions and the best division associated with the highest improvement in LMN's performance is considered. Then, two new validity functions are constructed for these regions and parameters of their LLMs are estimated by Equation (5).

In each iteration of the HBT algorithm, the validity functions are constructed by proper multiplication of sigmoid splitting functions, ψ_i,

$$\psi_i = \frac{1}{1 + e^{-\sigma_i(w_{i,0} + w_{i,1}u_1 + \ldots + w_{i,p}u_p)}} \tag{8}$$

where direction vector $w_i = [w_{i,1} \ldots w_{i,px}]^T$ sets division direction, position parameter $w_{i,0}$ determines the position of the split and the smoothness parameter σ_i determines the smoothness of the split. A detailed description of the HBT algorithm can be found in [22,23].

4. Design of Local Linear Control

While the DC-DC system model identification is developed/ derived, the LLC is designed to meet the requirements of the power converter. The inverse error dynamics control pursued in this paper includes precise tracking control.

In the exact tracking controller, the error equation is solved to define the control input which is essential for the next error value, i.e.,

$$v_{ref}(k+1) - v_{out}(k+1) = 0 \tag{9}$$

Considering Equations (1) and (2), the estimated converter output voltage is as follows,

$$\hat{v}_{out}(k+1) = \sum_{i=1}^{M} \Phi_i \times \left(\begin{array}{l} \theta_{i,0} + \theta_{i,1}v_{out}(k-1) \\ +\theta_{i,2}v_{out}(k) + \theta_{i,3}i_L(k-1) \\ +\theta_{i,4}i_L(k) \end{array} \right) + \tag{10}$$

$$+ \sum_{i=1}^{M} \Phi_i(\theta_{i,5}d(k-1)) + \sum_{i=1}^{M} \Phi_i(\theta_{i,6}d(k))$$

If we set,

$$\underline{\theta_i} = [\theta_{i,0}\theta_{i,1} \ldots \theta_{i,5}], \tag{11}$$

$$\underline{x}(k) = [1 \; v_{out}(k-1) \; v_{out}(k)i_L(k-1)i_L(k)d(k-1)]^T$$

$$c_i = \theta_{i,6}$$

Then, Equation (10) becomes,

$$\hat{v}_{out}(k+1) = \sum_{i=1}^{M} \Phi_i \times \left(\underline{\theta_i} \times \underline{x}(k)\right) + \sum_{i=1}^{M} \Phi_i \times (c_i \times d(k)) \tag{12}$$

By replacing Equation (12) into Equation (9),

$$d(k) = \frac{v_{ref}(k+1) - \sum_{i=1}^{M} \Phi_i \times \left(\underline{\theta}_i \times \underline{x}(k) \right)}{\sum_{i=1}^{M} \Phi_i c_i} \tag{13}$$

Note that, by defining,

$$d_i(k) = \frac{v_{ref}(k+1) - \underline{\theta}_i \times \underline{x}(k)}{c_i} \tag{14}$$

The control input recommended by the *i*th local linear model, the control input of Equation (12), is as follows,

$$d(k) = \frac{\sum_{i=1}^{M} \Phi_i c_i d_i(k)}{\sum_{i=1}^{M} \Phi_i c_i} \tag{15a}$$

Or

$$d(k) = \sum_{i=1}^{M} \alpha_i d_i(k) \tag{15b}$$

where $\alpha_i = \frac{\Phi_i c_i}{\sum_{i=1}^{M} \Phi_i c_i}$ and $\sum_{i=1}^{M} \alpha_i = 1$.

Therefore, based on Equation (15), the control input in Equation (13) is the same as the weighted average of the control inputs of all local linear models, stated in Equation (14). Therefore, a local linear controller is obtained. The schematic diagram of the LLC is illustrated in Figure 4.

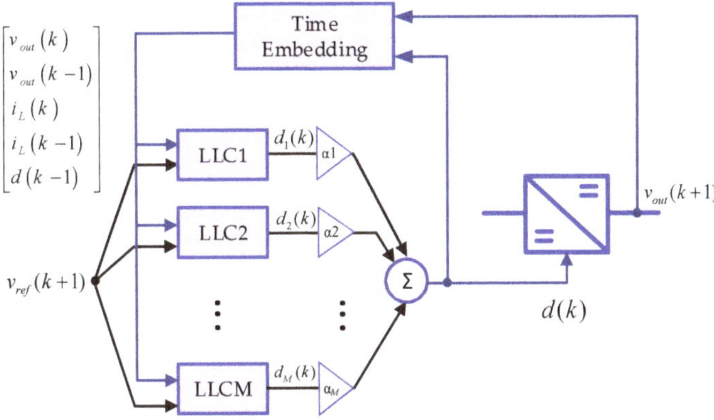

Figure 4. Schematic diagram of the LLC.

5. Simulation Results for a Standalone Boost Converter

This section reports the identification and voltage regulation results for the boost converter of Figure 2.

5.1. Results of Converter Identification

To assess performance of the LMN in the identification of boost converter dynamic behavior, two error criteria, namely root mean square error (RMSE) and mean absolute percentage error (MAPE), were used

$$RMSE = \sqrt{\frac{1}{T} \sum_{k=1}^{T} (v_{out}(k) - \hat{v}_{out}(k))^2} \tag{16}$$

$$MAPE = \left(\frac{1}{T}\right) \cdot \sum_{k=1}^{T} \left\{ \frac{\left|v_{out}(k) - \hat{v}_{out}(k)\right|}{v_{out}(k)} \right\} \tag{17}$$

where T is the number of test samples.

Figure 5 presents the actual and estimated converter output voltage for the test samples. Clearly, the LMN successfully captured the converter dynamic behaviors since the estimated voltages flawlessly matched the actual values. The RMSE and MAPE for estimation of test data are listed in Table 1. The identified system's robustness against measurement noise is also shown in Table 1.

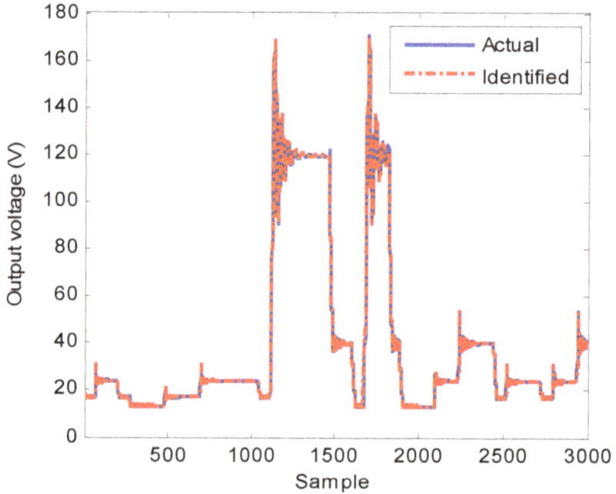

Figure 5. Identification results for the test dataset.

Table 1. Clean and noisy measurement data comparison of the LMN performance.

Case Study	RMSE (Volt)	MAPE (Percent)
Clean data	0.5504	1.0553
Noise contaminated	0.5806	1.0905

5.2. Voltage Regulation

The switching frequency of 100 kHz was adopted for converter control. Using the LLC scheme, the step response of the converter was evaluated for the input voltage of 12 V and the reference voltage of 24 V. The converter step response is illustrated in Figure 6a and the numerical evaluations of the step response are presented in Table 2.

To evaluate the controller performance against the source voltage variations, two-step changes from 12 to 15 V and from 15 to 12 V were applied to the input at time instants 1 ms, 2 ms and 3 ms, respectively. Robustness of the proposed controller against fast variations in the source voltage is noticeable in Figure 6b.

Simulations were performed to evaluate robustness of the controller with respect to load changes. The load changed from 20 Ω to 40 Ω at time instant 5 ms and back again to 20 Ω at 6 ms. The erformance of the proposed controller is depicted in Figure 6c. As can be seen, the LLC almost kept the output voltage unchanged during the load variations.

For more investigation of the proposed LLC, its performance in reference to output voltage tracking was analyzed. Figure 6d shows the LLC performance is tracking various reference voltages including 24 V, 26 V, 28 V and 25 V. Interestingly, the proposed control approach quickly followed variations in reference voltage.

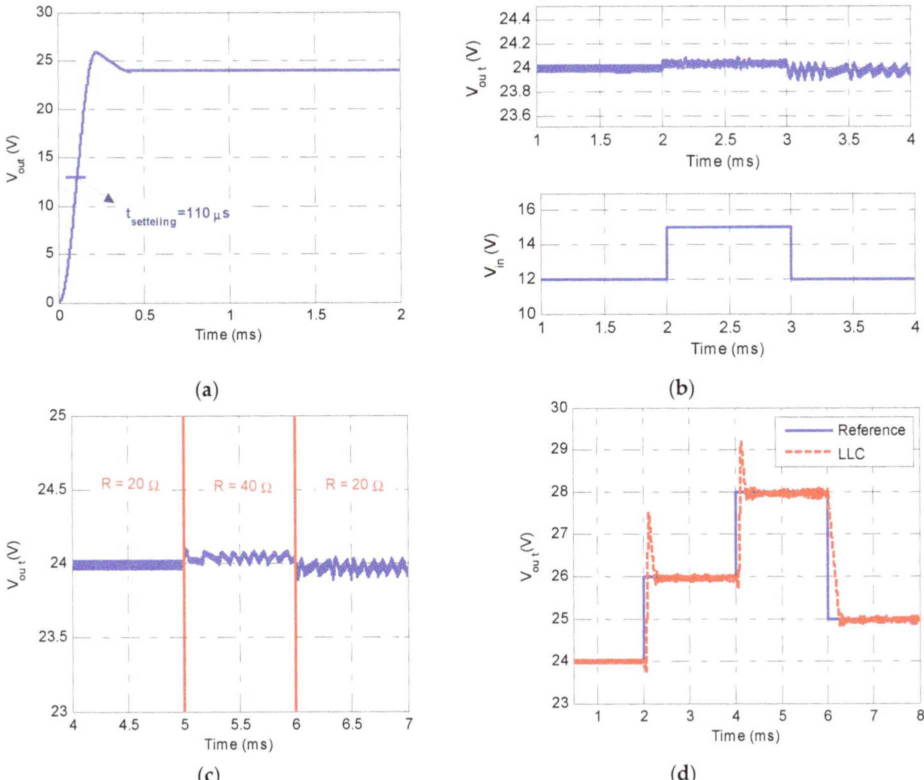

Figure 6. (**a**) Boost converter step response for 24 V reference output voltage; (**b**) response to variations in source voltage; (**c**) response to load variations; and (**d**) ability of the proposed LLC to reference voltage tracking.

Table 2. Numerical evaluations of the step response achieved by the proposed method.

Controller	Criterion		
	Rise Time	**Settling Time**	**Overshoot**
PI	340 µs	850 µs	12%
LLC	110 µs	350 µs	6.2%

Apart from the proposed control performance of the approach against variations in the source voltage, load resistance or reference voltage, its robustness against variations in converter's parameters is also of importance. To assess this capability, converter's parameters (the value of inductance and capacitance in Figure 2a) were varied between −10% to +10% with 5% steps. Then, in each case, the step response was simulated and the integral of time multiplied by the absolute error (ITAE), which is defined by Equation (18), was computed,

$$ITAE = \int_0^{0.002} t.\left|v_{out}(t) - \hat{v}_{out}(t)\right| dt \tag{18}$$

Table 3 compares the values obtained for ITAE in each simulation case. It is clear that ITAEs for all simulations were of the same order. Therefore, it can be concluded that the proposed control approach exhibited great robustness against variations in converter parameters.

Table 3. LLC robustness against variations in parameters of the boost converter.

Parameters Variation	ITAE
−10%	3.12×10^{-7}
−5%	3.31×10^{-7}
Base case	3.51×10^{-7}
5%	5.17×10^{-7}
10%	7.66×10^{-7}

6. Application to a test DC Microgrid

To better investigate the effectiveness of the proposed approach, a DC microgrid was considered. The DC microgrid featured a photovoltaic (PV) panel as well as battery storage and resistive load, as illustrated in Figure 7. The network voltage was 685 V. A DC-DC boost converter interfaced the PV panel to the microgrid. The duty ratio of this converter (d_{MPPT}) was tuned by the incremental conductance algorithm [24] for the maximum power point tracking (MPPT). On the other hand, energy storage was connected to the DC microgrid via a bidirectional DC-DC converter. The bidirectional converter regulated the grid's voltage based on the DDS, inside the microgrid. The DC microgrid specifications are provided in Table 4.

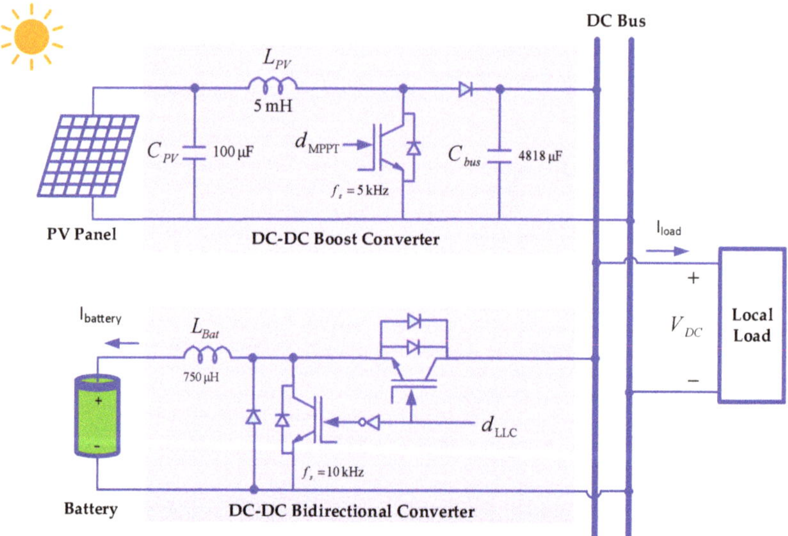

Figure 7. Test DC microgrid structure.

Table 4. Microgrid Specifications.

Device	Specification
PV panel converter with MPPT	20 kW
Bidirectional converter	20 kW
Battery	500 Ah
Load	15 kW

6.1. Identification of Bidirectional Converter Dynamics

Similar to the previous simulation study, the dynamic model of the bidirectional converter was derived from the measurement data. Hence, an APRBS considering 55 duty cycle levels was utilized to produce measurement data. The sampling time was considered 5 µs. Totally, 3000 input–output data sample were produced, 2000 of them served as the training data and the rest were used to validate the identified model.

For modeling of the bidirectional converter, 5-LMN was generated as the best-performing model. The actual and estimated output voltage of the bidirectional converter are shown in Figure 8. Next, the LLC for the bidirectional converter was designed based on the derived LMN.

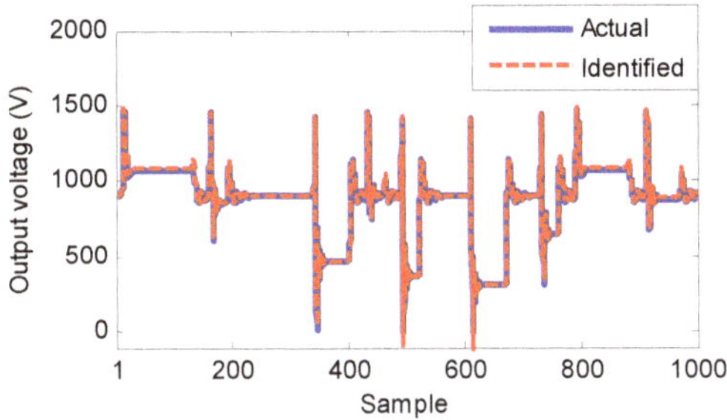

Figure 8. Profile of actual and estimated output voltage for bidirectional converter.

6.2. LLC of the Bidirectional Converter

The design of the local linear controller was carried on based on Equation (15). In this study, three case studies, namely step response, load resistance variation and solar irradiance variation, were applied. Furthermore, PI-based control scheme was designed for comparison. The PI controller comprised the voltage control loop (VCL) and current control loop (CCL), as proposed in [23–26]:

$$TF_{VCL} = \frac{4.26S + 400}{S} \tag{19}$$

$$TF_{CCL} = \frac{0.0026S + 0.4938}{S + 1} \tag{20}$$

6.2.1. Response to Step Change

The response to step change for the LLC is shown in Figure 9a. The achieved response featured speed with negligible overshoot. Table 5 presents the comparison between LLC and PI step responses. Evidently, the proposed LLC provided lower rise time and settling time as well as less overshoot compared to PI.

Table 5. Comparison of the Step Response for the Bidirectional Converter.

Controller	Criterion		
	Rise Time (ms)	Settling Time (ms)	Overshoot (%)
PI	14.15	35.66	6.12
LLC	7.01	11.10	2.62

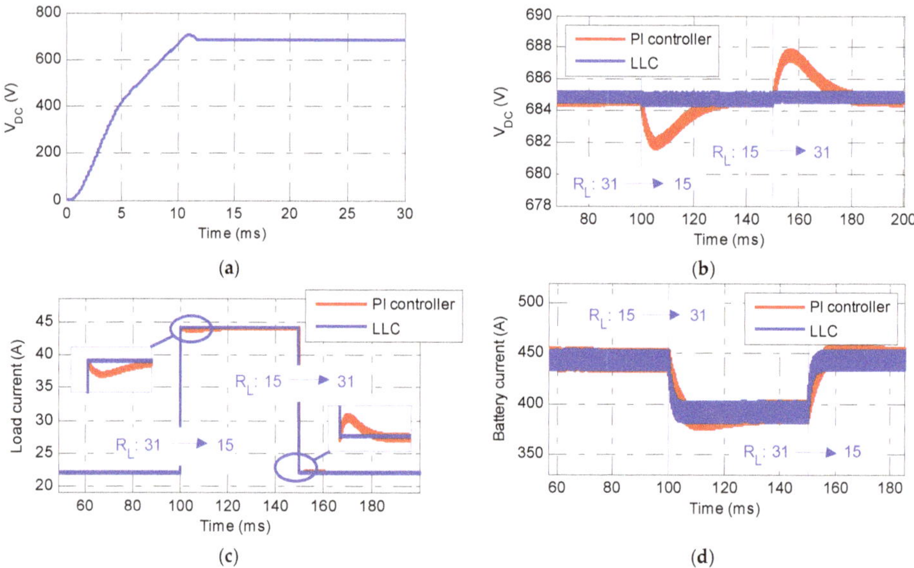

Figure 9. (**a**) Step response of the proposed LLC for the bidirectional converter; (**b**) response to load resistance variations; (**c**) load current during load resistance variation; and (**d**) battery current during load resistance variations.

6.2.2. Load variations

The load variation scenario included load changing from 31 Ω to 15 Ω and back again. The performance of the PI controller and LLC against the load variation is depicted in Figure 9b. As can be seen, the LLC showed outstanding performance.

To present a more detailed comparison, the load and battery currents achieved by both controllers are demonstrated in Figure 9c,d.

6.2.3. Variations in Solar Irradiance

The last study on the DC microgrid was dedicated to the irradiance variations. Based on the control system action, it is expected that the load side voltage is maintained at the reference value during variations in irradiance. The solar irradiation was changed from 1000 to 0 W/m^2 at 130 ms.

Based on the results illustrated in Figure 10a, it is evident that the LLC demonstrated more satisfactory robustness against the variations in the solar irradiance, compared to the PI controller.

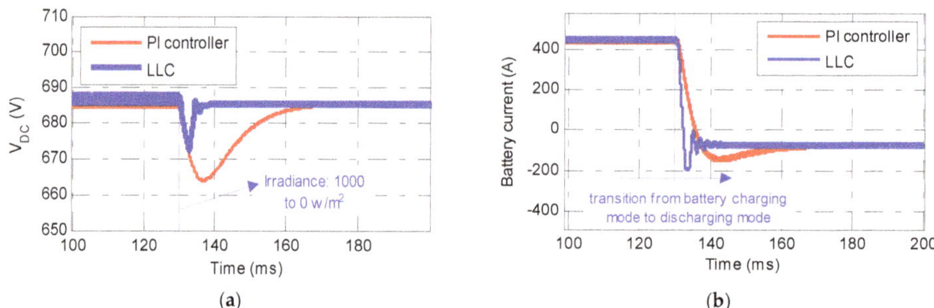

Figure 10. (**a**) Response to variations in irradiance; and (**b**) profile of battery current during irradiance variations.

The PI controller yielded around 3% of pre-specified voltage with slowly-damped undershoot, while the LLC restored the voltage back to the pre-specified value rapidly with around 1.5% undershoot.

The profile of the battery current in Figure 10b shows the charge to discharge mode transition, due to the absence of PV generation. It is clear that the achieved results by the LLC reached the steady-state more quickly than the PI controller.

7. Conclusions

This paper develops a data-driven identification strategy by use of LMNs for DC-DC converters identification and voltage control. The proposed LMNs employ input–output measurement data to identify converter dynamics at the first step. Then, the identified model was used to derive the control law for the voltage regulation of the converter. The proposed approach was used for identification and voltage regulation of converters in both standalone mode as well as grid-connected operation.

For the standalone converter, step response as well as response to load and source variations confirmed the pleasing performance of the LLC. Furthermore, for the DC-DC converters in a DC microgrid, simulations were performed under various scenarios indicating satisfactory performance with the proposed control technique. In both test scenarios, comparisons to conventionally tuned PI control demonstrated the superiority of the proposed approach.

Author Contributions: All authors have contributed equally to this work and all authors have read and approved the final manuscript.

Funding: This research received no external funding.

Conflicts of Interest: The authors declare no conflict of interest.

References

1. Rouzbehi, K.; Candela, J.I.; Luna, A.; Gharehpetian, G.B.; Rodriguez, P. Flexible Control of Power Flow in Multiterminal DC Grids using DC–DC Converter. *IEEE J. Emerg. Sel. Top. Power Electron.* **2016**, *4*, 1135–1144. [CrossRef]
2. Rouzbehi, K.; Yazdi, S.S.H.; Moghadam, N.S. Power Flow Control in Multi-Terminal HVDC Grids using a Serial-Parallel DC Power Flow Controller. *IEEE Access.* **2018**, *6*, 56934–56944. [CrossRef]
3. Yazdi, S.S.H.; Milimonfared, J.; Fathi, S.H.; Rouzbehi, K. Optimal Placement and Control Variable Setting of Power Flow Controllers in Multi-terminal HVDC Grids for Enhancing Static Security. *Int. J. Electr. Power Energy Syst.* **2018**, *102*, 272–286. [CrossRef]
4. Rouzbehi, K.; Gharehpetian, G.B.; Candela, J.; Luna, A.; Harnefors, L.; Rodriguez, P. Multi-terminal DC grids: Operating analogies to ac power systems. *Renew. Sustain. Energy Rev.* **2017**, *70*, 886–895. [CrossRef]
5. Rodriguez, P.; Rouzbehi, K. Multi-Terminal DC grids: Challenges and Prospects. *Mod. Power Syst. Clean Energy* **2017**, *5*, 515–523. [CrossRef]
6. Rouzbehi, K.; Miranian, A.; Luna, A.; Rodriguez, P. Towards fully controllable multi-terminal DC grids using flexible DC transmission systems. Energy Convers. *Congr. Expo. (ECCE)* **2014**, 5312–5316. [CrossRef]
7. Rouzbehi, K.; Miranian, A.; Candela, J.I.; Luna, A.; Rodriguez, P. A hybrid power flow controller for flexible operation of multi-terminal DC grids. In Proceedings of the International Conference on Renewable Energy Research and Application (ICRERA), Milwaukee, WI, USA, 19–22 October 2014; pp. 550–555. [CrossRef]
8. Rouzbehi, K.; Miranian, A.; Candela, J.I.; Luna, A.; Rodriguez, P. Proposals for flexible operation of multi-terminal dc grids: Introducing flexible dc transmission system (FDCTS). In Proceedings of the International Conference on Renewable Energy Research and Application (ICRERA), Milwaukee, WI, USA, 19–22 October 2014; pp. 180–184.
9. De Medeiros, R.L.P.; Barra, W.; De Bessa, I.V.; Filho, J.E.C.; Ayres, F.A.D.C.; Das Neves, C.C. Robust Decentralized Controller for Minimizing Coupling Effect in Single Inductor Multiple Output DC-DC Converter Operating in Continuous Conduction. *Mode. ISA Trans.* **2018**, *73*, 112–129. [CrossRef] [PubMed]
10. Restrepo, C.; Calvente, J.; Romero, A.; Vidal-Idiarte, E.; Giral, R. Current-Mode Control of A Coupled-Inductor Buck–Boost DC–DC Switching Converter. *IEEE Trans. Power Electron.* **2012**, *27*, 2536–2549. [CrossRef]

11. Shen, L.; Li, C.; Lu, D.D.C. Adaptive Sliding Mode Control Method for DC–DC Converters. *IET Power Electron.* **2015**, *8*, 1723–1732. [CrossRef]

12. Elshaer, M.; Mohamed, A.; Mohammed, O. Smart Optimal Control of DC-DC Boost Converter in PV Systems. In Proceedings of the Transmission and Distribution Conference and Exposition, Sao Paulo, Brazil, 8–10 November 2010.

13. Cheng, K.H.; Hsu, C.F.; Lin, C.M.; Lee, T.T.; Li, C. Fuzzy–Neural Sliding-Mode Control for DC-DC Converters Using Asymmetric Gaussian Membership Functions. *IEEE Trans. Ind. Electron.* **2007**, *54*, 1528–1536. [CrossRef]

14. Rouzbehi, K.; Miranian, A.; Citro, C.; Luna, A.; Rodriguez, P. Enhanced Average Current-Mode Control for DC-DC Converters Based on An Optimized Fuzzy Logic Controller. In Proceedings of the IECON 2012—38th Annual Conference of IEEE Industrial Electronics Society, Montreal, QC, Canada, 25–28 October 2012; pp. 382–387.

15. Bastos, R.F.; Aguiar, C.R.; Gonçalves, A.F.Q.; Machado, R.Q. An Intelligent Control System Used to Improve Energy Production From Alternative Sources With DC/DC Integration. *IEEE Trans. Smart Grid.* **2014**, *5*, 2486–2495. [CrossRef]

16. Naik, B.; Mehta, A. Sliding mode controller with modified sliding function for DC-DC Buck Converter. *ISA Trans.* **2017**, *70*, 279–287. [CrossRef]

17. Nelles, O. *Nonlinear System Identification: From Classical Approaches to Neural Networks and Fuzzy Models*; Springer-Verlag: Berlin, Germany, 2001.

18. Hametner, C.; Jakubek, S. Nonlinear Identification with Local Model Networks using GTLS Techniques and Equality Constraints. *IEEE Trans. Neural Networks* **2011**, *22*, 1406–1418. [CrossRef] [PubMed]

19. Kazemi, R.; Abdollahzade, M. Introducing an Evolving Local Neuro-Fuzzy Model—Application to modeling of car-following behavior. *ISA Trans.* **2015**, *59*, 375–384. [CrossRef] [PubMed]

20. Oysal, Y.; Yilmaz, S. Fuzzy Wavelet Neural Network Models for Prediction and Identification of Dynamical Systems. *IEEE Trans. Neural Networks.* **2010**, *21*, 1599–1609.

21. Takagi, T.; Sugeno, M. Fuzzy Identification of Systems and Its Applications to Modeling and Control. *IEEE Trans. Syst. Man Cybern.* **1985**, *15*, 116–132. [CrossRef]

22. Miranian, A.; Abdollahzade, M. Developing a Local LSSVM-Based Neuro-Fuzzy Model for Nonlinear and Chaotic Time Series Prediction. *IEEE Trans. Neural Netw.* **2013**, *24*, 207–218. [CrossRef]

23. Miranian, A.; Rouzbehi, K. Nonlinear Power System Load Identification Using Local Model Networks. *IEEE Trans. Power Syst.* **2013**, *28*, 2872–2881. [CrossRef]

24. Narendra, K.; Lin, Y.-H.; Valavani, L. Stable Adaptive Controller Design, Part II: Proof of Stability. *IEEE Trans. Autom.* **1980**, *25*, 440–448. [CrossRef]

25. Alonge, F.; Rabbeni, R.; Pucci, M.; Vitale, G. Identification and Robust Control of a Quadratic DC/DC Boost Converter by Hammerstein Model. *IEEE Trans. Ind. Appl.* **2015**, *51*, 3975–3985. [CrossRef]

26. Xiao, W.; Dunford, W.; Palmer, P.; Capel, A. Application of Centered Differentiation and Steepest Descent to Maximum Power Point Tracking. *IEEE Trans. Ind. Electron.* **2007**, *54*, 2539–2549. [CrossRef]

Article

Design and Realization of a Bidirectional Full Bridge Converter with Improved Modulation Strategies

Filippo Pellitteri [1],*, Rosario Miceli [1], Giuseppe Schettino [1], Fabio Viola [1] and Luigi Schirone [2]

[1] Department of Engineering, University of Palermo, 90128 Palermo, Italy; rosario.miceli@unipa.it (R.M.); giuseppe.schettino@unipa.it (G.S.); fabio.viola@unipa.it (F.V.)
[2] School of Aerospace Engineering, Sapienza University of Rome, 00138 Rome, Italy; luigi.schirone@uniroma1.it
* Correspondence: filippo.pellitteri@unipa.it

Received: 8 April 2020; Accepted: 26 April 2020; Published: 28 April 2020

Abstract: In this paper a Full-Bridge Converter (FBC) for bidirectional power transfer is presented. The proposed FBC is an isolated DC-DC bidirectional converter, connected to a double voltage source—a voltage bus on one side and a Stack of Super-Capacitors (SOSC) on the other side. The control law aims at the regulation either of the bus current (when the load requires power) or of the SOSC current (when the stack requires a recharge). Analysis and design of the proposed FBC are discussed. A Phase Shift Modulation (PSM) scheme is proposed, along with an improved modulation variant for the efficiency optimization, through a proper reduction of the transformer power losses. The realized prototype, compliant with automotive applications, is presented and experimental results are highlighted. The target power level is 2 kW.

Keywords: full-bridge converter; phase shift modulation; supercapacitors; isolated DC-DC bidirectional converter

1. Introduction

Energy Storage Systems allow the durable collection of electrical energy arising from different sources. With respect to the electrical grid, alternative ways of obtaining power are renewable sources and energy harvesting [1–3]. As far as storage systems are concerned, their hybridization is gathering momentum in several fields, such as automotive or aerospace, due to the opportunity to integrate different features in the same storage system. Among the possible electrical storage elements, batteries and supercapacitors offer significant benefits in terms of energy density and power density respectively, so that their combination can lead to improvements in terms of total cost and efficiency [4–9]. Investigation of possible power system architectures and power converter topologies in order to properly manage the electrical energy inside a hybrid storage system is therefore an attractive research topic [10–12].

According to specific application and power level, different types of power system architectures are possible. A common DC voltage bus is generally used to supply the existing loads and this bus could be either directly connected to a battery or connected to it through a DC-DC converter. The supercapacitor instead shall be properly managed in order to supply the load or to recharge the battery itself—a bidirectional DC-DC converter can be used for this purpose. The mentioned DC-DC bidirectional converter shall be able to control the charge/discharge of the supercapacitor from/to the bus.

State of art and future trends concerning DC-DC converters for automotive applications are widely provided in Reference [13], whereas different investigations on DC-DC bidirectional converters for both supercapacitors and battery charging applications are provided in References [14–17], with particular focus on reliability, ease of control, sizing and robustness. Due to the potentially high voltage range of a stack of supercapacitors, proper DC-DC converter architectures could be used—in References [18–21] high-voltage-ratio topologies are described, feasible for several applications such as automotive, DC microgrids and renewable energy sources. At the same time, such high voltage levels could suggest the use of galvanic isolation in order to guarantee safety—in References [22–27] different applications of isolated DC-DC converters for energy storage management are reported.

In this paper, an insulated DC-DC bidirectional converter is proposed for the management of a hybrid storage system based on battery and supercapacitor. The proposed topology is a full-bridge converter (FBC), which has been realized and experimentally tested with two DC sources emulating a Stack Of Supercapacitors (SOSC) whose maximum voltage is 70 V and a 28 V bus. The rated power level is 2 kW. In order to avoid expensive clamping networks and maintain a high efficiency target, the conventional Phase Shift Modulation (PSM) has been modified through a novel variant aiming at reducing undesirable voltage spikes.

The proposed power converter could find its application whenever a battery-supercapacitor hybrid storage system is present—the battery provides average power, whereas the supercapacitance is able to provide the peak power pulses. Typical application examples of power system architectures are those inside an Electric Vehicle (EV), based on an automotive DC voltage bus or inside a space launcher, based on an avionic DC voltage bus.

This paper is organized as follows—Section 2 provides a description of the proposed power system architecture and analyzes the proposed bidirectional DC-DC converter; Section 3 focuses on the Full Bridge converter design, accordingly, comparing different modulation strategies; Section 4 reports experimental results concerning a Full Bridge converter prototype, able to transfer a 2 kW power; in Section 5 conclusions are given.

2. Analysis of the Full Bridge Converter (FBC)

The aim of the proposed DC-DC converter is to manage the energy flow between two energy sources, being in the specific case a Stack Of Supercapacitors (SOSC) and a DC voltage bus, as highlighted in Figure 1, showing the power system architecture. A typical current profile, required from the bus section, is highlighted as well: the positive current values correspond to a SOSC discharge, whereas the negative ones correspond to a SOSC recharge. The maximum required current value is 70 A, which is equivalent to a maximum rated power of 2 kW for a nominal DC bus voltage equal to 28 V.

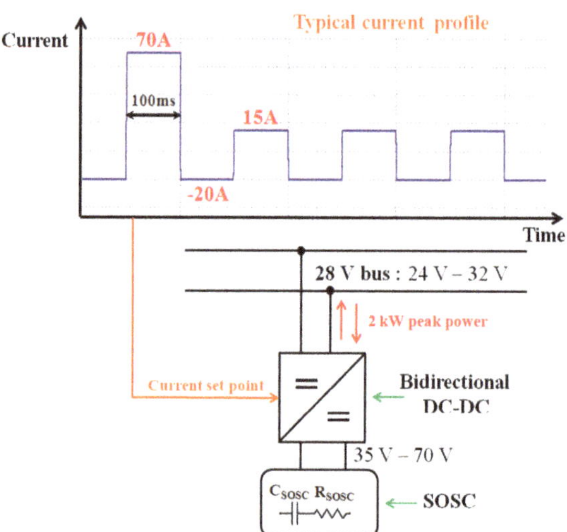

Figure 1. Power system architecture.

2.1. Energy Sources Model

The SOSC voltage lies in the range 35 V–70 V, whereas the nominal DC bus voltage is 28 V. In Table 1 the SOSC parameters and voltage range are reported, where C_{SOSC} and R_{SOSC} are the SOSC equivalent capacitance and resistance. The reported values arise from the considered specific stack, which is supposed to be a 26s11p (26 series–11 parallel) connection of 2.7 V 10 F 35 mΩ ultracapacitor cells (Maxwell Technologies, San Diego, CA, USA).

Table 1. Stack of Super-Capacitors (SOSC) parameters and voltage range.

C_{SOSC}	R_{SOSC}	V_{min}	V_{max}
4 F	70 mΩ	35 V	70 V

Being equal to the product between C_{SOSC} and R_{SOSC}, the SOSC charge/discharge time constant is equivalent to hundreds of ms, so that the SOSC can be considered as a DC power source with respect to a converter switching frequency in the order of tens-of-kHz. For this reason the proposed converter can be analyzed and designed as an actual DC-DC converter.

In Table 2 the DC bus specifications concerning the voltage level are reported.

Table 2. DC bus voltage minimum, nominal and maximum level.

V_{min}	V_{nom}	V_{maz}
24 V	28 V	32 V

2.2. Full Bridge Converter Analysis

In Figure 2, the schematic of the proposed Full Bridge Converter (FBC) is shown. The FBC is an insulated topology, featuring a transformer between the sections connected to the DC voltages V_1 and V_2, representing the SOSC and the bus respectively. As far as the transformer is concerned, L_1 and L_m are the equivalent primary leakage and magnetizing inductances respectively, whereas $n{:}m$ is the turns ratio. Each of the H-bridges in the primary and in the secondary side of the transformer consists of four switches, in this specific case four enhancement n-channel MOSFETs—the primary side H-bridge consists of M1-M2-M3-M4, whereas the secondary side one consists of M5-M6-M7-M8. In this network therefore energy can flow in both directions, either from V_1 to V_2 or from V_2 to V_1—in the first case the primary H-bridge acts as an inverter and the secondary one as a rectifier, in the second case the H-bridges play the opposite roles.

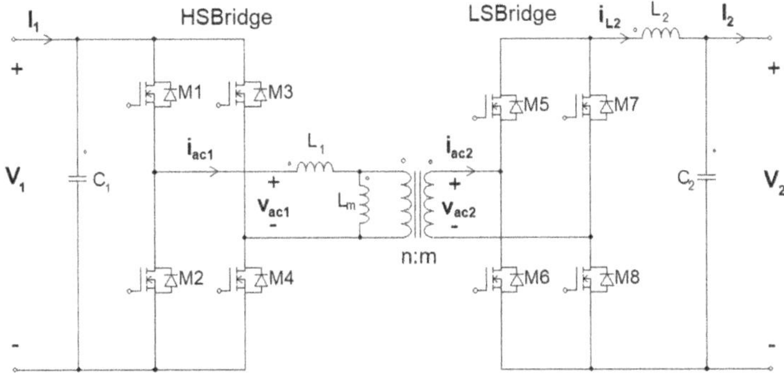

Figure 2. Schematic of the Full Bridge Converter (FBC).

If a voltage source is connected to V_1 port and a load is connected to V_2 port, the proposed converter implements a step-down operation.

The primary and secondary H-bridges can therefore be referenced as High-Side Bridge (HSBridge) and Low-Side Bridge (LSBridge).

L_2 is the converter inductor, placed in series with the LSBridge.

2.3. Possible Modulation Techniques

Two possible modulation schemes have been investigated on the proposed FBC, as shown in Figures 3 and 4, *Vgn* being the logic level applied to the gate-source voltage of the MOSFET Mn—the Pulse Width Modulation (PWM) and the Phase Shift Modulation (PSM).

Figure 3 shows the PWM scheme—M1 and M4 gate signals are in phase, as well as M2 and M3 gate signals and a phase difference occurs between the diagonals M1–M4 and M2–M3; the secondary-side gate signals are obtained by logical negation (NOT) of the primary-side signals, as highlighted in the figure. The on-time of each HSBridge switch is *DTs*, where *D* is the duty-cycle and *Ts* the switching period, being the duty-cycle limited to less than 50% in order to avoid short-circuit at the primary side.

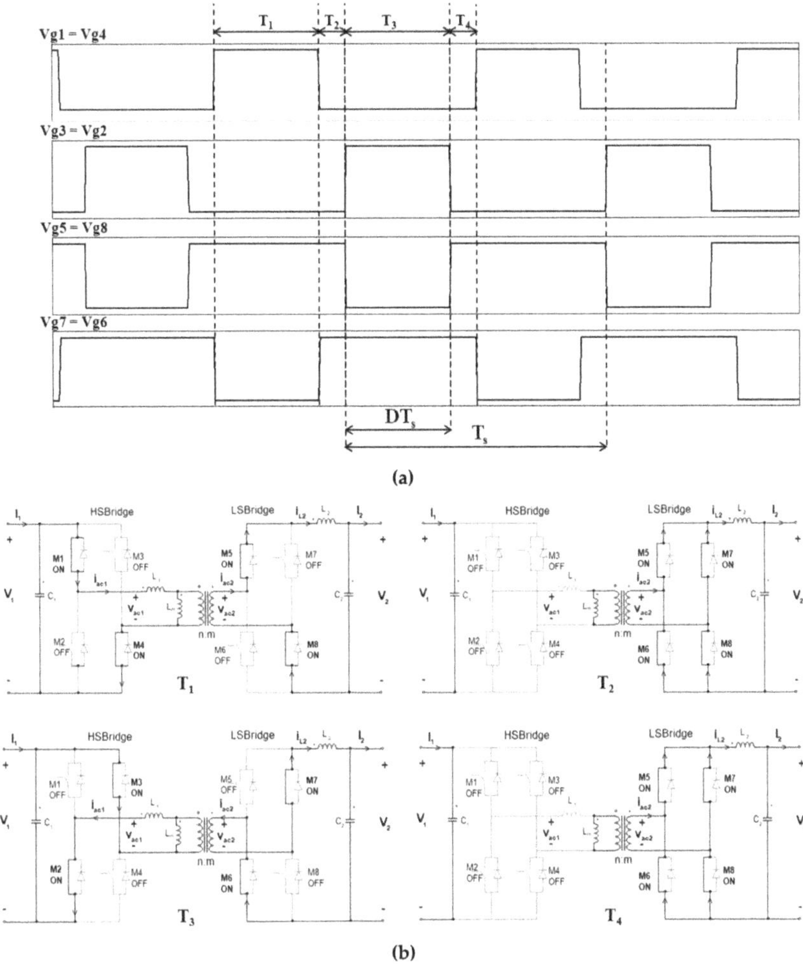

Figure 3. Pulse Width Modulation (PWM) case: (**a**) gate signals; (**b**) modes of operation.

As highlighted by the modes of operation, if D goes higher than 50%, the time windows T_2 and T_4, corresponding to the L_2 discharge towards short circuit, would be deleted, thus avoiding a proper converter working; moreover, the time windows T_1 and T_3 would be in overlap, thus leading to an undesirable short circuit on V_1.

Figure 4 shows the PSM scheme—M1 and M2 are always in phase opposition, as well as M3 and M4; a phase difference, corresponding to the duty-cycle D, occurs between them. The PWM, therefore, is converted into a phase shift modulation.

At the same way as in the PWM, the duty-cycle D is limited to less than 50% in order to avoid open-circuit at the secondary side, which could lead to voltage spikes due to the inductance L_2.

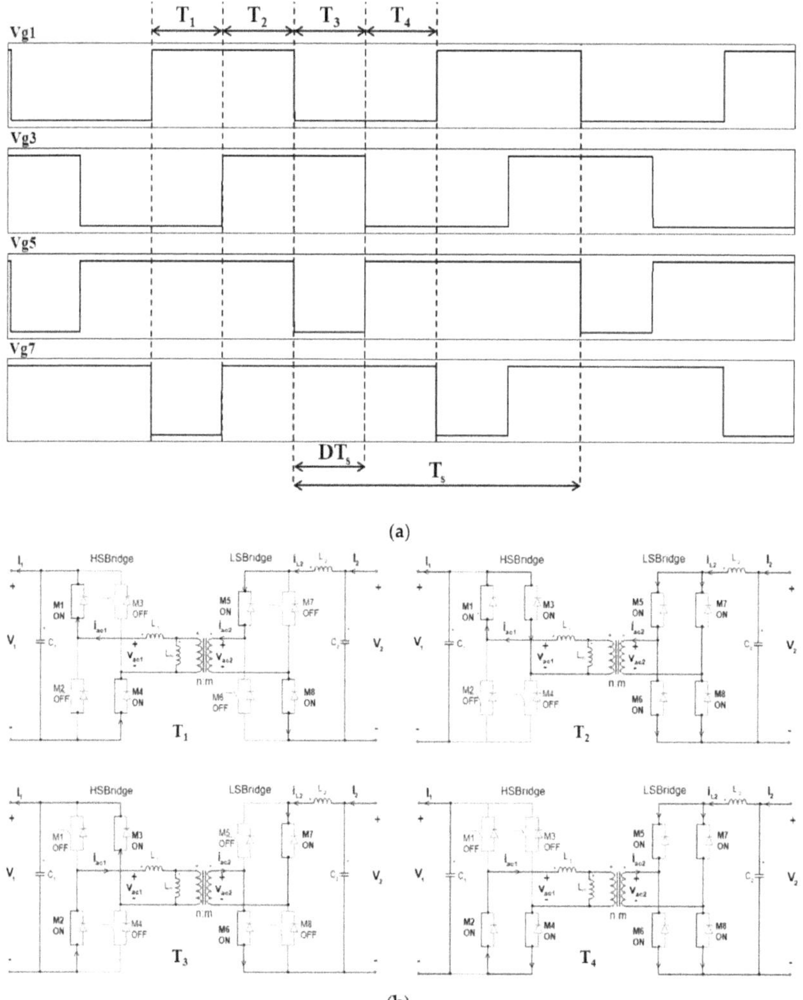

Figure 4. Phase Shift Modulation (PSM) case: (**a**) gate signals; (**b**) modes of operation.

As highlighted by the modes of operation, if D goes higher than 50%, the time windows T_2 and T_4, corresponding to the L_2 discharge towards short circuit, would be deleted, thus avoiding a proper converter working, such as in the PWM mode; moreover, the time windows T_1 and T_3 would be in overlap, thus leading to an undesirable open circuit on L_2.

The PSM scheme has been preferred to the PWM, since the PWM scheme involves higher switching power losses, due to the Zero Voltage Switching (ZVS) condition which is reached in the PSM.

The proposed converter aims at the regulation of the power in terms of both amount and direction.

Considering that V_1 and V_2 are two DC voltage sources—a Stack Of Supercapacitors and a DC bus respectively—the power regulation is therefore consisting in a current regulation.

3. Design and Modulation Strategies for the Proposed Converter

3.1. Choice of the n:m Transformer Ratio

In a conventional full-bridge converter, with a resistive load at its output, the ratio of the output voltage V_2 to the input voltage V_1 is equal to:

$$\frac{V_2}{V_1} = \frac{2mD}{n}.$$ (1)

This means that for the designed power converter, where voltage sources are applied at both ports (the V_1 SOSC source at the input and the V_2 bus source at the output), this condition corresponds to a "current balance," meaning that no DC current is flowing.

In order to produce a current flow, this condition shall be modified. If the SOSC has to discharge into the bus, the duty-cycle, representing the control parameter, shall be increased towards the maximum limit, that is $D = 0.5$.

The most critical condition is represented by the minimum voltage difference between SOSC and bus, that is when the SOSC is at its minimum voltage ($V_1 = 35$ V) and the bus is at its maximum voltage ($V_2 = 32$ V). In this condition, the V_2-to-V_1 ratio is at its maximum value, so that, considering that D must be less than 0.5, the secondary-to-primary ratio m:n shall be higher than 1 according to (1), especially considering the case of a positive bus current I_2. For this reason, the selected n and m have been chosen according to the following ratio:

$$n : m = 2 : 3.$$ (2)

A higher-than-1 turns ratio is even more required considering the voltage drop in the primary side, due to the R_{SOSC}, possibly leading the voltage V_1 from 35 V to less than 32 V and all the voltage drops concerning the different components.

If the leakage inductance effect is neglected, the L_2 average current $I_{L2,av}$, corresponding to the bus current, is the following:

$$I_{L2,av} = \frac{2\frac{m}{n}DV_1 - V_2}{R_{eq} + \left(2\frac{m}{n}D\right)^2 R_{bosc}} = I_2,$$ (3)

where R_{eq} refers to all the resistive losses in the converter.

3.2. Modulation Strategies

In Figure 5, all the eight gate signals in the case of the previously described Phase Shift Modulation (PSM) are shown in simulation.

As expected by the modulation law and highlighted by the figure, each LSbridge gate signal event is simultaneous with an HSbridge gate signal event.

Figure 5. Gate signals in conventional PSM.

In Figure 6, the resulting secondary-side waveforms are highlighted, where V_{L2} is the voltage at the output of the LSbridge and T_s is the switching period.

The considered case concerns the SOSC recharge, that is when I_2 is negative and $I_{L2,av}$ as well. In this case, I_{L2} is divided into the drain-source currents of M5 and M7. Every $T_s/2$, whenever Vg5–Vg8 or Vg6–Vg7 goes low, that is when M5 or M7 is opened, there is no conduction path through the opened MOSFET, due to the reverse direction of the MOSFET-connected diode. Therefore, if before the switch opening the drain-source current on the other switch was negative, a voltage spike on V_{L2} is provoked since the inductor energy is not free to circulate. This is better highlighted in Figure 7.

A useless energy waste is therefore shown in case of energy flow from V_2 to V_1 with the conventional PSM.

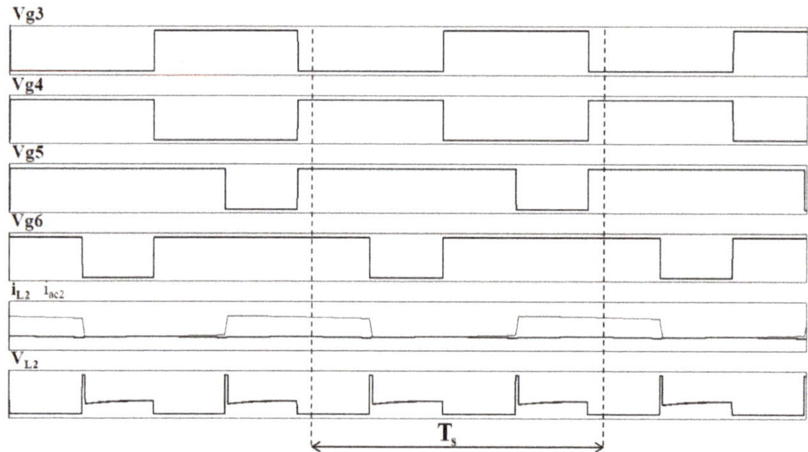

Figure 6. Secondary-side waveforms for the conventional PSM.

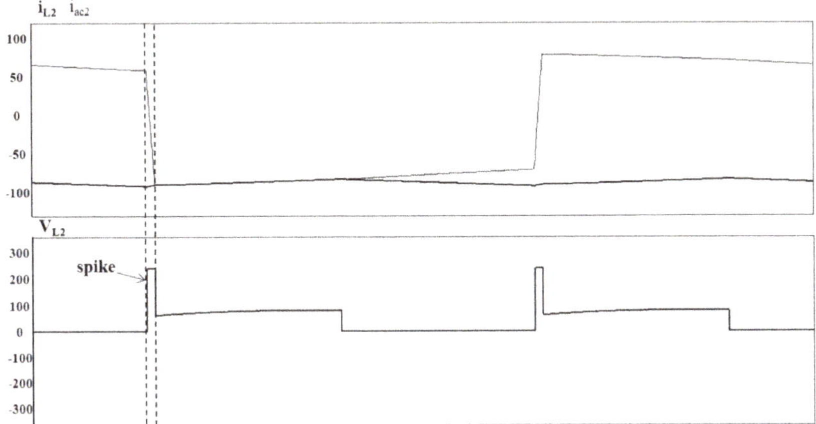

Figure 7. Zoom on the secondary-side waveforms for the conventional PSM.

In order to avoid the mentioned voltage spikes, thus avoiding to limit the overall power system efficiency or to add an expensive clamping network, an improved modulation strategy is proposed in the following.

In Figure 8 the gate signals concerning the proposed modified Phase Shift Modulation (PSM) are shown in simulation.

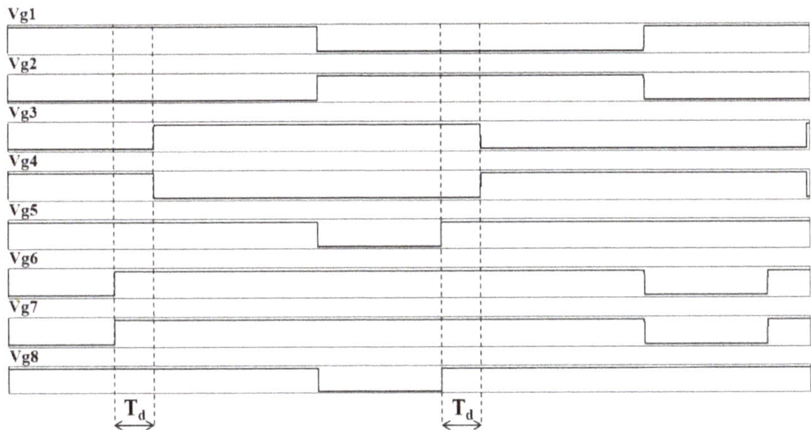

Figure 8. Gate signals in modified PSM.

The aim of the improved control law is to open M5–M8 (M6–M7) when the drain-source current on M6–M7 (M5–M8) is positive (negative), that is when i_{ac2} is positive (negative). In order to reach this goal, the positive events of *Vgn* concerning the LSbridge are anticipated by a time window, referenced as T_d, so that i_{ac2} has the time to reverse its sign.

In Figure 9, the resulting secondary-side waveforms are highlighted—this time V_{L2} presents the ideal shape, as better highlighted by Figure 10.

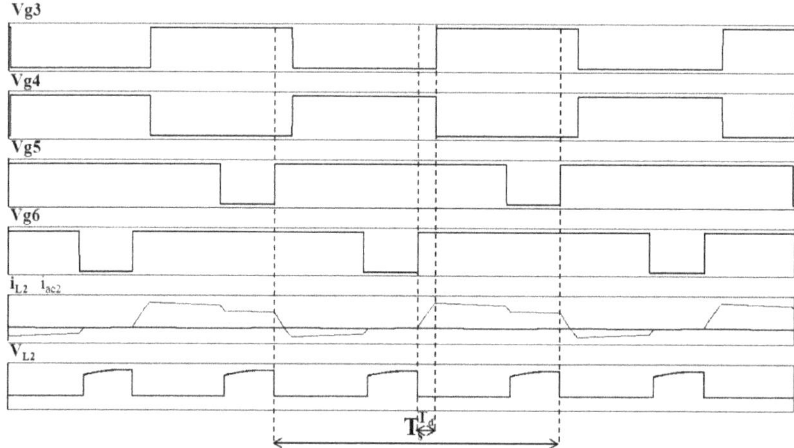

Figure 9. Improved secondary-side waveforms for the modified PSM.

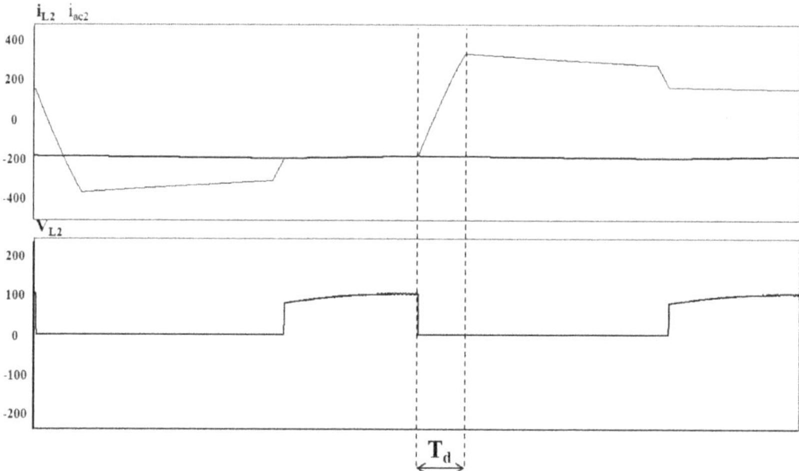

Figure 10. Zoom on the secondary-side waveforms for the modified PSM.

A power efficiency increase is therefore achievable by means of this improved modulation strategy in case of SOSC recharge, as highlighted in Table 3, reporting system efficiency η for conventional and improved PSM, with respect to the extreme values of V_1 and V_2, at full power.

Table 3. Efficiency at full power in case of SOSC recharge, as resulting from simulation.

Conventional PSM			Improved PSM		
V_1	V_2	η	V_1	V_2	η
70 V	32 V	0.72	70 V	32 V	0.84
70 V	24 V	0.62	70 V	24 V	0.80
35 V	32 V	0.73	35 V	32 V	0.84
35 V	24 V	0.67	35 V	24 V	0.85

4. Experimental Tests

A FBC has been realized aiming at a 2 kW power target and compliant with the specifications concerning the DC voltage sources and the current requirements. The mounted converter consists of the following parts—the power board, including the actual FBC, filtering networks, protection switches and transducers for telemetries; a control board, including signal conditioning, input/output interfaces, controller and gate signals generation; an auxiliary power supply, including power converters to generate all internal supply lines; a mechanical frame/heatsink for thermal dissipation. The control loop aims at the regulation of the bus current value according to its desired value and it has been entirely implemented through analog components.

In Figure 11 the prototypal breadboard is shown. Auxiliary power supply board and control board on the bottom-left corner and bottom-right position can be noted.

Figure 11. The prototype of the bidirectional FBC, as built and mounted.

4.1. H-bridges Realization

Each power switch in the power board consists of four parallel MOSFETs. The selected components are compliant with Automotive applications and show a TO-247 package. Though-hole type was found convenient as it natively provides connection to multiple PCB layers. In the SOSC-side, AUIRFP4568 nMOSFETs have been used, rated for a maximum 171 A drain current $I_{d,max}$ and for a maximum 150 V source-to-drain voltage V_{DSS}, with a maximum 5.9 mΩ on resistance; in the bus-side, IXFH140N20 $\times$ 3 nMSOFETs have been used, rated for a maximum 140 A drain current $I_{d,max}$ and for a maximum 200 V source-to-drain voltage V_{DSS}, with a maximum 9.6 mΩ on resistance.

The bus-side switches have been selected with a higher maximum voltage due to the transformer ratio of the secondary side (bus-side) turns to the primary side (SOSC-side) turns, which is higher than one.

All MOSFETs are driven by means of isolated gate drivers, useful to maintain galvanic insulation between controller and power circuits. The selected component is UCC21521, providing dual independent channels, useful for managing the low-side and the high-side MOSFETs and guaranteeing a 4–6 A (source/sink) current capability and a 16 ns time rise on a 2 nF load capacitance.

4.2. Reactive Components Realization

As far as the used capacitors are concerned, Multi-Layer Ceramic Capacitances (MLCC) are preferred to provide the highest frequency current peaks, since MLCC present equivalent impedances with higher cut frequencies with respect to the electrolytic and Metallized Polyester (PET) ones, which instead contribute to the rms component of the pulsed currents.

The magnetic components with pulsed currents in the tens of A range (the converter inductor L_2 and the transformer) have been implemented with EE-type-cores. For the transformer, two magnetic structures have been used, as highlighted by the figure. Copper foils have been used for the windings and multiple wires for the connections with the PCB, in order to have a large current capability. The material used for the transformer cores is 3C95, a power ferrite with high saturation levels and low losses. For L_2 a powder core with distributed air gap is used, whose material is Kool Mµ® 40 from (Magnetics, Phoenix, AZ, USA).

In Tables 4 and 5 the main characteristics of the designed and realized transformer and L_2 are respectively reported.

Table 4. Characteristics of the realized transformer.

Core	Material	n	m	Measured L_m	Measured L_1
EE80/38/20	3C95	4	6	60 µH	140 nH

Table 5. Characteristics of the realized inductor.

Core	Material	Turns	Measured L_2
00K7228E040	Kool Mµ® 40	13.5	24 µH

In Table 6, the values of the selected converter switching frequency f_{sw} and reactive components are reported.

For filtering inductors, with DC currents and small ripple, toroidal powder cores are used, with multi-wire windings. The used toroidal core materials are MolyPermalloy Powder (MPP) and High Flux (Magnetics, Phoenix, AZ, USA).

Table 6. Selected values of switching frequency and reactive components for the designed converter.

Parameter	Value
f_{sw}	50 kHz
L_2	24 µH
C_1	130 µF
C_2	200 µH

The choice of L_m is based on specifications concerning the desired magnetizing current, whereas the reactive elements selection is based on specifications concerning the desired current ripple at the input and output sections. Details regarding these specifications are beyond the purpose of this paper.

4.3. Experimental Setup

A schematic of the test setup, proposed for full power (2 kW) transfer, is shown in Figure 12.

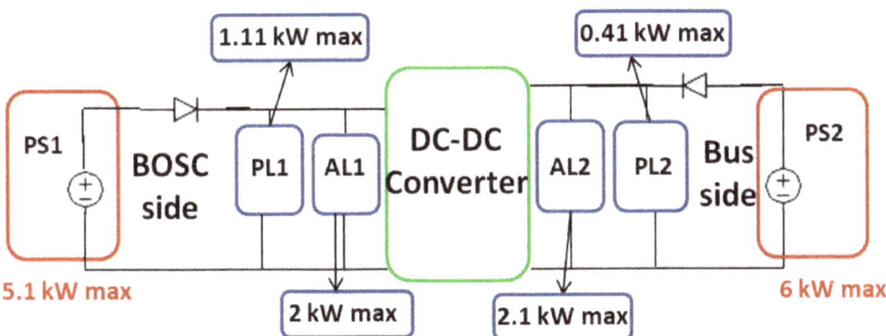

Figure 12. Schematic of the test setup.

Each side (SOSC or bus) is emulated by a 1Q (One Quadrant) Power Supply Stack—PS1 and PS2 for SOSC and bus respectively—and a Load Stack, so that a bi-directional power flow can be tested—when the power flows from the SOSC to the bus, PS1 provides the required energy to the bus-side Load Stack; when the power flows from the bus to the SOSC, the SOSC-side Load Stack adsorbs energy arising from PS2.

In order to avoid a negative current flowing through the power supplies, a protection diode is used between the supply and the load. Therefore, 2 protection diodes are used, one on the SOSC side and one on the bus side. Each of them is able to withstand a maximum 200 A current, considering that a maximum 70 A current is supposed to flow in the converter and a maximum 80 A current is needed to continuously supply the Passive Loads. Therefore, in order to guarantee that both the protection diodes are continuously polarized in direct way, a minimum direct current (supposed to be equal to 5 A) shall flow through each of them. Some Passive Loads are needed for this purpose, in order to guarantee the required power absorption.

4.3.1. SOSC-Side Setup

The goal of the SOSC test setup is to guarantee—(35 V–70 V) voltage range; (−2 kW–+2 kW) power range. The SOSC-side Power Supply (PS1) Stack can provide a total maximum power of 5.1 kW. The SOSC-side Load Stack consists of an Active Load (AL1) and a Passive Load (PL1), for a total maximum power of 3.11 kW.

4.3.2. Bus-Side Setup

The goal of the bus test setup is to guarantee—(24 V–32 V) voltage range; (−2 kW–+2 kW) power range. The bus-side Power Supply (PS2) Stack can provide a total maximum power of 6 kW. The bus-side Load Stack consists of an Active Load (AL2) and a Passive Load (PL2), for a total maximum power of 2.51 kW.

4.4. Experimental Results

In Figure 13 a picture of the arranged test setup is shown, along with the highlighted parts.

Figure 13. The arranged test setup.

The improved PSM was implemented for the built prototype. The proper behavior of the converter is shown in Figure 14, showing the proper bus current response to a square wave command. The transduction factor in the command and monitoring signals is equal to 1 V/10% of full power, where "full power" corresponds to a 70 A bus current. Therefore, the image refers to a step between two levels corresponding to about 70% of the full negative power (during the SOSC recharge) and of the full positive power (during the SOSC discharge). There are no overshoot phenomena in the transients.

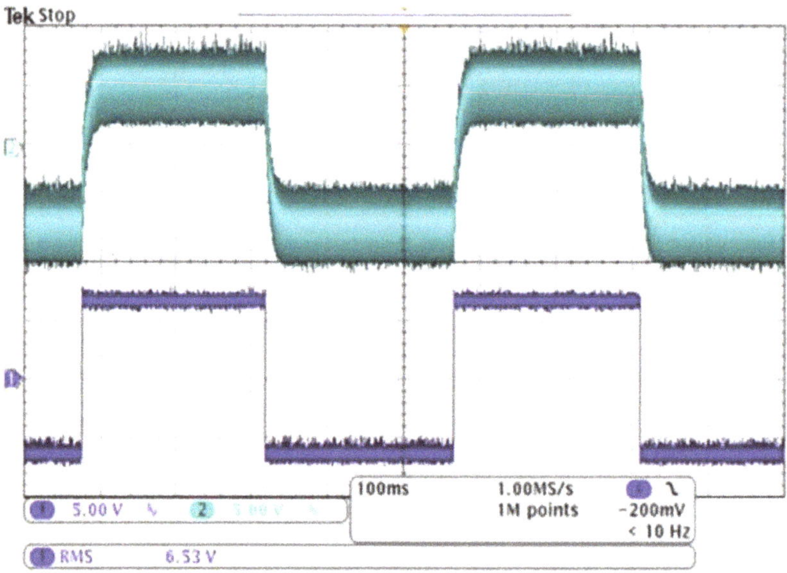

Figure 14. The proposed behavior of the bus current (in light blue), as experimentally tested, corresponding to a given power profile (in blue). For the monitoring signal (light blues) and the command signal (blue) 1 V is equivalent to the 10% of the full power.

Figures 15 and 16 show a zoom of the positive and negative transients respectively. Response time is compliant with the requirements, according to typical applications where 100 ms maximum response times are allowed. A further improvement of the response time is possible through a slight refinement of the control loop

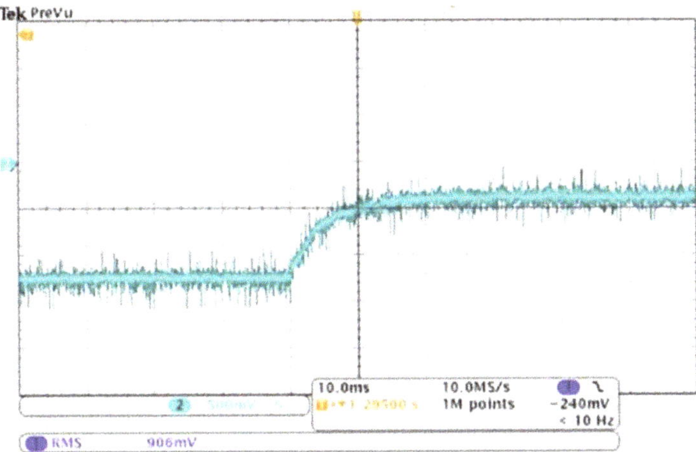

Figure 15. Zoom on a positive bus current transient, responding to an instantaneous positive command.

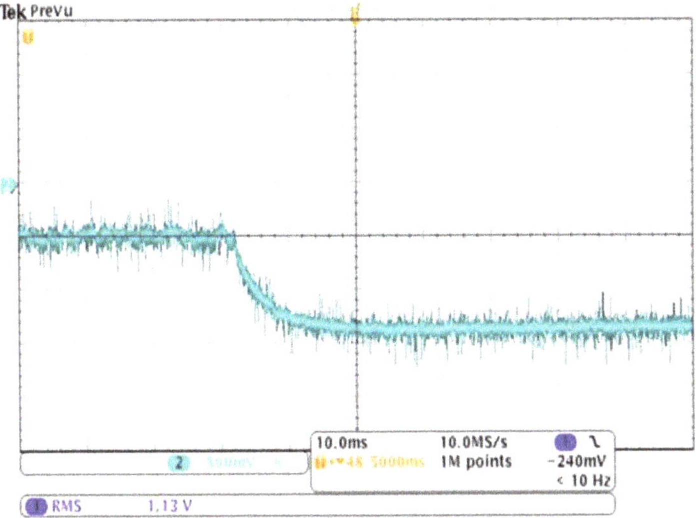

Figure 16. Zoom on a negative bus current transient, responding to an instantaneous negative command.

5. Conclusions

In this paper, an insulated bidirectional DC-DC converter for the management of a storage system is proposed. Electrical Storage Systems find different possible application fields—automotive, zero-energy buildings, aerospace and so forth. The described topology is a Full-Bridge Converter (FBC), connected to a double voltage source—a voltage bus on one side and a Stack of Super-Capacitors (SOSC) on the other side. Analysis, design and modulation strategies of the proposed FBC are discussed. An improved modulation strategy has been proposed by authors, aiming at avoiding expensive clamping networks and at the power losses reduction. A 2 kW converter prototype, including also control and auxiliary power supply boards and compliant with automotive applications, has been realized and the related experimental results have been presented as well, proving the proper behavior of the converter.

Author Contributions: Conceptualization, F.P. and L.S.; Data curation, F.P. and G.S.; Investigation, F.V.; Methodology, L.S.; Project administration, R.M.; Software, G.S. and F.V.; Supervision, R.M.; Validation, F.P.; Visualization, G.S., F.V. and L.S.; Writing–original draft, F.P.; Writing–review & editing, R.M. All authors have read and agreed to the published version of the manuscript.

Funding: This work was financially supported by: MIUR-Ministero dell'Istruzione, dell'Università e della Ricerca (Italian Ministry of Education, University and Research), PON R&I 2014-2020-AIM (Attraction and International Mobility), project AIM1851228-1; AEROSPOWER Lab (Aerospace Power Laboratory) of Scuola di Ingegneria Aerospaziale, "Sapienza" University of Rome; ESA (European Space Agency), Contract No. 4000116941/16/NL/MH; Airbus Safran Launchers SAS; PON R&I 2015-2020 PROpulsione e Sistemi IBridi per velivoli ad ala fissa e rotante PROSIB, CUP no:B66C18000290005; PRIN 2017, Advanced power-trains and -systems for full electric aircrafts, prot. no.: 2017MS9F49; project REACTION (first and euRopEAn siC eighT Inches pilOt liNe), co-funded by the ECSEL Joint Undertaking under grant agreement No 783158; SDESLab (Sustainable Development and Energy Saving Laboratory) of University of Palermo; RPLab (Rapid Prototyping Laboratory-University of Palermo); LEAP (Laboratory of Electrical APplications) of University of Palermo.

Conflicts of Interest: The authors declare no conflict of interest.

References

1. Livreri, P.; Caruso, M.; Castiglia, V.; Pellitteri, F.; Schettino, G. Dynamic reconfiguration of electrical connections for partially shaded PV modules: Technical and economical performances of an Arduino-based prototype. *Int. J. Renew. Energy Res.* **2018**, *8*, 336–344.

2. Di Carlo, C.A.; Di Donato, L.; Mauro, G.S.; La Rosa, R.; Livreri, P.; Sorbello, G. A circularly polarized wideband high gain patch antenna for wireless power transfer. *Microw. Opt. Technol. Lett.* **2018**, *60*, 620–625. [CrossRef]

3. La Rosa, R.; Zoppi, G.; Finocchiaro, A.; Papotto, G.; Di Donato, L.; Sorbello, G.; Bellomo, F.; Di Carlo, C.A.; Livreri, P. An over-the-distance wireless battery charger based on RF energy harvesting. In Proceedings of the 14th International Conference on Synthesis, Modeling, Analysis and Simulation Methods and Applications to Circuit Design (SMACD), Giardini Naxos, Italy, 12–15 June 2017; pp. 1–4.

4. Yoo, H.; Sul, S.-K.; Park, Y.; Jeong, J. System integration and power-flow management for a series hybrid electric vehicle using supercapacitors and batteries. *IEEE Trans. Ind. Appl.* **2008**, *44*, 108–114. [CrossRef]

5. Abdelhedi, R.; Chiheb Ammari, A.; Sari, A.; Lahyani, A.; Venet, P. Optimal power sharing between batteries and supercapacitors in Electric vehicles. In Proceedings of the 7th International Conference on Sciences of Electronics, Technologies of Information and Telecommunications (SETIT), Hammamet, Tunisia, 18–20 December 2016; pp. 97–103.

6. Angerer, C.; Krapf, S.; Wassiliadis, N.; Lienkamp, M. Reduction of aging-effects by supporting a conventional battery pack with ultracapacitors. In Proceedings of the Twelfth International Conference on Ecological Vehicles and Renewable Energies (EVER), Monte Carlo, Monaco, 11–13 April 2017; pp. 1–12.

7. Shah, N.; Czarkowski, D. Supercapacitors in Tandem with Batteries to Prolong the Range of UGV Systems. *Electronics* **2018**, *7*, 6. [CrossRef]

8. Zhang, Q.; Li, G. Experimental Study on a Semi-Active Battery-Supercapacitor Hybrid Energy Storage System for Electric Vehicle Application. *IEEE Trans. Power Electron.* **2019**, *35*, 1014–1021. [CrossRef]

9. Yaïci, W.; Kouchachvili, L.; Entchev, E.; Longo, M. Dynamic Simulation of Battery/Supercapacitor Hybrid Energy Storage System for the Electric Vehicles. In Proceedings of the International Conference on Renewable Energy Research and Applications (ICRERA), Brasov, Romania, 3–6 November 2019; pp. 460–465.

10. Pellitteri, F.; Castiglia, V.; Livreri, P.; Miceli, R. Analysis and design of bi-directional DC-DC converters for ultracapacitors management in EVs. In Proceedings of the International Conference on Ecological Vehicles and Renewable Energies (EVER), Monte-Carlo, Monaco, 10–12 April 2018; pp. 1–6.

11. Livreri, P.; Castiglia, V.; Pellitteri, F.; Miceli, R. Design of a Battery/Ultracapacitor Energy Storage System for Electric Vehicle Applications. In Proceedings of the 4th International Forum on Research and Technologies for Society and Industry (RTSI), Palermo, Italy, 10–13 September 2018; pp. 1–5.

12. Castiglia, V.; Livreri, P.; Miceli, R.; Pellitteri, F.; Schettino, G.; Viola, F. Power Management of a Battery/Supercapacitor System for E-Mobility Applications. In Proceedings of the AEIT International Conference of Electrical and Electronic Technologies for Automotive (AEIT AUTOMOTIVE), Torino, Italy, 2–4 July 2019; pp. 1–5.

13. Chakraborty, S.; Vu, H.N.; Hasan, M.M.; Tran, D.D.; Baghdadi, M.E.; Hegazy, O. DC-DC Converter Topologies for Electric Vehicles, Plug-in Hybrid Electric Vehicles and Fast Charging Stations: State of the Art and Future Trends. *Energies* **2019**, *12*, 1569. [CrossRef]

14. Karbozov, A.; Ibanez, F.M. Optimal Design Methodology for High-Power Interleaved Bidirectional Buck-Boost Converters for Supercapacitors in Vehicular Applications. In Proceedings of the 8th International Conference on Renewable Energy Research and Applications (ICRERA), Brasov, Romania, 3–6 November 2019; pp. 152–157.

15. Sadoun, R.; Rizoug, N.; Bartholumeus, P.; Barbedette, B.; LeMoigne, P. Sizing of hybrid supply (battery-supercapacitor) for electric vehicle taking into account the weight of the additional buck-boost chopper. In Proceedings of the First International Conference on Renewable Energies and Vehicular Technology, Hammamet, Tunisia, 26–28 March 2012; pp. 8–14.

16. Camara, M.B.; Gualous, H.; Gustin, F.; Berthon, A. Design and new control of DC/DC converters to share energy between supercapacitors and batteries in hybrid vehicles. *IEEE Trans. Veh. Technol.* **2013**, *57*, 2721–2735. [CrossRef]

17. Xue, L.K.; Wang, P.; Wang, Y.F.; Bei, T.Z.; Yan, H.Y. A four-phase high voltage conversion ratio bidirectional DC-DC converter for battery applications. *Energies* **2015**, *8*, 6399–6426. [CrossRef]

18. Zhang, H.; Chen, Y.; Park, S.J.; Kim, D.H. A Family of Bidirectional DC–DC Converters for Battery Storage System with High Voltage Gain. *Energies* **2019**, *12*, 1289. [CrossRef]

19. Lai, C.M. Development of a novel bidirectional DC/DC converter topology with high voltage conversion ratio for electric vehicles and DC-microgrids. *Energies* **2016**, *9*, 410. [CrossRef]

20. Ikeda, S.; Kajiwara, K.; Tsuji, K.; Kurokawa, F. Efficiency improvement of isolated bidirectional boost full bridge dc-dc converter. In Proceedings of the 7th International Conference on Renewable Energy Research and Applications (ICRERA), Paris, France, 14–17 October 2018; pp. 673–676.

21. Lee, I.O.; Lee, J.Y. A High-Power DC-DC Converter Topology for Battery Charging Applications. *Energies* **2017**, *10*, 871. [CrossRef]

22. Shen, C.L.; Shen, Y.S.; Tsai, C.T. Isolated DC-DC Converter for Bidirectional Power Flow Controlling with Soft-Switching Feature and High Step-Up/Down Voltage Conversion. *Energies* **2017**, *10*, 296. [CrossRef]

23. Pellitteri, F.; Boscaino, V.; Di Tommaso, A.O.; Miceli, R.; Capponi, G. Experimental test on a Contactless Power Transfer system. In Proceedings of the Ninth International Conference on Ecological Vehicles and Renewable Energies (EVER), Monte-Carlo, Monaco, 25–27 March 2014; pp. 1–6.

24. Inoue, S.; Akagi, H. A bidirectional DC–DC converter for an energy storage system with galvanic isolation. *IEEE Trans. Power Electron.* **2007**, *22*, 2299–2306. [CrossRef]

25. Pellitteri, F.; Caruso, M.; Castiglia, V.; Di Tommaso, A.O.; Miceli, R.; Schirone, L. An inductive charger for automotive applications. In Proceedings of the 42nd Annual Conference of the IEEE Industrial Electronics Society (IECON), Florence, Italy, 23–26 October 2016; pp. 4482–4486.

26. Caruso, M.; Castiglia, V.; Di Tommaso, A.O.; Miceli, R.; Pellitteri, F.; Schirone, L. Efficient contactless power transfer system for EVs. In Proceedings of the 51st International Universities Power Engineering Conference (UPEC), Coimbra, Portugal, 6–9 September 2016; pp. 1–6.

27. Pellitteri, F.; Boscaino, V.; Di Tommaso, A.O.; Miceli, R.; Capponi, G. Control subsystem design for wireless power transfer. In Proceedings of the International Conference on Renewable Energy Research and Application (ICRERA), Milwaukee, WI, USA, 19–22 October 2014; pp. 980–984.

Article

Single DC Source Multilevel Inverter with Changeable Gains and Levels for Low-Power Loads

Emad Samadaei [1,*]**, Aidin Salehi** [2]**, Mina Iranian** [3] **and Edris Pouresmaeil** [4,*]

[1]   Electromobility Department, Volvo Group Technology, 41875 Gothenburg, Sweden
[2]   Department of Engineering, Mazandaran University of Science and Technology, Babol 47166-85635, Iran; Aidin_salehi_98@yahoo.com
[3]   Department of Engineering, Atlas Danesh Co, Ghaemshahr 47658-37449, Iran; iranian_mina@yahoo.com
[4]   Department of Electrical Engineering and Automation, Aalto University, 02150 Espoo, Finland
*   Correspondence: emad.samadaei@volvo.com (E.S.); edris.pouresmaeil@aalto.fi (E.P.); Tel.: +358-50-598-4479 (E.P.)

Received: 12 April 2020; Accepted: 3 June 2020; Published: 4 June 2020

Abstract: Different types of multilevel converters have been used to convert DC voltage to AC voltage for different applications. It is, however, desirable for flexible AC output voltage to be created from a single DC source at lower cost. This paper presents a new type of single DC source multilevel inverter which has the ability to create a different number of levels with low rate components. This device is also able boost the magnitude of the output voltage by means of a variable gain with one DC source. The advantages of the multilevel inverter are rapid stepping among levels, and its ability to produce different ranges of levels (seven, nine and eleven) and gains (two, four and eight). The proposed multilevel inverter includes six semiconductor switches, eight diodes, two capacitors and two inductors, making it suitable for low-power applications. A simulation in MATLAB and experimental tests on a prototype setup showed good performance for different modulations, with THD% down to 2.29%, which are meets the IEEE standard (IEEE 519).

Keywords: boost inverter; multilevel inverter; power electronics; single source; low components

---

1. Introduction

Multilevel inverters (MLIs) have been used in various applications as voltage source converters (VSC) to connect DC power systems to AC power systems. MLIs include an array of semiconductors and DC sources to generate different output voltage levels with high quality waveforms. Low harmonic contents, high resolution on the output voltage and scalability make this kind of inverter more suitable in comparison with traditional two-level inverters for energy conversion applications such as photovoltaic systems [1], HVDC [2,3], wind turbines [4], active power filers [5], drives systems [6,7], electrical vehicles [8] and power grids [9]. Multilevel inverters are classified into three well-known topologies that have been used in various industries: Neutral Point Clamped (NPC) [10], Flying Capacitor (FC) [11] and Cascade H-bridge (CHB) [12]. CHB topologies are gaining attention due to certain aspects, such as simpler control strategy, scalability, the number of levels, the number of semiconductors, etc. However, classical topologies require a large number of components to increase the number of output voltage levels. Circuit complexity, complicated voltage control strategies and higher manufacturing costs are expected when the number of components increases. Nevertheless, creating more voltage levels with fewer components and DC sources is an important objective for newer multilevel inverter topologies [13–16]. In [17], a structural review and comparative study of all categories of Multilevel Inverters are presented.

In [18], a series connection of modules consisting of one DC source and two switches was introduced. The output voltage is generated by the sum of each module's output voltage. The module can only generate zero and positive voltage levels. An H-bridge was added at the end of the series module in [19] to create both negative and positive levels. Also, the module has a simple circuit and control strategy, and the stress on the switches is low compared with conventional CHBs. On the other hand, a large number of cascade connections to generate higher output voltage levels can increase the complexity and construction cost. A similar structure with an H-bridge at the end of the module was introduced in [20]. It is used a diode instead of a switch by [21], as the number of switches decreased in this way. The number of required electronic switches and DC sources are a key factor in designing MLIs, because the circuit size, cost, installation area and control complexity depend on them. Compared with other modules, the device described in [22] can generate more voltage levels with fewer switches. However, the cost of using several DC sources still remains high. Another type of MLI using crossing switches to oppose the polarity of DC links was introduced in [23,24] to generate more levels. The authors of [25,26] introduced modules with few semiconductors in order to achieve maximum levels from four DC sources. Recently, different storage elements such as capacitors have been mixed with DC sources to provide a sinusoidal voltage source in various multilevel inverter structures and increase the output voltage levels with the same number of DC sources. The authors of [27,28] redesigned the configuration of [11,23], replacing some DC sources with capacitors and increasing the number of voltage levels. Some configurations with mixed DC sources and capacitors were proposed in [29,30]. A new topology with only one DC source is presented in [31]. The authors of [32] redesigned the configuration of [31], removing some switches and replacing them with diodes. In some modules that use a single source to generate output levels, the number of semiconductors increased, and the charging/discharging of capacitors and switch driving become complicated [31,33]. Also, the authors of [34] replaced half of the DC sources used in [25,26] with capacitors, and rearranged the semiconductor configuration to present a new multilevel inverter with fewer DC sources.

In this paper, a flexible multilevel inverter for one DC source is proposed which can create a different number of levels and voltages at the output for low-power applications. The main core of the proposed module is a new type of DC–DC converter to change the levels very quickly by a variable duty cycle, in order to create a staircase waveform. The main converter should be rapid with a tiny transient time, and have the ability to create all voltage levels; otherwise, it can produce high harmonics and destroy the staircase waveform. The other part of the proposed inverter is a full-bridge converter (H-bridge) to evenly produce the positive and negative half-cycles of a sinusoidal waveform. The proposed multilevel inverter has the ability to create up to 11 levels and boost the output voltage up to 8 times compared to the DC source. The proposed multilevel inverter is illustrated in Section 2. The modular configuration, switching patterns, equations, and a comparative study with conventional inverters are included in this section. In Section 3, the calculations for the switching pattern and duty cycle for a different number of levels and different gains are presented. Finally, the simulation and experimental results are presented for nine different cases (number of levels = 7, 9 and 11 versus gain = 2, 4 and 8) in Sections 4 and 5.

2. The Proposed Multilevel Inverter

Figure 1 shows the proposed multilevel inverter, which is divided into two parts: the main converter and the full-bridge converter. The main converter provides the levels, and the full-bridge converter changes the polarity of output. The main converter is very fast and comprises a high gain DC–DC converter with good dynamic reactivity to changes in the duty cycle. Each level of output can be achieved in the specified amount of the duty cycle. On the other hand, the duty cycle must be set to a certain number, which is calculated in lookup tables (Tables 1–9). Nevertheless, the duty cycle was changed in one cycle continuously to create different levels in the multilevel inverter. Then, the full-bridge converter changes the polarity for each half-cycle from the positive or negative half-cycles. Figure 2 illustrates this issue with an example for eleven levels on the output.

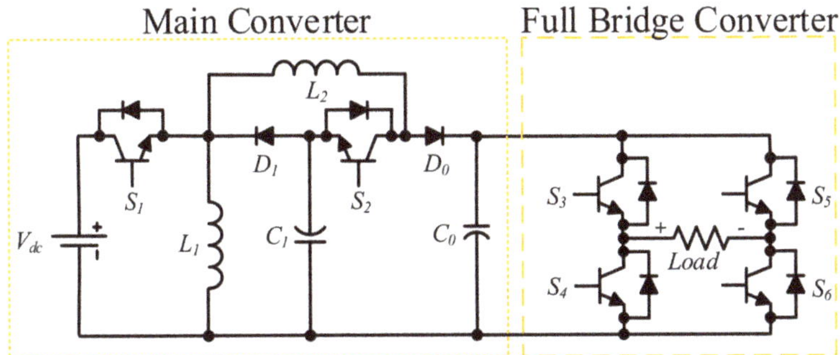

Figure 1. The proposed single source multilevel inverter.

Table 1. Switching table level = 11, gain = 8.

Voltage Level		Duty Cycle S_1 (%)	Duty Cycle S_2 (%)	S_3	S_4	S_5	S_6
Positive Level	$8\,V_{DC}$	74	74	1	0	0	1
	$6.66\,V_{DC}$	72	72	1	0	0	1
	$5.66\,V_{DC}$	69	69	1	0	0	1
	$4\,V_{DC}$	66	66	1	0	0	1
	$1.66\,V_{DC}$	56	56	1	0	0	1
Zero Level	0	0	0	1	0	1	0
	0	0	0	0	1	0	1
Negative Level	$-1.66\,V_{DC}$	56	56	0	1	1	0
	$-4\,V_{DC}$	66	66	0	1	1	0
	$-5.66\,V_{DC}$	69	69	0	1	1	0
	$-6.66\,V_{DC}$	72	72	0	1	1	0
	$-8\,V_{DC}$	74	74	0	1	1	0
Num. of turning on per 1-cycle		20,000	20,000	1	1	2	2

Table 2. Switching table level = 11, gain = 4.

Voltage Level		Duty Cycle S_1 (%)	Duty Cycle S_2 (%)	S_3	S_4	S_5	S_6
Positive Level	$4\,V_{DC}$	66	66	1	0	0	1
	$3.33\,V_{DC}$	64	64	1	0	0	1
	$2.83\,V_{DC}$	62	62	1	0	0	1
	$2\,V_{DC}$	58	58	1	0	0	1
	$0.84\,V_{DC}$	47	47	1	0	0	1
Zero Level	0	0	0	1	0	1	0
	0	0	0	0	1	0	1
Negative Level	$-0.84\,V_{DC}$	47	47	0	1	1	0
	$-2\,V_{DC}$	58	58	0	1	1	0
	$-2.83\,V_{DC}$	63	63	0	1	1	0
	$-3.33\,V_{DC}$	72	72	0	1	1	0
	$-4\,V_{DC}$	66	66	0	1	1	0
Num. of turning on per 1-cycle		20,000	20,000	1	1	2	2

Table 3. Switching table level = 11, gain = 2.

Voltage Level		Duty Cycle S_1 (%)	Duty Cycle S_2 (%)	S_3	S_4	S_5	S_6
Positive Level	$2\,V_{DC}$	58	58	1	0	0	1
	$1.66\,V_{DC}$	56	56	1	0	0	1
	$1.41\,V_{DC}$	54	54	1	0	0	1
	$1\,V_{DC}$	50	50	1	0	0	1
	$0.42\,V_{DC}$	39	39	1	0	0	1
Zero Level	0	0	0	1	0	1	0
	0	0	0	0	1	0	1
Negative Level	$-0.42\,V_{DC}$	39	39	0	1	1	0
	$-1\,V_{DC}$	50	50	0	1	1	0
	$-1.41\,V_{DC}$	54	54	0	1	1	0
	$-1.66\,V_{DC}$	56	56	0	1	1	0
	$-2\,V_{DC}$	58	58	0	1	1	0
Num. of turning on per 1-cycle		20,000	20,000	1	1	2	2

Table 4. Switching table level = 9, gain = 8.

Voltage Level		Duty Cycle S_1 (%)	Duty Cycle S_2 (%)	S_3	S_4	S_5	S_6
Positive Level	$8\,V_{DC}$	74	74	1	0	0	1
	$5.83\,V_{DC}$	70	70	1	0	0	1
	$3.33\,V_{DC}$	64	64	1	0	0	1
	$1.66\,V_{DC}$	56	56	1	0	0	1
Zero Level	0	0	0	1	0	1	0
	0	0	0	0	1	0	1
Negative Level	$-1.66\,V_{DC}$	56	56	0	1	1	0
	$-3.33\,V_{DC}$	64	64	0	1	1	0
	$-5.83\,V_{DC}$	70	70	0	1	1	0
	$-8\,V_{DC}$	74	74	0	1	1	0
Num. of turning on per 1-cycle		20,000	20,000	1	1	2	2

Table 5. Switching table level = 9, gain = 4.

Voltage Level		Duty Cycle S_1 (%)	Duty Cycle S_2 (%)	S_3	S_4	S_5	S_6
Positive Level	$4\,V_{DC}$	66	66	1	0	0	1
	$2.91\,V_{DC}$	63	63	1	0	0	1
	$1.66\,V_{DC}$	56	56	1	0	0	1
	$0.84\,V_{DC}$	47	47	1	0	0	1
Zero Level	0	0	0	1	0	1	0
	0	0	0	0	1	0	1
Negative Level	$-0.84\,V_{DC}$	47	47	0	1	1	0
	$-1.66\,V_{DC}$	56	56	0	1	1	0
	$-2.91\,V_{DC}$	63	63	0	1	1	0
	$-4\,V_{DC}$	66	66	0	1	1	0
Num. of turning on per 1-cycle		20,000	20,000	1	1	2	2

Table 6. Switching table level = 9, gain = 2.

Voltage Level		Duty Cycle S_1 (%)	Duty Cycle S_2 (%)	S_3	S_4	S_5	S_6
Positive Level	$2\,V_{DC}$	58	58	1	0	0	1
	$1.45\,V_{DC}$	55	55	1	0	0	1
	$0.84\,V_{DC}$	47	47	1	0	0	1
	$0.42\,V_{DC}$	39	39	1	0	0	1
Zero Level	0	0	0	1	0	1	0
	0	0	0	0	1	0	1
Negative Level	$-0.42\,V_{DC}$	39	39	0	1	1	0
	$-0.84\,V_{DC}$	47	47	0	1	1	0
	$-1.45\,V_{DC}$	55	55	0	1	1	0
	$-2\,V_{DC}$	58	58	0	1	1	0
Num. of turning on per 1-cycle		20,000	20,000	1	1	2	2

Table 7. Switching table level = 7, gain = 8.

Voltage Level		Duty Cycle S_1 (%)	Duty Cycle S_2 (%)	S_3	S_4	S_5	S_6
Positive Level	$8\,V_{DC}$	74	74	1	0	0	1
	$5.41\,V_{DC}$	69	69	1	0	0	1
	$2.5\,V_{DC}$	61	61	1	0	0	1
Zero Level	0	0	0	1	0	1	0
	0	0	0	0	1	0	1
Negative Level	$-2.5\,V_{DC}$	61	61	0	1	1	0
	$-5.41\,V_{DC}$	69	69	0	1	1	0
	$-8\,V_{DC}$	74	74	0	1	1	0
Num. of turning on per 1-cycle		20,000	20,000	1	1	2	2

Table 8. Switching table level = 7, gain = 4.

Voltage Level		Duty Cycle S_1 (%)	Duty Cycle S_2 (%)	S_3	S_4	S_5	S_6
Positive Level	$4\,V_{DC}$	66	66	1	0	0	1
	$2.70\,V_{DC}$	62	62	1	0	0	1
	$1.25\,V_{DC}$	52	52	1	0	0	1
Zero Level	0	0	0	1	0	1	0
	0	0	0	0	1	0	1
Negative Level	$-1.25\,V_{DC}$	52	52	0	1	1	0
	$-2.70\,V_{DC}$	62	62	0	1	1	0
	$-4\,V_{DC}$	66	66	0	1	1	0
Num. of turning on per 1-cycle		20,000	20,000	1	1	2	2

Table 9. Switching table level = 7, gain = 2.

Voltage Level		Duty Cycle S_1 (%)	Duty Cycle S_2 (%)	S_3	S_4	S_5	S_6
Positive Level	$2\,V_{DC}$	58	58	1	0	0	1
	$1.35\,V_{DC}$	53	53	1	0	0	1
	$0.62\,V_{DC}$	44	44	1	0	0	1
Zero Level	0	0	0	1	0	1	0
	0	0	0	0	1	0	1
Negative Level	$-0.62\,V_{DC}$	44	44	0	1	1	0
	$-1.35\,V_{DC}$	53	53	0	1	1	0
	$-2\,V_{DC}$	58	58	0	1	1	0
Num. of turning on per 1-cycle		20,000	20,000	1	1	2	2

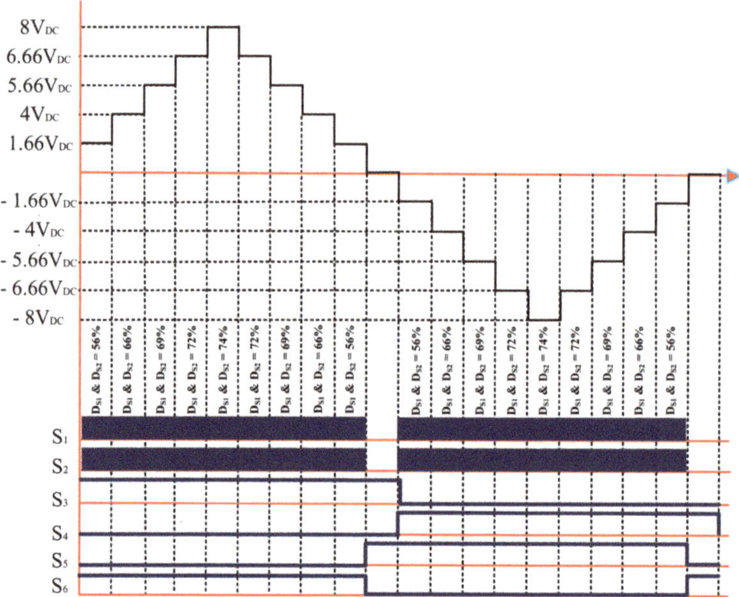

Figure 2. Switching pattern and duty cycle of switches in one-cycle for the proposed multilevel inverter (ex: the number of levels = 11 and gain = 8).

The duty cycle of each level for S1 and S2 are calculated (according to Section 3) and shown in Figure 2. The states for S3, S4, S5, and S6 to create the polarity of the output are also shown.

In order to demonstrate the detail and performance, Figure 3 depicts the two modes (on/off) for the main converter part. It is noticeable that S1 and S2 switch on and off, simultaneously.

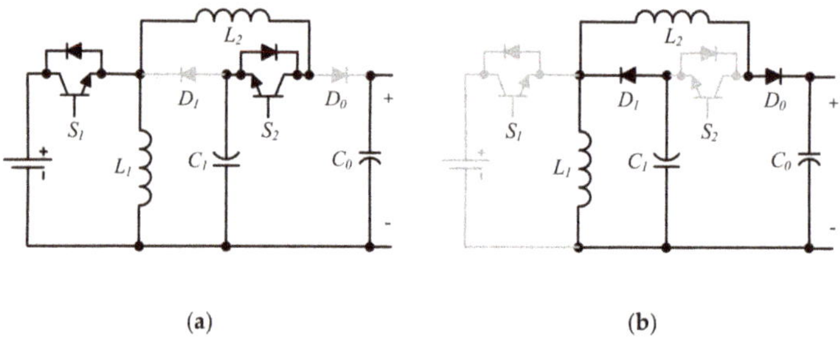

Figure 3. The mode of the main converter: (**a**) on mode; (**b**) off mode.

On mode: S1 and S2 are in an on state, and D0 and D1 are in reverse bias (Figure 3a). On the other hand, switches S1 and S2 conduct the current to magnetize L1 and L2, respectively. And C0 and C1 are discharged to the output. Then:

$$V_{L1} = V_{in} \tag{1}$$

$$V_{L2} = V_{in} + V_{C1} \tag{2}$$

Off mode: S1 and S2 are turned off, and D0 and D1 are conducting in forward bias (Figure 3b). L1 and L2 demagnetize to charge C0 and C1, respectively, and supply output. Thus:

$$V_{L1} = -V_{C1} \tag{3}$$

$$V_{L2} = -(V_{C1} + V_0) \tag{4}$$

Figure 4 demonstrates the behavior of the components in the two modes.

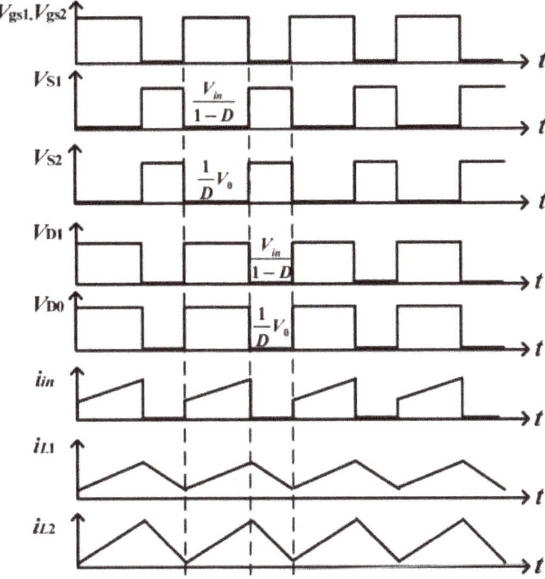

Figure 4. The behaviors of the components for two modes.

From (1) and (3) and balancing in L1:

$$DV_{in} + (1-D)(-DV_{C1}) = V_{L1} = 0 \tag{5}$$

Then

$$V_{C1} = \frac{D}{1-D} V_{in} \tag{6}$$

Also, from (2) and (4),

$$D(V_{in} + V_{C1}) + (1-D)(-(V_{C1} + V_o) = V_{L2} = 0 \tag{7}$$

Consequently,

$$\frac{V_0}{V_{in}} = \left(\frac{D}{1-D}\right)^2 \tag{8}$$

Equation (8) provides the formula to calculate the duty cycle which creates the levels. The duty cycles for each level and each gain according to (8) will be presented in Section 3.

A comparative study between the proposed multilevel inverter and some other configurations in cases of producing 11 levels of output voltage is shown in Figure 5. In order to make a suitable comparison, various aspects, such as the number of switches, the number of diodes, the number of DC sources and TSV, were considered.

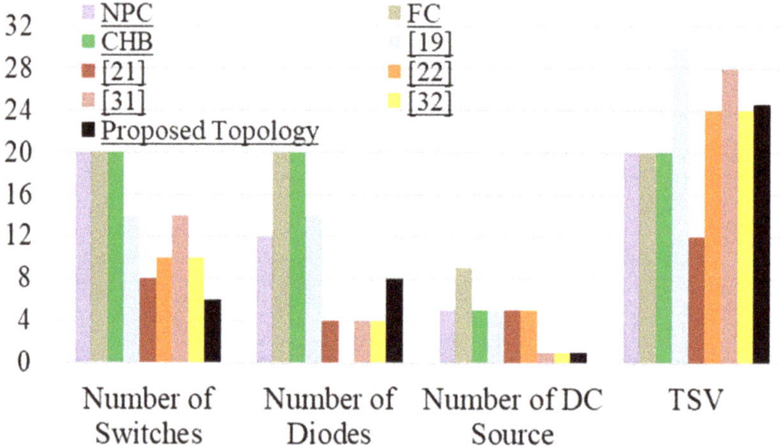

Figure 5. Comparative studies: The number of switches, number of diodes, number of sources and TSV, in terms of the number of levels.

According to Figure 5, the proposed module can get maximum levels with fewer DC sources. As a comparison only in the case of the number of DC sources, [31,32] require the same number of DC sources, but the number of switches should be considered. One of the major advantages of the proposed module is that it uses fewer switches. On the other hand, the module can be considered as being in the group with a low number of diodes. The complexity of the module is reduced by using diodes instead of switches. Since the module is intended for use in low-power applications, the range of TSV is reasonable, as can be seen in Figure 5.

3. Calculation of the Switching Pattern and Duty Cycle

The proposed multilevel inverter is divided into two parts, as mentioned above: the DC–DC converter and the full-bridge. In the DC–DC converter, the number of levels can be achieved by controlling the duty cycles. Thus, the duty cycle should be changed continuously to create each level and a staircase waveform in one cycle of a sinusoidal waveform. In the full-bridge, the switching modes should be set to alternatively change the polarity of the half-cycle. The duty cycle and switching pattern are calculated for levels 7, 9 and 11 and gains 2, 4 and 8. Tables 1–9 illustrate the calculation of the switching pattern and duty cycle according to Equation (8). S3 and S6 are for positive levels and S4 and S5 are for negative levels, which have a low frequency (up to 100 Hz); S1 and S2 are used to set a duty cycle with 20 kHz.

The nearest level control method (NLC) was also considered as a switching technique for level timing, as presented in [35]. This technique is a very simple means by which to reduce the number of calculations in the processor. The modulation scheme and the control diagram are shown in Figure 6. According to Figure 6a, the controller gets a sample from the reference voltage (Vref) and then rounds it to the nearest voltage level (VaN). Each voltage level has a switching table according to the information in Tables 1–9 to change switch modes and duty cycles (Figure 6b). The sampling is repetitive for each sample time (Ts).

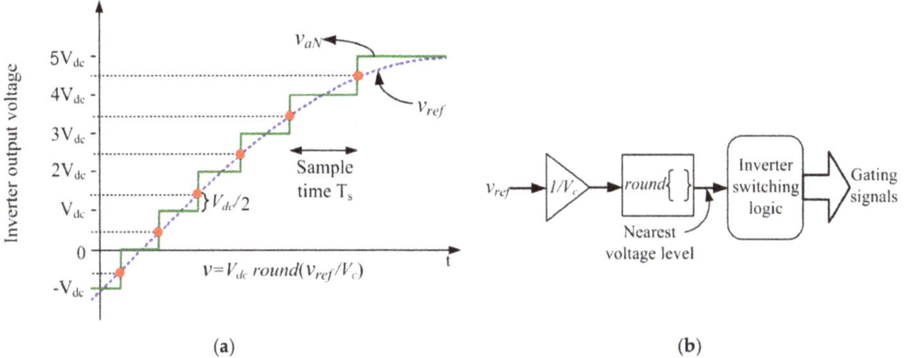

(a)   (b)

Figure 6. Nearest level control: (**a**) Waveform synthesis; (**b**) Control diagram.

4. Simulation Results

A simulation was carried out to evaluate the proposed multilevel inverter at different outputs. The specifications of the simulation are shown in Table 10. The aim of the simulation test was to produce a sinusoidal waveform of 50 Hz from one DC source at 36 volts.

Table 10. Simulation configuration.

Switches		IGBT
Diodes		Diode
Inductors	$L_1 = 50\ \mu H$	$L_2 = 250\ \mu H$
Capacitors	$C_0 = 2\ \mu F$	$C_1 = 1\ \mu F$
DC Source	36 V	
Load	500 Ω	

Nine output conditions were considered for the waveforms. The output levels were set for seven, nine and eleven, along with three different gains (output voltage amount/input voltage amount), i.e., two, four and eight. Figure 7 summarizes the nine simulation setups. The simulation results showed the ability of the proposed circuit to create different waveforms with sinusoidal output voltages.

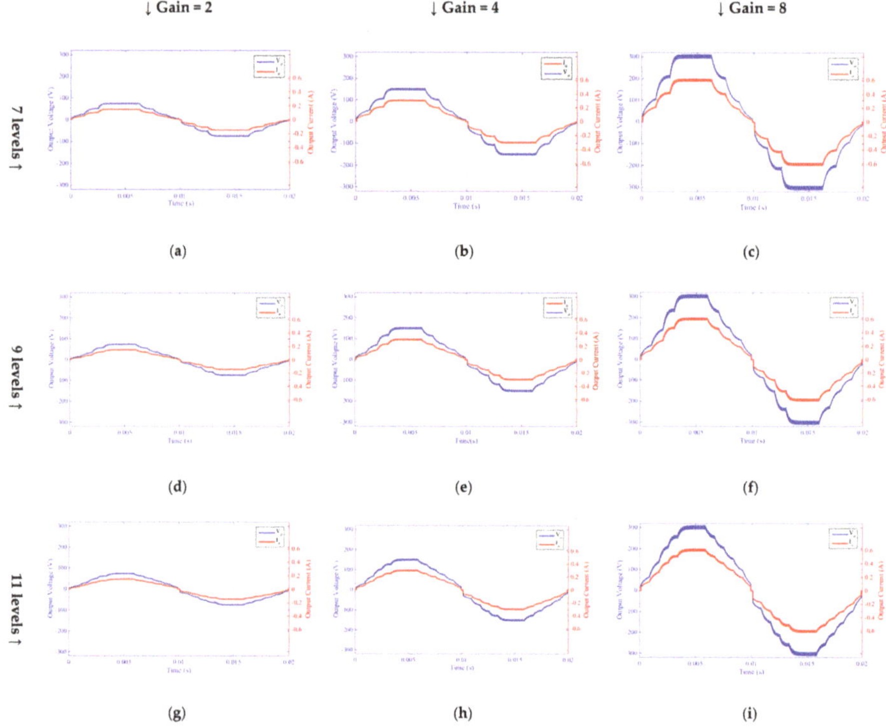

Figure 7. Simulation waveform of output voltages and currents for the proposed multilevel inverter: (**a**) levels = 7, gain = 2; (**b**) levels = 7, gain = 4; (**c**) levels = 7, gain = 8; (**d**) levels = 9, gain = 2; (**e**) levels = 9, gain = 4; (**f**) levels = 9, gain = 8; (**g**) levels = 11, gain = 2; (**h**) levels = 11, gain = 4; (**i**) levels = 11, gain = 8.

5. Experimental Results

A prototype setup was implemented in the laboratory to verify the performance of the proposed single source multilevel inverter with flexible levels and gains. Figure 8 provides a picture of the experimental setup in the laboratory. The specifications of the prototype setup are given in Table 11. An experimental test was also carried out on nine cases (the number of levels for 7, 9 and 11 versus gains for 2, 4 and 8) to generate a sinusoidal waveform of 50 Hz from one DC source at 12 volts.

Table 11. Specifications of the prototype setup configuration.

Switches	IGBT FGH60N60SFD (S_1,S_2) (S_3,S_4,S_5,S_6) IGBT IKW40N120H3	
Diodes	RHRP15120	
Inductors	L_1 = 50 μH	L_2 = 250 μH
Capacitors	C_0 = 2 μF	C_1 = 1 μF
DC Source	12 V	
Optocoupler	HCPL3120	
Microcontroller	ATmega32A	
Buffer	7404	
Load	500 Ω	

Figure 8. Experimental setup in the laboratory.

Figure 9 shows the voltages and currents of the proposed inverter with nine different outputs. It should be mentioned that the THD% of all the waveforms meets the IEEE standard (IEEE 519), in that they are under 8%. The THD% is shown in Table 12. The worst THD% was 6.82% and the best was 2.08%.

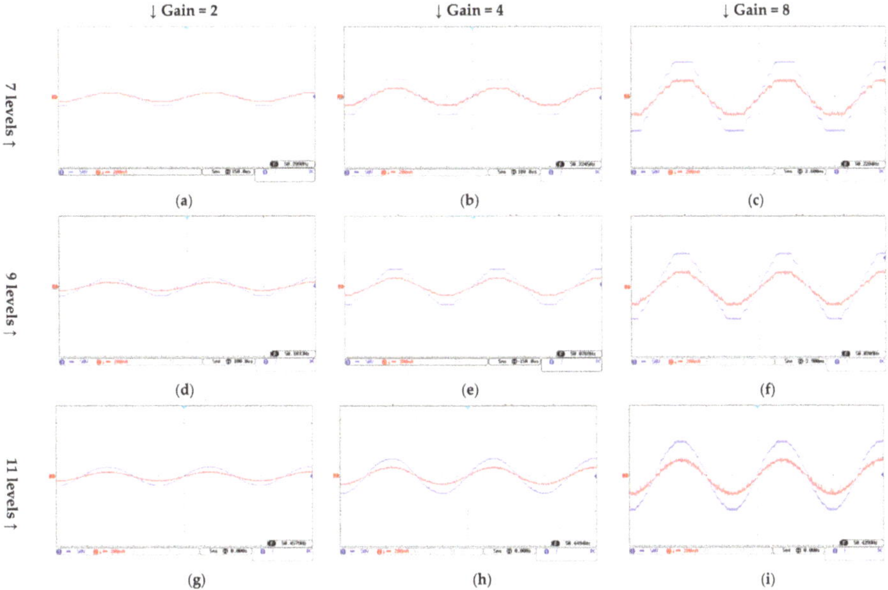

Figure 9. The experimental test waveform of output voltages and currents for the proposed multilevel inverter: (**a**) levels = 7, gain = 2; (**b**) levels = 7, gain = 4; (**c**) levels = 7, gain = 8; (**d**) levels = 9, gain = 2; (**e**) levels = 9, gain = 4; (**f**) levels = 9, gain = 8; (**g**) levels = 11, gain = 2; (**h**) levels = 11, gain = 4; (**i**) levels = 11, gain = 8.

Table 12. THD% of voltages for nine outputs.

7Levels, Gain 2	6.61
7Levels, Gain 4	6.70
7Levels, Gain 8	6.82
9Levels, Gain 2	4.88
9Levels, Gain 4	4.77
9Levels, Gain 8	5.27
11Levels, Gain 2	2.08
11Levels, Gain 4	2.28
11Levels, Gain 8	2.29

The proposed multilevel inverter (with few components and one DC source) is economical, with good output voltage for low-power applications. It can create different voltage levels for different voltages with just one DC source. Also, a low number of harmonics make it especially promising.

Figure 10 illustrates the voltages and currents of the switches and diodes in the proposed inverter.

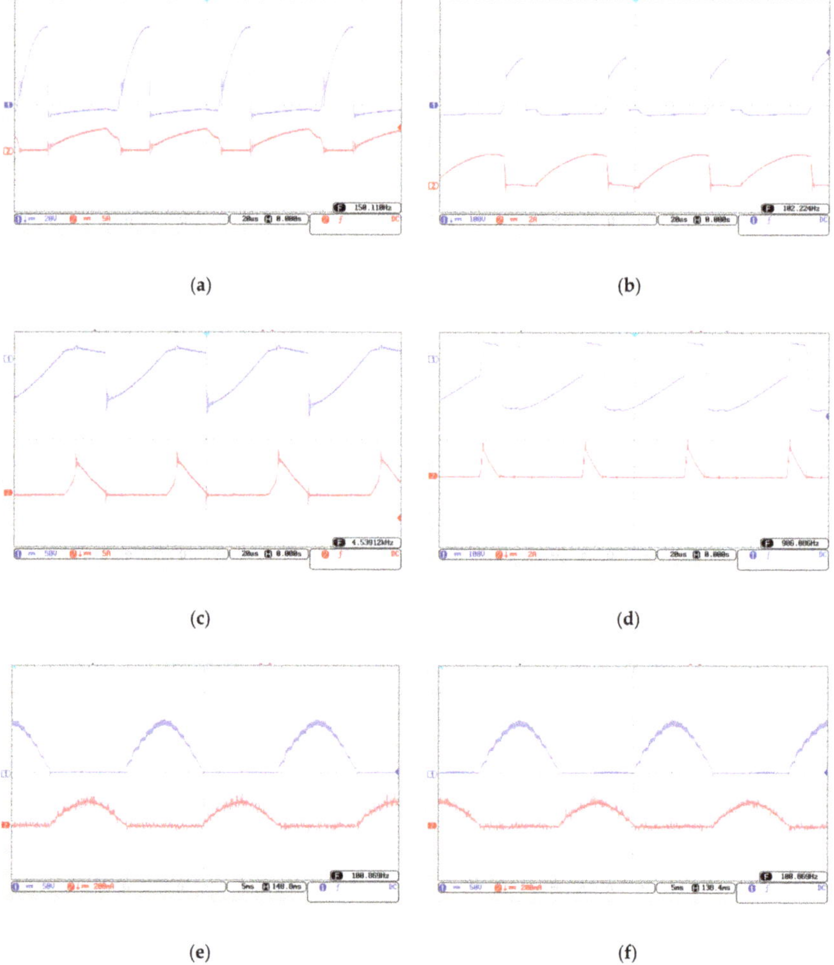

(a)

(b)

(c)

(d)

(e)

(f)

Figure 10. *Cont.*

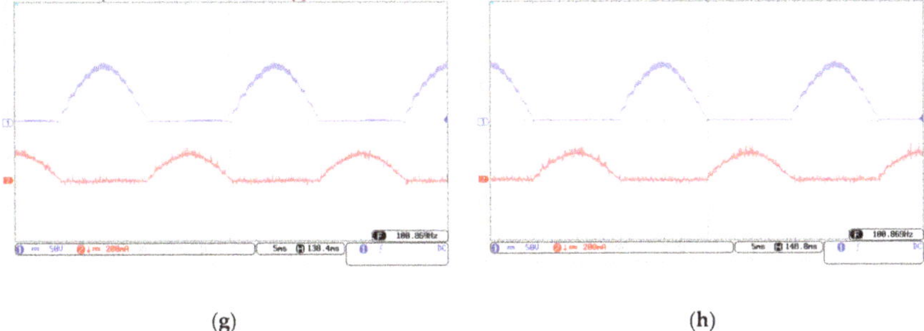

(g)   (h)

Figure 10. Experimental test waveform of output voltages and currents for the proposed multilevel inverter: (**a**) the voltage and current of S1; (**b**) the voltage and current of S2; (**c**) the voltage and current of D1; (**d**) the voltage and current of D0; (**e**) the voltage and current of S3; (**f**) the voltage and current of S4; (**g**) the voltage and current of S5; (**h**) the voltage and current of S6.

Finally, the voltage and current of the capacitors and inductors are depicted in Figure 11.

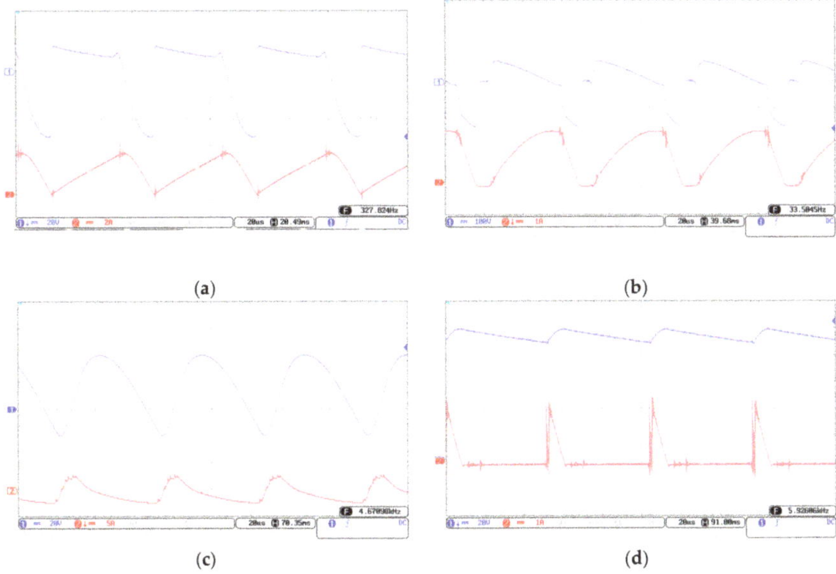

(a)   (b)

(c)   (d)

Figure 11. Experimental test waveform of output voltages and currents for the proposed multilevel inverter: (**a**) the voltage and current of L1; (**b**) the voltage and current of L2; (**c**) the voltage and current of C1; (**d**) the voltage and current of C0.

6. Conclusions

A new kind of multilevel inverter for low-power applications, operating from one DC source, was introduced in this paper. The proposed inverter has the ability to produce different levels of output, i.e., up to 11 levels, and to increase the output voltage by up to eight times compared to the input voltage. It also has flexible levels and gains. The circuit of the proposed inverter comprises six semiconductor switches, eight diodes, two capacitors and two inductors, i.e., a low number of components for a multilevel inverter. These components are organized into two parts: the main converter and the full-bridge converter. The main converter is involved in fast DC–DC conversion, with a tiny transient time to create levels and gains by changing the duty cycle according to different strategies. Meanwhile, the full-bridge changes the polarities for the positive and negative half-cycles. Verification of the proposed inverter was carried out in a simulation and in experimental tests using nine different modulations. The device was shown to produce three level modes for 7, 9 and 11, and three gain modes for 2, 4 and 8. The simulation and experimental results showed a high level of performance for all the test cases; the THD% was shown to be between 2.08% and 6.82% (in the permitted range of the IEEE 519 standard). The illustrated features of the proposed inverter make it suitable for single DC sources in low-power applications.

Author Contributions: All authors contributed equally to this work and all authors have read and approved the final manuscript. All authors have read and agreed to the published version of the manuscript.

Funding: This research received no external funding.

Conflicts of Interest: The authors declare no conflict of interest.

References

1. Essakiappan, S.; Krishnamoorthy, H.S.; Enjeti, P.; Balog, R.; Ahmed, S. Multilevel Medium-Frequency Link Inverter for Utility Scale Photovoltaic Integration. *IEEE Trans. Power Electron.* **2015**, *30*, 3674–3684. [CrossRef]
2. Nami, A.; Liang, J.; Dijkhuizen, F.; Demetriades, G.D. Modular Multilevel Converters for HVDC Applications: Review on Converter Cells and Functionalities. *IEEE Trans. Power Electron.* **2014**, *30*, 18–36. [CrossRef]
3. Gowaid, I.A.; Adam, G.; Massoud, A.; Ahmed, S.; Williams, B.W. Hybrid and Modular Multilevel Converter Designs for Isolated HVDC–DC Converters. *IEEE J. Emerg. Sel. Top. Power Electron.* **2017**, *6*, 188–202. [CrossRef]
4. Yuan, X. A Set of Multilevel Modular Medium-Voltage High Power Converters for 10-MW Wind Turbines. *IEEE Trans. Sustain. Energy* **2014**, *5*, 524–534. [CrossRef]
5. Ahmadi, D.; Wang, J. Online Selective Harmonic Compensation and Power Generation with Distributed Energy Resources. *IEEE Trans. Power Electron.* **2013**, *29*, 3738–3747. [CrossRef]
6. Mathew, J.; Rajeevan, P.P.; Mathew, K.; Azeez, N.A.; Gopakumar, K. A Multilevel Inverter Scheme with Dodecagonal Voltage Space Vectors Based on Flying Capacitor Topology for Induction Motor Drives. *IEEE Trans. Power Electron.* **2012**, *28*, 516–525. [CrossRef]
7. Nair, V.; Pramanick, S.; Gopakumar, K.; Franquelo, L.G. Novel Symmetric Six-Phase Induction Motor Drive Using Stacked Multilevel Inverters with a Single DC Link and Neutral Point Voltage Balancing. *IEEE Trans. Ind. Electron.* **2017**, *64*, 2663–2670. [CrossRef]
8. Zheng, Z.; Wang, K.; Xu, L.; Li, Y. A Hybrid Cascaded Multilevel Converter for Battery Energy Management Applied in Electric Vehicles. *IEEE Trans. Power Electron.* **2013**, *29*, 3537–3546. [CrossRef]
9. Haw, L.K.; Dahidah, M.S.A.; Almurib, H. A New Reactive Current Reference Algorithm for the STATCOM System Based on Cascaded Multilevel Inverters. *IEEE Trans. Power Electron.* **2015**, *30*, 3577–3588. [CrossRef]
10. Nabae, A.; Takahashi, I.; Akagi, H. A New Neutral-Point-Clamped PWM Inverter. *IEEE Trans. Ind. Appl.* **1981**, *17*, 518–523. [CrossRef]
11. Meynard, T.A.; Foch, H. Multi-Level Choppers for High Voltage Applications. *EPE J.* **1992**, *2*, 45–50. [CrossRef]
12. Peng, F.Z.; Lai, J.-S.; McKeever, J.; VanCoevering, J. A multilevel voltage-source inverter with separate DC sources for static VAr generation. *IEEE Trans. Ind. Appl.* **1996**, *32*, 1130–1138. [CrossRef]
13. Zare, F. *Power Electronics Education Electronic-Book*; School of Engineering Systems, Queensland University of Technology: Brisbane, SEQ, Australia, 2008.
14. Gupta, K.K.; Ranjan, A.; Bhatnagar, P.; Sahu, L.K.; Jain, S.; Ranjan, A. Multilevel Inverter Topologies with Reduced Device Count: A Review. *IEEE Trans. Power Electron.* **2015**, *31*, 1. [CrossRef]
15. Debnath, S.; Qin, J.; Bahrani, B.; Saeedifard, M.; Barbosa, P. Operation, Control, and Applications of the Modular Multilevel Converter: A Review. *IEEE Trans. Power Electron.* **2015**, *30*, 37–53. [CrossRef]
16. Venkataramanaiah, J.; Suresh, Y.; Panda, A.K. A review on symmetric, asymmetric, hybrid and single DC sources based multilevel inverter topologies. *Renew. Sustain. Energy Rev.* **2017**, *76*, 788–812. [CrossRef]
17. Vijeh, M.; Rezanejad, M.; Samadaei, E.; Bertilsson, K. A General Review of Multilevel Inverters Based on Main Submodules: Structural Point of View. *IEEE Trans. Power Electron.* **2019**, *34*, 9479–9502. [CrossRef]

18. Babaei, E.; Hosseini, S.H. New cascaded multilevel inverter topology with minimum number of switches. *Energy Convers. Manag.* **2009**, *50*, 2761–2767. [CrossRef]

19. Babaei, E.; Farhadi-Kangarlu, M.; Sabahi, M. Extended multilevel converters: An attempt to reduce the number of independent DC voltage sources in cascaded multilevel converters. *IET Power Electron.* **2014**, *7*, 157–166. [CrossRef]

20. Najafi, E.; Yatim, A.H. Design and Implementation of a New Multilevel Inverter Topology. *IEEE Trans. Ind. Electron.* **2011**, *59*, 4148–4154. [CrossRef]

21. Alishah, R.S.; Nazarpour, D.; Sabahi, M.; Hosseini, S.H. New hybrid structure for multilevel inverter with fewer number of components for high-voltage levels. *IET Power Electron.* **2014**, *7*, 96–104. [CrossRef]

22. Malathy, S.; Ramaprabha, R. A new single phase multilevel inverter topology with reduced number of switches. In Proceedings of the 2016 3rd International Conference on Electrical Energy Systems (ICEES), Chennai, India, 17–19 March 2016; pp. 139–144.

23. Gupta, K.; Jain, S. Topology for multilevel inverters to attain maximum number of levels from given DC sources. *IET Power Electron.* **2012**, *5*, 435–446. [CrossRef]

24. Farhadi-Kangarlu, M.; Babaei, E. Cross-switched multilevel inverter: An innovative topology. *IET Power Electron.* **2013**, *6*, 642–651. [CrossRef]

25. Samadaei, E.; Gholamian, S.A.; Sheikholeslami, A.; Adabi, J. An Envelope Type (E-Type) Module: Asymmetric Multilevel Inverters with Reduced Components. *IEEE Trans. Ind. Electron.* **2016**, *63*, 7148–7156. [CrossRef]

26. Samadaei, E.; Sheikholeslami, A.; Gholamian, S.A.; Adabi, J. A Square T-Type (ST-Type) Module for Asymmetrical Multilevel Inverters. *IEEE Trans. Power Electron.* **2018**, *33*, 987–996. [CrossRef]

27. Metri, J.I.; Vahedi, H.; Kanaan, H.Y.; Al-Haddad, K. Real-Time Implementation of Model-Predictive Control on Seven-Level Packed U-Cell Inverter. *IEEE Trans. Ind. Electron.* **2016**, *63*, 4180–4186. [CrossRef]

28. Vahedi, H.; Al-Haddad, K. Real-Time Implementation of a Seven-Level Packed U-Cell Inverter with a Low-Switching-Frequency Voltage Regulator. *IEEE Trans. Power Electron.* **2015**, *31*, 5967–5973. [CrossRef]

29. Zamiri, E.; Vosoughi, N.; Hosseini, S.H.; Barzegarkhoo, R.; Sabahi, M. A New Cascaded Switched-Capacitor Multilevel Inverter Based on Improved Series–Parallel Conversion with Less Number of Components. *IEEE Trans. Ind. Electron.* **2016**, *63*, 3582–3594. [CrossRef]

30. Liu, J.; Wu, J.; Zeng, J.; Guo, H. A Novel Nine-Level Inverter Employing One Voltage Source and Reduced Components as High-Frequency AC Power Source. *IEEE Trans. Power Electron.* **2017**, *32*, 2939–2947. [CrossRef]

31. Taghvaie, A.; Adabi, J.; Rezanejad, M. Circuit Topology and Operation of a Step-Up Multilevel Inverter with a Single DC Source. *IEEE Trans. Ind. Electron.* **2016**, *63*, 6643–6652. [CrossRef]

32. Tang, Y.; Ran, L.; Alatise, O.; Mawby, P. Capacitor Selection for Modular Multilevel Converter. *IEEE Trans. Ind. Appl.* **2016**, *52*, 3279–3293. [CrossRef]

33. Sun, X.; Wang, B.; Zhou, Y.; Wang, W.; Du, H.; Lu, Z. A Single DC Source Cascaded Seven-Level Inverter Integrating Switched-Capacitor Techniques. *IEEE Trans. Ind. Electron.* **2016**, *63*, 7184–7194. [CrossRef]

34. Samadaei, E.; Kaviani, M.; Bertilsson, K. A 13-Levels Module (K-Type) with Two DC Sources for Multilevel Inverters. *IEEE Trans. Ind. Electron.* **2018**, *66*, 5186–5196. [CrossRef]

35. Meshram, P.; Borghate, V.B. A Simplified Nearest Level Control (NLC) Voltage Balancing Method for Modular Multilevel Converter (MMC). *IEEE Trans. Power Electron.* **2014**, *30*, 450–462. [CrossRef]

Article

The P-Type Module with Virtual DC Links to Increase Levels in Multilevel Inverters

Emad Samadaei [1,*], Mohammad Kaviani [2], Mina Iranian [3] and Edris Pouresmaeil [4,*]

[1]   Department of Electromobility, Volvo Truck Group Technology, SE- 417 15 Gothenburg, Sweden
[2]   Department of Engineering, Mazandaran University of Science and Technology, Babol 47166-85635, Iran;
      mohammad.kaviani70@gmail.com
[3]   Department of Engineering, Atlas Danesh Co, Ghaemshahr 47658-37449, Iran; iranian_mina@yahoo.com
[4]   Department of Electrical Engineering and Automation, Aalto University, 00076 Aalto, Finland
*   Correspondence: emad.samadaei@volvo.com (E.S.); edris.pouresmaeil@aalto.fi (E.P.);
      Tel.: +358-50-598-4479 (E.P.)

Received: 11 October 2019; Accepted: 20 November 2019; Published: 2 December 2019

Abstract: There has been an active interest in the evolution of newer multilevel inverter topologies in which the highest operation of DC sources become an important subject. In the paper, a new structure module presented a seventeen levels asymmetrical multilevel inverter by using two unequal DC sources (with the ratio 3:1). The configuration was focused on creating virtual DC links by two chargeable capacitors. The module had a simple inherent charging for capacitors without any additional circuit. The proposed multilevel inverter could produce higher voltage levels by a lower number of components; therefore, it is suitable for a wide range of applications. Also, the cascade connection of the module led to a modular topology with more voltage levels at higher voltages. The capability of the inherent negative voltage was involved. The simulation results obtained in MATLAB/Simulink, as well as the experimental results, verified the proposed topology.

Keywords: asymmetric; capacitors; multilevel inverter; power electronics; self-charging; virtual DC links

---

1. Introduction

Multilevel inverters (MLIs) have obtained more attention in recent years against two-level inverters because of their abilities in medium to high power applications, such as wind turbine [1], HVDC (High Voltage Direct Current) for transmission line [2,3], photovoltaic systems [4], drives systems [5,6], active power filer [7], power grid [8], and electrical vehicle [9]. MLIs synthesize the desired stepped output waveform from several DC voltage sources by the proper arrangement of the semiconductor switches. One of the important advantages of MLIs is using fewer components to create higher levels of the output voltage. Besides increasing the number of output voltage levels, high resolution on the output voltage and low harmonic components will be expected. Also, scalability, modularity, and lower switches stress are some of the other MLIs outstanding features due to the ability of cascade connection. Various topologies are introduced for MLIs. They can be categorized in three main types: NPC (Neutral Point Clamps) [10], FC (Flying Capacitor) [11], CHB (Cascade H-Bridge) [12]. Some disadvantages in NPC and FC, such as bulky capacitors, unbalanced DC links, and high switch stress, make CHB topologies more interesting. CHB has some comparable aspects: the number of semiconductors and DC sources and levels; total standing voltage (TSV) on switches; the inherent polarity levels, etc. Some conventional and vanguard topologies for the last decade were investigated in [13–16]. In [17], the module generated each level from one DC source by two switches. A series connection of the module can create more levels. The module can only generate positive polarity, and it requires an extra circuit for negative polarity. Full-bridge was added to the series module in [18] to create both negative and positive levels. By adding full-bridge circuits, negative voltage polarity is generated by the penalty of high switching stress on the semiconductors in the additional circuit and increasing the number of components. The enhancement of multilevel inverters' performances depends on creating higher output voltage levels by using a lower number of switches and DC voltage sources. Recently, asymmetric multilevel inverters with unequal DC sources have been addressed to increase the output levels without any complexity to the power circuit. Modules are designed based on the optimal combination of DC links and semiconductors. On another side, unequal DC links in asymmetric multilevel inverters may influence the stress on switches. The stress on switches is indexed with total standing voltage (TSV), which is the sum of the highest voltage stress on each switch. In [19,20], crossing switches were introduced as a solution to dividing of stress on switches and generating more levels. [21,22] presented extended H-bridge with different amounts of DC links. As the number of voltage levels increases, H-bridge switches tolerate higher stress. So, these topologies need higher rate semiconductors. Another kind of MLIs was proposed in [23] that is well-known as hybrid type topologies, although stress on switches is still obvious. Using full-bridge for negative voltages increases stress on switching and total standing voltages (TSV) on semiconductors. [24,25] introduced modules with inherent negative levels based on the maximum levels of achievement with four DC sources and low semiconductors to overcome these disadvantages. Different energy sources or storage elements, such as capacitors, can be applied instead of some DC sources to form a sinusoidal waveform in various multilevel inverter structures [26]. As a result, output voltage levels are increased with the same number of DC sources. [27–29] redesigned the structure of [11,19] to replace capacitors with some DC sources. In order to decrease the number of sources, [30–33] used a single source. In these configurations, they needed more semiconductors for the charging/discharging of capacitors. Some stair modular configurations with diverse DC sources and capacitors were proposed in [34–36], although full-bridge was applied in the circuits for producing the negative levels.

In the proposed inverter module, a new arrangement of semiconductors was introduced, which used only two DC sources. Two DC sources were unequal with $1 \times V_{DC}$ and $3 \times V_{DC}$ (one/three times scale of base voltage, respectively). The proposed inverter generated 17 output voltage levels. On the other hand, the proposed asymmetric multilevel module could produce eight positive levels, eight negative levels, and zero level (total 17 levels). Also, it did not need any extra circuit (such as a full-bridge) for negative voltage levels. The cascade connections of the module were described to create more levels for high voltage applications. The proposed multilevel inverter was illustrated in Section 2. Module introduction, switching patterns, charging and discharging of capacitors, cascade connection, and comparison study were included in this section. In Section 3, the nearest level control (NLC) scheme as a switching modulation was described. The voltage ripple on capacitors was investigated in Section 4. Finally, simulation and experimental results were presented in Sections 5 and 6, respectively.

2. Proposed Module

Figure 1 illustrates a general concept diagram of the proposed multilevel inverter with two DC sources. In order to achieve maximum output levels from sources, capacitors could be added to the configurations. Some extra DC links were created by capacitors to get more levels with the same DC sources. As shown in Figure 1, two DC sources with ratio 3:1 ($V_{DC1} = 3V_{DC2}$) could create nine voltage levels; and it could be redesigned with two capacitors and suitable arrangements to create seventeen voltage levels in output. The charging paths of capacitors should be considered as well. The charging paths of capacitors could be provided by a suitable designing of semiconductors arrangement to achieve the output levels paths without using an additional circuit.

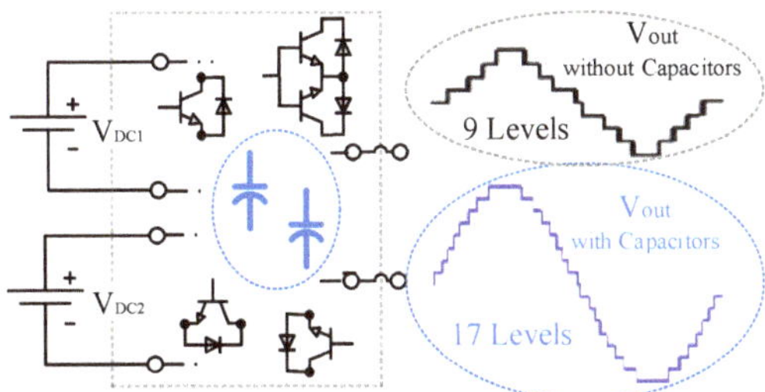

Figure 1. The general concept diagram of the proposed multilevel inverter (MLI) with capacitors.

2.1. Module Configuration

Asymmetric multilevel inverters could produce a different number of output voltage levels by using a fewer number of semiconductors in which it caused lower harmonic components as well. This promotion could be achieved by using two DC sources with different amounts as $1 \times V_{DC}$ and $3 \times V_{DC}$. It means the amount of one source was three times greater than the other one, and they were rewritten as $1V_{DC}$ and $3V_{DC}$ to simplify for the rest of the paper. In order to increase the number of DC links without any change in the number of DC sources, capacitors could be used. This idea gave four DC links involving 2 DC sources and 2 capacitors. Figure 2 shows the proposed module with a new arrangement of the components that contained 18 switches (8 unidirectional switches and 5 bidirectional switches) and 18 diodes in combination with 2 unequal DC sources and 2 capacitors. This configuration produced 17 levels of voltage at the output, including eight positive levels, eight negative levels, and zero level. This means that this module had an inherent negative level ability by connecting each DC link to other ones through different paths from different sides of a DC link. The structure of the proposed topology was figured to polygon, so it is named "P-Type" (Polygon-Type). The proper designing of the proposed module made that DC source with $1V_{DC}$ to charge the capacitor with $1V_{DC}$, and DC source with $3V_{DC}$ to charge the capacitor with $3V_{DC}$ without any additional circuit. Figure 3 draws the switching paths of all output voltage levels in the presented structure, and the state of switches in each level is listed in Table 1. The proposed module and their switching paths were designed accurately, as well as the positive terminals of DC links were not connected to the anode of diodes to cause the shortcuts. On the other hand, it was protected from short currents in which Figure 3 shows that the switching paths did not form any closed loop for DC links. Thus, diodes and bidirectional switches guaranteed that short-circuiting would not occur in the module.

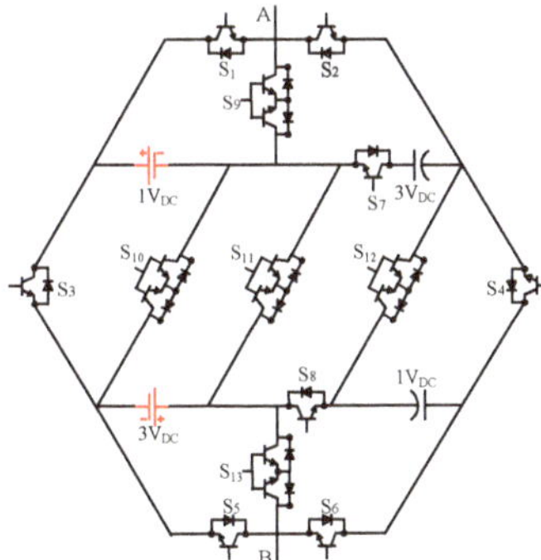

Figure 2. The proposed module (P-Type, Polygon-Type) for the multilevel inverter.

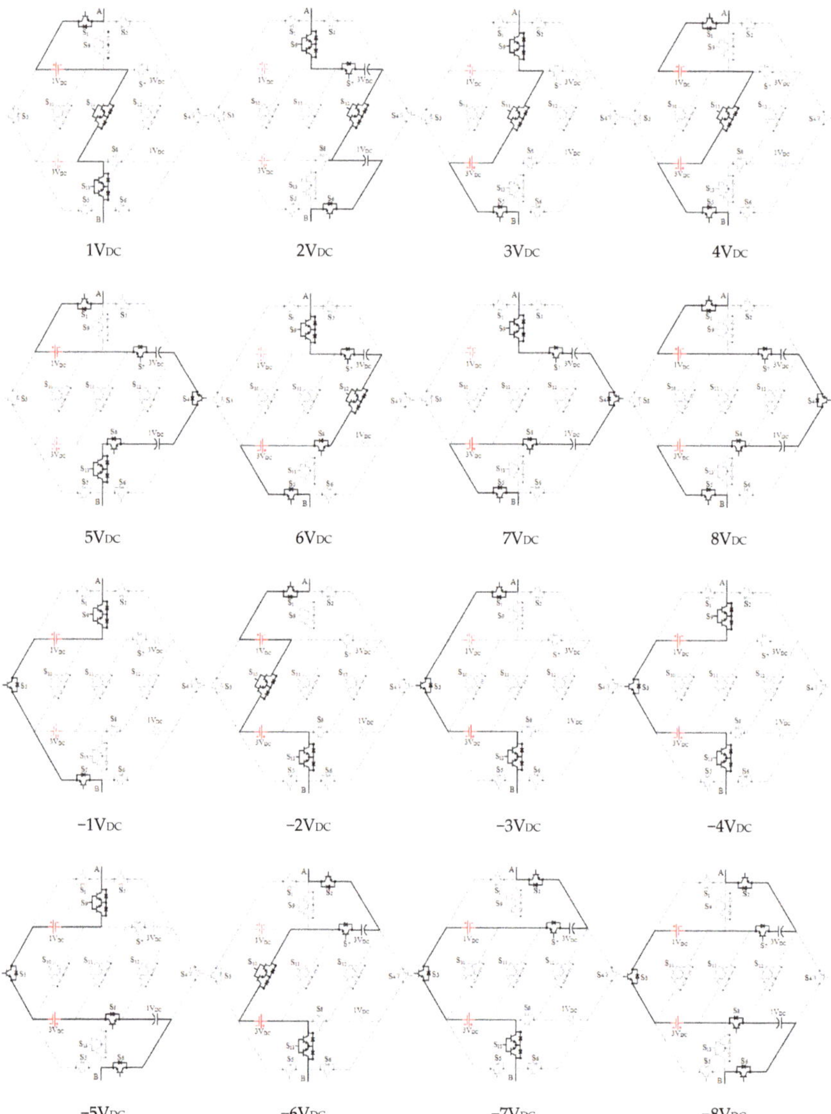

Figure 3. The switching paths of the proposed module.

Table 1. Sweating table.

		S_1	S_2	S_3	S_4	S_5	S_6	S_7	S_8	S_9	S_{10}	S_{11}	S_{12}	S_{13}
Positive level	$8V_{DC}$	1	0	0	1	1	0	1	1	0	0	0	0	0
	$7V_{DC}$	0	0	0	1	1	0	1	1	1	0	0	0	0
	$6V_{DC}$	0	0	0	0	1	0	1	1	1	0	0	1	0
	$5V_{DC}$	1	0	0	1	0	0	1	1	0	0	0	0	1
	$4V_{DC}$	1	0	0	0	1	0	0	0	0	0	1	0	0
	$3V_{DC}$	0	0	0	0	1	0	0	0	1	0	1	0	0
	$2V_{DC}$	0	0	0	0	0	1	1	0	1	0	0	1	0
	$1V_{DC}$	1	0	0	0	0	0	0	0	0	0	1	0	1
Negative level	$-1V_{DC}$	0	0	1	0	1	0	0	0	1	0	0	0	0
	$-2V_{DC}$	1	0	0	0	0	0	0	0	0	1	0	0	1
	$-3V_{DC}$	1	0	1	0	0	0	0	0	0	0	0	0	1
	$-4V_{DC}$	0	0	1	0	0	0	0	0	1	0	0	0	1
	$-5V_{DC}$	0	0	1	0	0	1	0	1	1	0	0	0	0
	$-6V_{DC}$	0	1	0	0	0	0	1	0	0	1	0	0	1
	$-7V_{DC}$	0	1	1	0	0	0	1	0	0	0	0	0	1
	$-8V_{DC}$	0	1	1	0	0	1	1	1	0	0	0	0	0
Num. of turning on per 1-cycle		6	2	5	4	4	6	5	5	7	4	4	4	6

Table 1 shows the on and off states of the switches at each level. It was clear that some pair switches could not be turned on simultaneously, such as (S_1, S_9); otherwise, the short circuit on DC sources was expectable. Additionally, the number of turning on per one cycle for each switch is shown in Table 1. As can be seen in the last row of Table 1, all switches had low operation frequency. In order to show this fact, S_2 and S_9 were selected as the switches with the lowest and highest number of turning on in one cycle to calculate the operation frequency. According to Table 1, S_2 and S_9 would be turning on 2 and 7 times in one cycle, respectively. By considering the fundamental frequency as 50 Hz, the operation frequency of a microprocessor for 32 steps would be calculated 1600 Hz. However, the operation frequency of S_2 and S_9 in one cycle was 100 Hz and 350 Hz, respectively. It was clear that these switches worked even lower than the overall microprocessor frequency (1600 Hz). It proved that all switches in the proposed module tolerated low-frequency stress.

Schematic output voltage levels associated with different switching states of the proposed module in one cycle are illustrated in Figure 4. Figure 4 also shows the schedule of the DC sources and capacitors, which were used for each level. Consequently, the proposed multilevel inverters, along with two unequal DC sources, could create 17-levels.

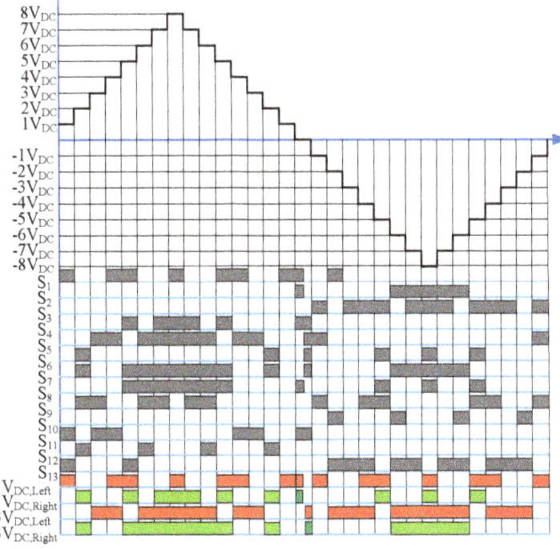

Figure 4. Switching pattern/schedule of DC links for the proposed inverter in one-cycle.

This module did not require any additional circuit to charge capacitors. According to Figure 4, capacitors were charged at level "zero". The module was designed based on the charging paths consisting of two loops for the charging of DC links that are shown in Figure 5. DC source with $1V_{DC}$ was charging $1Vc$ (Figure 5a), and DC source with $3V_{DC}$ was charging $3Vc$ (Figure 5b).

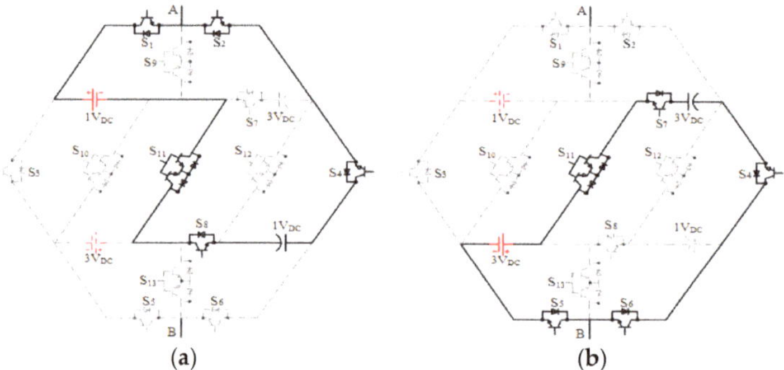

Figure 5. The charging paths of capacitors in the P-Type module, (**a**) charging for 1Vc; (**b**) charging for 3Vc.

Table 2 shows the equations of the module. The number of the semiconductor, DC sources, capacitors, drivers, and TSV (total standing voltages) based on the number of module units (n) were determined in the middle column and the number of output levels (NL), according to the mentioned variable parameters, were calculated in the last column. The symbol "[]" represents floor function.

Table 2. The equations of the proposed module.

	Based on the Number of Module Units	Based on the Number of Desired Levels
Levels	16n+1	N_L
Number of Switches	18n	$18\left[\frac{N_L-2}{16}+1\right]$
Number of Diodes	18n	$18\left[\frac{N_L-2}{16}+1\right]$
Driver	13n	$13\left[\frac{N_L-2}{16}+1\right]$
DC sources	2n	$13\left[\frac{N_L-2}{16}+1\right]$
Capacitors	2n	$2\left[\frac{N_L-2}{16}+1\right]$
TSV	57n	$57\left[\frac{N_L-2}{16}+1\right]$

According to Figure 3, the maximum magnitude of the blocking voltage was considered for each power switch. The total of all switch blocking voltages was introduced as TSV. The voltage standing on the switches in each level and the circuit study are presented in Figures 6 and 7, respectively. For each level, the voltage standing on each switch was separated by different colors, as shown in Figure 6. Low purple parts in Figure 6 confirms that the voltage stresses on switches were rare, and most of the area had low switch stress in the levels. Figure 7 demonstrates the voltage of switches on the circle graph, showing that the voltage standing in comparison with the total standing voltages was low in S_1, S_2, S_5, S_6, S_7, S_8, S_9, S_{11}, S_{12}, and S_{13}.

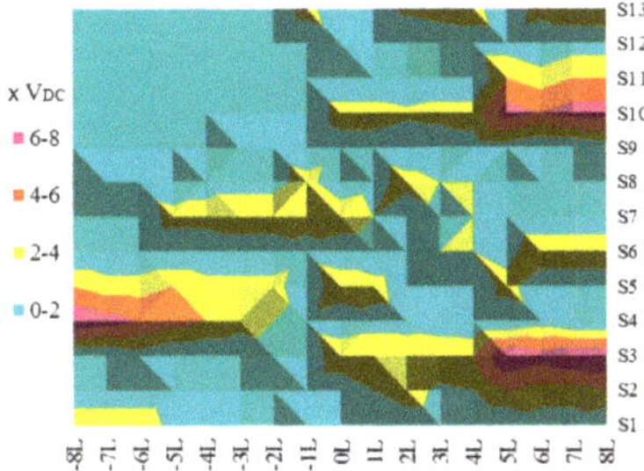

Figure 6. The surface graph of the voltage restrictions on the circuit for each level.

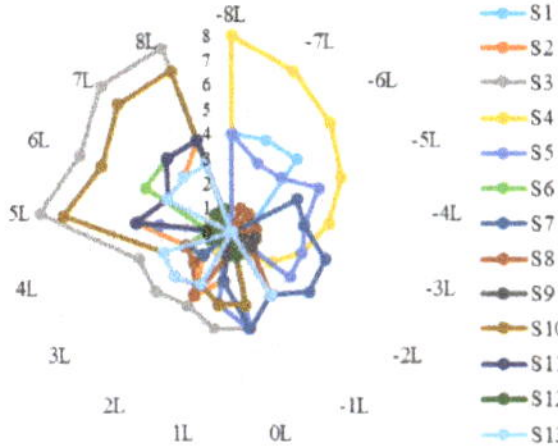

Figure 7. The circle graph of the voltage restrictions on each switch.

2.2. Module Extension

The modularity of the proposed model led to achieving more voltage levels. The cascade configuration was attractive for the medium and high voltage applications with cumulative DC links, such as solar Photovoltaic farms. The cascade connection of the two sequential units is shown in Figure 8. In this configuration, the unit produced 0, $\pm 1V_{DC}$, $\pm 2V_{DC}$, $\pm 3V_{DC}$, $\pm 4V_{DC}$, $\pm 5V_{DC}$, $\pm 6V_{DC}$, $\pm 7V_{DC}$, and $\pm 8V_{DC}$.

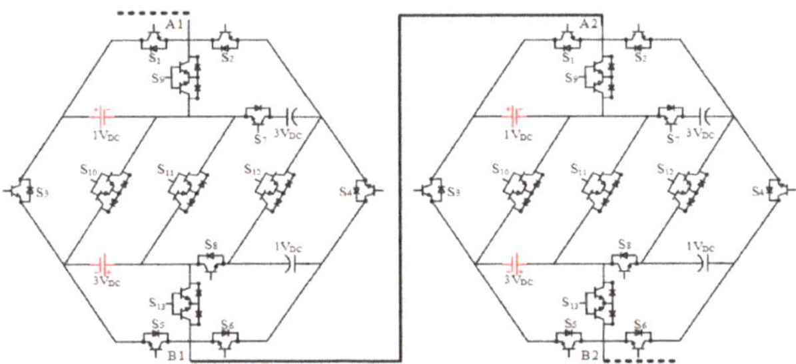

Figure 8. The cascade connection of the modular proposed multilevel.

Table 3 demonstrates that the combination of unit 1 and unit 2 created 16 positive levels, 16 negative levels, and zero level (total 33 levels). In Table 3, $u1=\frac{V_{A1B1}}{V_{DC}}$, $u2=\frac{V_{A2B2}}{V_{DC}}$. The extending of the proposed module as several units in series could make some redundant paths.

Table 3. Output levels for two modules to create 33 levels.

$u2=$ \ $u1=$	−8	−7	−6	−5	−4	−3	−2	−1	0	1	2	3	4	5	6	7	8
−8	−16	−15	−14	−13	−12	−11	−10	−9	−8	−7	−6	−5	−4	−3	−2	−1	0
−7	−15	−14	−13	−12	−11	−10	−9	−8	−7	−6	−5	−4	−3	−2	−1	0	1
−6	−14	−13	−12	−11	−10	−9	−8	−7	−6	−5	−4	−3	−2	−1	0	1	2
−5	−13	−12	−11	−10	−9	−8	−7	−6	−5	−4	−3	−2	−1	0	1	2	3
−4	−12	−11	−10	−9	−8	−7	−6	−5	−4	−3	−2	−1	0	1	2	3	4
−3	−11	−10	−9	−8	−7	−6	−5	−4	−3	−2	−1	0	1	2	3	4	5
−2	−10	−9	−8	−7	−6	−5	−4	−3	−2	−1	0	1	2	3	4	5	6
−1	−9	−8	−7	−6	−5	−4	−3	−2	−1	0	1	2	3	4	5	6	7
0	−8	−7	−6	−5	−4	−3	−2	−1	0	1	2	3	4	5	6	7	8
1	−7	−6	−5	−4	−3	−2	−1	0	1	2	3	4	5	6	7	8	9
2	−6	−5	−4	−3	−2	−1	0	1	2	3	4	5	6	7	8	9	10
3	−5	−4	−3	−2	−1	0	1	2	3	4	5	6	7	8	9	10	11
4	−4	−3	−2	−1	0	1	2	3	4	5	6	7	8	9	10	11	12
5	−3	−2	−1	0	1	2	3	4	5	6	7	8	9	10	11	12	13
6	−2	−1	0	1	2	3	4	5	6	7	8	9	10	11	12	13	14
7	−1	0	1	2	3	4	5	6	7	8	9	10	11	12	13	14	15
8	0	1	2	3	4	5	6	7	8	9	10	11	12	13	14	15	16

2.3. Comparative Study

Getting maximum voltage levels from the two DC sources is the specialty of the P-Type. It should be mentioned that there are few configurations with the exact two sources to be compared with the proposed multilevel inverter. Table 4 shows some similar new multilevel inverter configurations, as well as the proposed module in case of producing 17 output voltage levels. Some of these configurations could create the same levels with the only use of DC sources without any capacitors [12,18,24,25,35,36], and the presented module in [34] with two DC sources and some capacitors had a close configuration to the P-Type.

Table 4. Comparison of some modular multilevel inverter topologies.

	CHB [12]	[18]	[34]	[35]	[36]	[24]	[25]	Proposed Module
Number of Switches	$4\left[\frac{N_L-2}{2}+1\right]$	$8\left[\frac{N_L-2}{4}+1\right]$	$22\left[\frac{N_L-3}{16}+1\right]$	$9\left[\frac{N_L-2}{6}+1\right]$	$8\left[\frac{N_L-2}{8}+1\right]$	$10\left[\frac{N_L-2}{12}+1\right]$	$12\left[\frac{N_L-3}{16}+1\right]$	$18\left[\frac{N_L-2}{16}+1\right]$
Number of Diodes	$4\left[\frac{N_L-2}{2}+1\right]$	$8\left[\frac{N_L-2}{4}+1\right]$	$28\left[\frac{N_L-3}{16}+1\right]$	$11\left[\frac{N_L-2}{6}+1\right]$	$14\left[\frac{N_L-2}{8}+1\right]$	$10\left[\frac{N_L-2}{12}+1\right]$	$12\left[\frac{N_L-3}{16}+1\right]$	$18\left[\frac{N_L-2}{16}+1\right]$
Number of DC sources	$\left[\frac{N_L-2}{2}+1\right]$	$\left[\frac{N_L-2}{4}+1\right]$	$2\left[\frac{N_L-3}{16}+1\right]$	$\left[\frac{N_L-2}{6}+1\right]$	$\left[\frac{N_L-2}{8}+1\right]$	$4\left[\frac{N_L-2}{12}+1\right]$	$4\left[\frac{N_L-3}{16}+1\right]$	$2\left[\frac{N_L-2}{16}+1\right]$
Number of capacitors	-	$2\left[\frac{N_L-2}{4}+1\right]$	$6\left[\frac{N_L-3}{16}+1\right]$	$2\left[\frac{N_L-2}{6}+1\right]$	$3\left[\frac{N_L-2}{8}+1\right]$	-	-	$2\left[\frac{N_L-2}{16}+1\right]$
TSV [1] (xV_{DC})	$\left[\frac{N_L-2}{2}+1\right]$	$12\left[\frac{N_L-2}{4}+1\right]$	$88\left[\frac{N_L-3}{16}+1\right]$	$17\left[\frac{N_L-2}{6}+1\right]$	$17\left[\frac{N_L-2}{8}+1\right]$	$28\left[\frac{N_L-2}{12}+1\right]$	$39\left[\frac{N_L-3}{16}+1\right]$	$57\left[\frac{N_L-2}{16}+1\right]$
Negative level	With H-Bridge	With H-Bridge	With H-Bridge	With H-Bridge	With H-Bridge	Inherent	Inherent	Inherent

[1] Total standing voltages.

As shown in Table 4, to make a suitable comparison, various aspects, such as the number of DC sources, the number of semiconductors, the number of capacitors, the ability to generate negative voltage level, and TSV, were considered in terms of a number of voltage levels (NL). The formula for the other configuration is referred to from [37].

Figure 9 depicts the parameters, as mentioned in Table 4, versus voltage levels in different topologies. It was noticeable that one module generated some ranges of levels with the constant components. If more levels were required, it should be a connected module with the module in cascade connection. This is why Figure 9 was a staircase form. According to Figure 9a,b, it was prominent that the proposed module could attain maximum voltage levels from two DC sources with a lower number of semiconductors. The number of switches versus voltage levels is an important factor for MLIs. As a comparison only in the case of a number of semiconductors, [24,25,36] required a lesser number of switches/diodes, but the number of DC sources should be considered. It was observed that the P-Type had significantly fewer semiconductors than [34] with the same number of DC sources. One of the promising advantages of the P-Type module was using lower DC sources except for the single source configurations (Figure 9c).

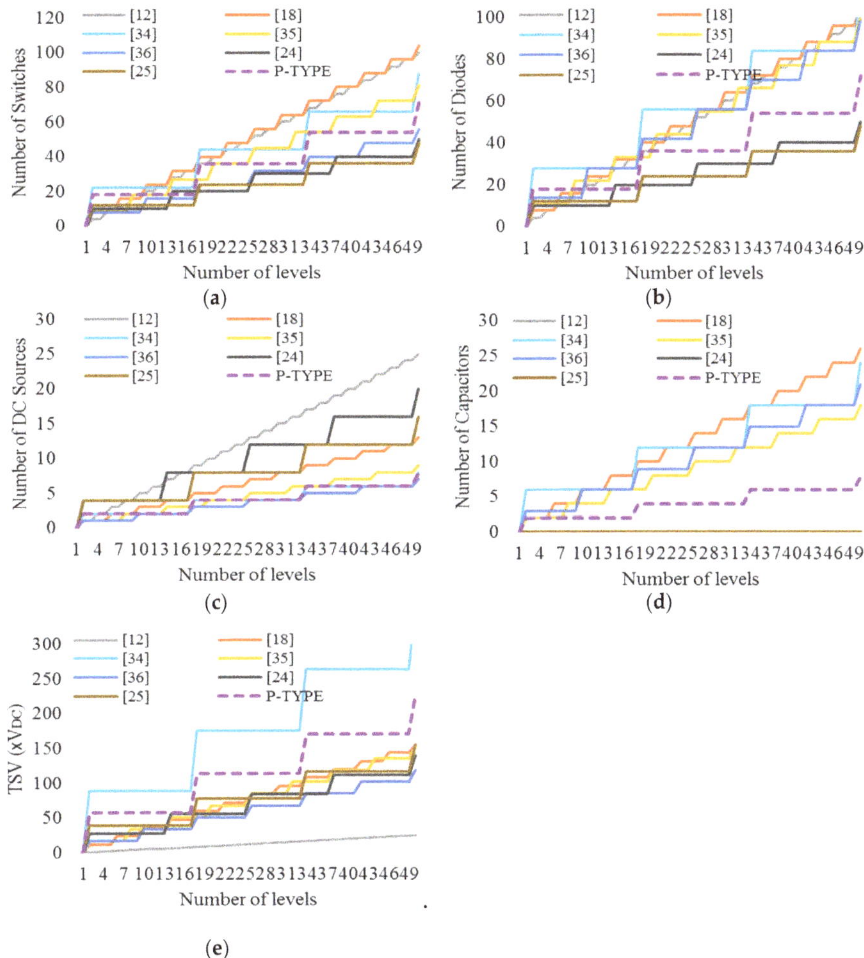

Figure 9. Comparative studies: The number of switches (**a**); The number of diodes (**b**); The number of sources (**c**); The number of capacitors (**d**); and TSV (**e**) in terms of the number of levels.

It is good to mention that P-Type needed the lowest number of capacitors in comparison with the module that used a capacitor as DC links (see Figure 9d). As shown in Figure 9e, the proposed module had a reasonable range of TSV. It could be referred to as Figure 6, which described the most of switches in the most of levels tolerating low switch stress (at the end of Section 2.1). It is noticeable that the presented topology and [24,25] had an inherent ability to generate negative voltage levels without any additional circuit. [12,18,34–36] could not have produced it without using full-bridge. This ability, along with lower components and switch stress, proved that the presented inverter could perform high in comparison with the other existing ones.

3. Nearest Level Control (NLC) Modulation Method

The nearest level control method (NLC) was used as a switching technique in the proposed multilevel [38]. This technique was applied in high voltage level converters to simplify and reduce the calculation of the processor. The modulation scheme and the control diagram are shown in Figure 10.

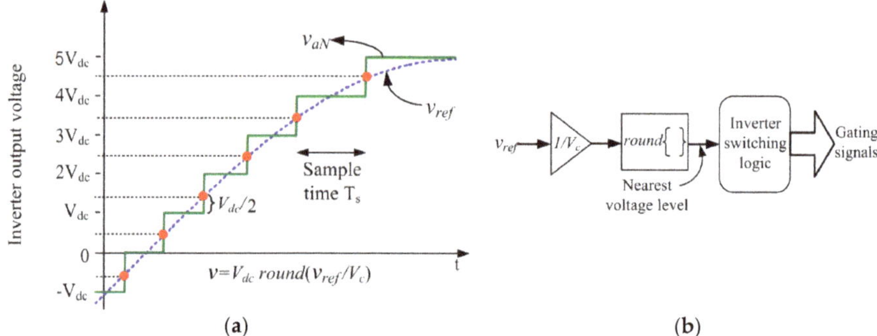

Figure 10. Nearest level control: (**a**) Waveform synthesis; (**b**) Control diagram.

According to Figure 10a, the controller sampled a point from the reference voltage (V_{ref}) and then rounded it to the nearest of the voltage level (V_{aN}). Each voltage level had a switching logic according to the switching table to change switches status (Figure 10b). The sampling was repeated for each sample time (T_s).

4. The Analysis of Capacitors Ripple

The capacitor voltage balancing is necessary to having constant voltage DC links in multilevel converters, which use capacitors as DC links. Since the capacitor voltage is kept constant, feeding the electrical load by MLIs would be guaranteed. Due to this fact, the voltage ripple of capacitors should be considered. To clarify this issue, ripple factor (RF) and figure factor (FF), as the main parameters in the ripple analysis of capacitors, are given by:

$$RF = \frac{V_{ac}}{V_{dc}} \tag{1}$$

$$FF = \frac{V_{rms}}{V_{dc}} \tag{2}$$

and,

$$V_{ac} = \sqrt{V_{rms}^2 - V_{dc}^2} \tag{3}$$

The ripple waveform is estimated to a sinusoidal waveform eight times bigger than the fundamental period. The maximum ripple is allowed to be less than 5% to obey this condition. Thus, the capacitor's drop voltages are not decreased than $0.95V_{max}$. The boundaries of the area of the integral are between 77° and 103°.

$$V_{dc} = \frac{1}{T}\int_{Start\ angle\ of\ ripple}^{End\ angle\ of\ ripple} f(\theta)\, d\theta = \frac{2 \times 1}{2\pi}\int_{77°}^{103°} V_{max} \sin\frac{\theta}{8}\, d\theta \tag{4}$$

$$V_{rms} = \sqrt{\frac{1}{T}\int_{Start\ angle\ of\ ripple}^{End\ angle\ of\ ripple} f^2(\theta)\, d\theta} = \sqrt{\frac{2 \times 1}{2\pi}\int_{77°}^{103°} V_{max}^2 \sin^2\frac{\theta}{8}\, d\theta} \tag{5}$$

As can be observed from the parallel charging in Figure 5, V_{max} for C_1 was 10 and for C_2 was 30. The following result could be calculated from Equations (1)–(5): Vdc,C_1 = 9.65, Vdc,C_2 = 28.95,

Vrms,C_1 = 9.75, Vrms,C_2 = 29.2, Vac,C_1 = 1.39, Vac,C_2 = 4.17, and therefore: FFC_1 = 1.01, RFC_1 = 0.14 and FFC_2 = 1.01, RFC_2 = 0.14.

The described analysis showed that the voltages of the capacitors were standard to use in the proposed multilevel inverter.

Taking into account that the amounts of capacitors directly depend on load application, the proper determination of the capacitors resulted in having enough energy to supply the load during each periodic cycle on their levels (see Figure 4). First, the typical AC electrical load was assumed to consume 60 Wh (or 0.333 mW for one cycle = 20 millisecond). Based on Figure 4, the C_1 as a DC link supplied levels −5, −8, 2, 5, 7, and 8, meaning 0.062 mW for each one cycle, and levels −8, −7, −6, 2, 5, 6, 7, and 8 were supplied by C_2, meaning 0.094 mW for each one cycle.

On the other hand, to keep the DC link voltages at the constant level, the drop voltage of capacitors must be less than 5%. This requires drop voltage to satisfy $\Delta V_{C1} < 0.5$ and $\Delta V_{C2} < 1.5$.

The energy stored in capacitors can be calculated by the following equation:

$$E_C = \frac{1}{2} C \Delta V^2 \tag{6}$$

According to the above equation and mentioned conditions, considering $C_1 \geq 496$ µF and $C_2 \geq 84$ µF would admit that the capacitor values were sufficient to keep their voltages constant with standard ripple.

It is obvious, the values of the capacitors were selected to limit the voltage ripples. Selecting a higher capacitor as a DC link led to a reduction in the voltage ripples correspondingly.

5. Simulation Results

The proposed 17 levels of multilevel inverter had been simulated by MATLAB/SIMULINK. The output voltage of P-Type is shown in Figure 11a. The magnitude of each level (V_{DC}) was 10 volts to create a 50 Hz sinusoidal waveform. Figure 11b despises the harmonics spectrums as well. The THD (Total Harmonic Distribution) was calculated as 3.12% by FFT analysis for the waveform of Figure 11a, which was lower than the acceptable amount in the IEEE519 standard (THD% $\leq$ 8% and each order $\leq$ 5%).

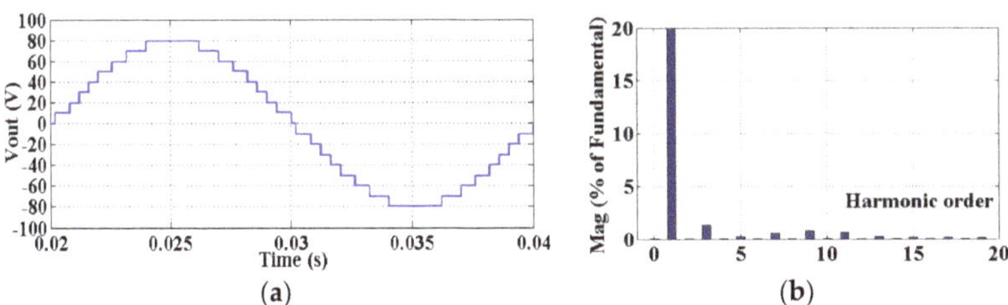

Figure 11. The simulated output voltage waveform of the proposed module: (**a**) Waveform; (**b**) Harmonics spectrums.

In order to indicate the performance of the modular mode, the cascade topology was simulated, and the results are depicted in Figure 12. The illustrated results confirmed the modular ability of proposed topology and its performance for creating 33 levels with THD = 1.54% for the cascade topology. IEEE519 was satisfied, as shown in Figure 12b, showing harmonic spectrums. Simulation results clarified the performance of the proposed module to create maximum output voltage waveforms with low harmonics.

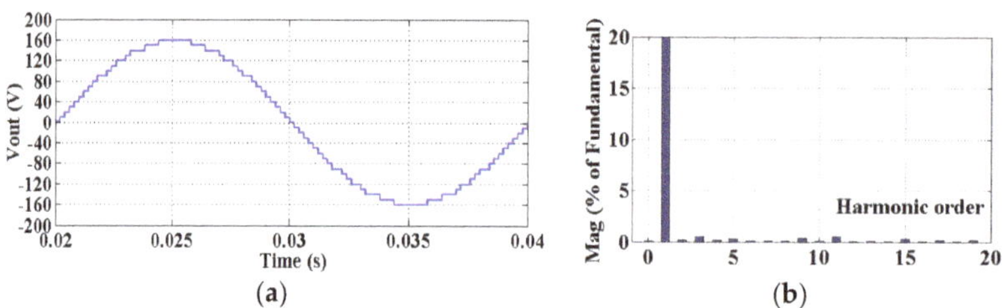

Figure 12. The simulated output voltage waveform of the first cascade topology (33 levels): (**a**) Waveform; (**b**) Harmonic spectrums.

6. Experimental Results

In order to verify the accurate performance of the proposed multilevel inverter and cascade topology connection for generating all output levels, an experimental prototype of the proposed module was built using IGBT12N60A4, Diode RHRP15120. The switching patterns of the different switches were generated by Microcontroller ATMEGA32, which provides on/off pulses for all switches. Based on Tables 1 and 3 and optocoupler-drivers (HCPL3120), the switches (MOSFET23N50E) were driven to create the sinusoidal waveform with a frequency of 50 Hz. Figure 13 depicts the experimental setup in the laboratory.

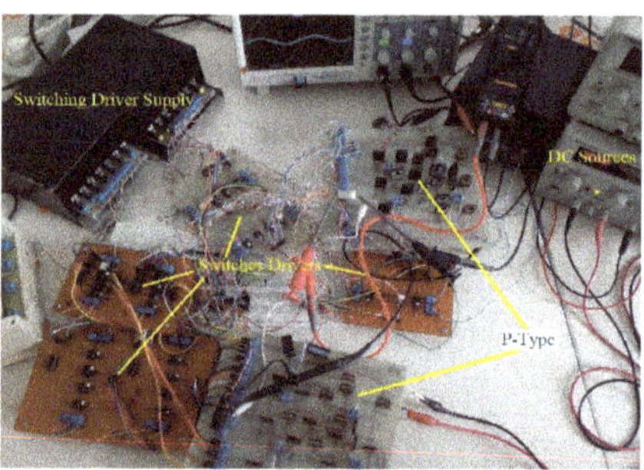

Figure 13. Experimental setup picture in the laboratory.

The experimental test on the setup system was performed, and each step of the voltage was considered 10 volts. Thus, the values of the used DC voltage sources were 10 (V) for $1V_{DC}$ and 30 (V) for $3V_{DC}$ in which $\pm 8xV_{DC}$ and $\pm 16xV_{DC}$ were generated for one module and cascade connection, respectively. The 17 levels and 33 levels MLI supplied the load with 40 Ω. Figure 14 shows the 50 Hz voltage and current sinusoidal waveforms of the proposed module for 17 levels. It should be mentioned that THD was 3.77% for 17 levels in the experimental test. Figure 15 contains results for 33 levels whose THD was 1.97%. The components were reduced directly affect the manufacturing cost. The new proposed multilevel inverter with a reduced number of DC sources was economical, and the smooth output voltage with low harmonic waveform made P-Type an interesting multilevel.

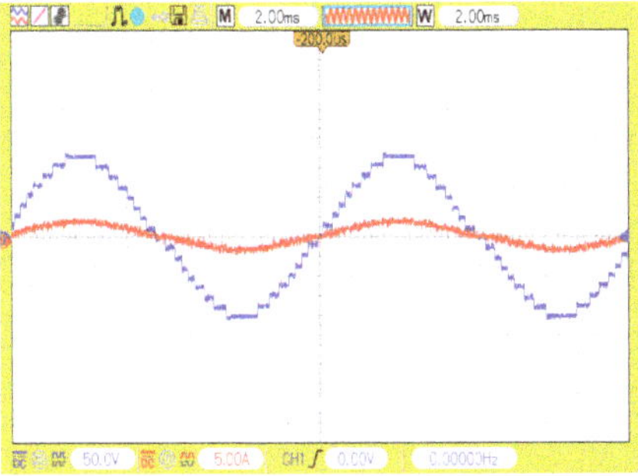

Figure 14. The output voltage of experimental results for 17 levels (case 1).

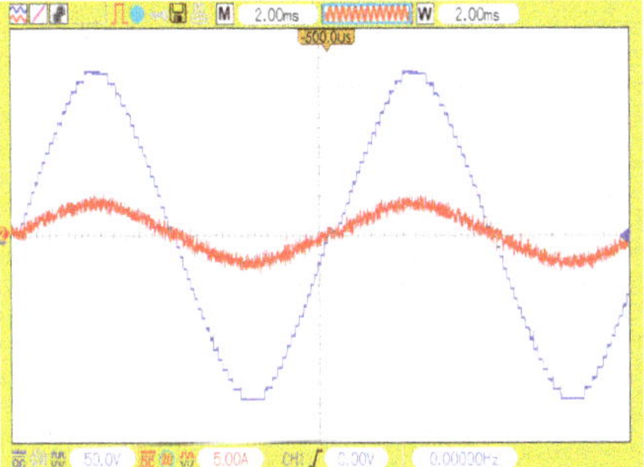

Figure 15. The output voltage of experimental results for 33 levels (case 2).

Finally, the voltages of V_{C1} and V_{C2} that were on 10 and 30 volts are demonstrated in Figures 16 and 17, respectively. It was objective that the voltages of the capacitors were constant during the experiment.

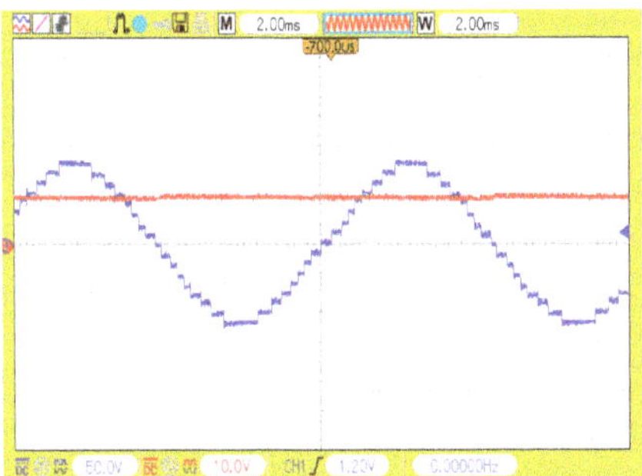

Figure 16. The voltage of V_{C1} experimental results.

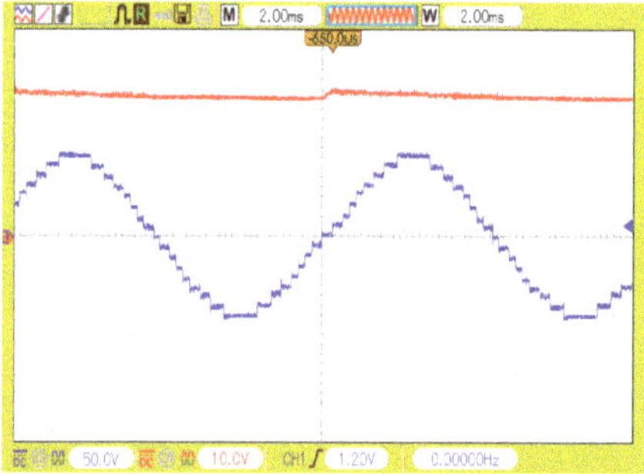

Figure 17. The voltage of V_{C2} experimental results.

7. Conclusions

In this paper, a new asymmetrical multilevel inverter module was introduced that is named P-Type. The configuration of P-Type produced 17 voltage levels by using only four DC links, including two DC sources and two capacitors. As a result, the maximum output voltage levels were produced at the output by the reduction of DC sources. By proper designing of the module, capacitors would be charged/discharged without any extra circuit. Modularity with low stress on semiconductors made the proposed module suitable for high power applications. The inherent negative voltage and low THDv were some main advantages of the proposed module. THDv% for one module was obtained as 3.12% and 3.77% in the simulation and experimental results, respectively, satisfying the harmonics standard (IEEE519). THDv% for cascade connection (two modules) was calculated to be 1.54% in simulation and 1.97% in experimental results. The experimental results proved the validity of the proposed module in producing the maximum output levels with a low amount of harmonics. The illustrated features of P-Type made it acceptable in power applications, which use unequal DC sources with ratio 3:1. The proposed module, with its all features, could be used in some applications with DC sources to supply AC loads. For example, it could be used in the solar farms with photovoltaic systems in which the unequal DC sources are accessible by the suitable connection of solar panels. Also, this system could be applied to other DC sources, such as fuel cells, batteries, etc.

Author Contributions: All authors contributed equally to this work and all authors have read and approved the final manuscript.

Funding: This research received no external funding.

Conflicts of Interest: The authors declare no conflict of interest.

References

1. Yuan, X. A Set of Multilevel Modular Medium-Voltage High Power Converters for 10-MW Wind Turbines. *IEEE Trans. Sustain. Energy* **2014**, *5*, 524–534. [CrossRef]
2. Nami, A.; Liang, J.; Dijkhuizen, F.; Demetriades, G.D. Modular Multilevel Converters for HVDC Applications: Review on Converter Cells and Functionalities. *IEEE Trans. Power Electron.* **2015**, *30*, 18–36. [CrossRef]
3. Gowaid, I.A.; Adam, G.P.; Massoud, A.M.; Ahmed, S.; Williams, B.W. Hybrid and Modular Multilevel Converter Designs for Isolated HVDC–DC Converters. *IEEE J. Emerg. Sel. Top. Power Electron.* **2018**, *6*, 188–202. [CrossRef]
4. Essakiappan, S.; Krishnamoorthy, H.S.; Enjeti, P.; Balog, R.S.; Ahmed, S. Multilevel Medium-Frequency Link Inverter for Utility Scale Photovoltaic integration. *IEEE Trans. Power Electron.* **2015**, *30*, 3674–3684. [CrossRef]
5. Mathew, J.; Rajeevan, P.P.; Mathew, K.; Azeez, N.A.; Gopakumar, K. A Multilevel Inverter Scheme with Dodecagonal Voltage Space Vectors Based on Flying Capacitor Topology for Induction Motor Drives. *IEEE Trans. Power Electron.* **2013**, *28*, 516–525. [CrossRef]

6. Viju Nair, R.; Arun Rahul, S.; Pramanick, S.; Gopakumar, K.; Franquelo, L.G. Novel Symmetric Six-Phase Induction Motor Drive Using Stacked Multilevel Inverters with a Single DC Link and Neutral Point Voltage Balancing. *IEEE Trans. Ind. Electron.* **2017**, *64*, 2663–2670. [CrossRef]

7. Ahmadi, D.; Wang, J. Online Selective Harmonic Compensation and Power Generation with Distributed Energy Resources. *IEEE Trans. Power Electron.* **2014**, *29*, 3738–3747. [CrossRef]

8. Haw, L.K.; Dahidah, M.S.A.; Almurib, H.A.F. A New Reactive Current Reference Algorithm for the STATCOM System Based on Cascaded Multilevel Inverters. *IEEE Trans. Power Electron.* **2015**, *30*, 3577–3588. [CrossRef]

9. Zheng, Z.; Wang, K.; Xu, L.; Li, Y. A Hybrid Cascaded Multilevel Converter for Battery Energy Management Applied in Electric Vehicles. *IEEE Trans. Power Electron.* **2014**, *29*, 35373546. [CrossRef]

10. Nabae, A.; Takahashi, I.; Akagi, H. A new neutral-point-clamped PWM inverter. *IEEE Trans. Ind. Appl.* **1981**, *IA-17*, 518–523. [CrossRef]

11. Meynard, T.A.; Foch, H. Multi-level choppers for high voltage applications. *EPE J.* **1992**, *2*, 45–50. [CrossRef]

12. Peng, F.Z.; Lai, J.-S.H.; McKeever, J.W.; VanCoevering, J. A multilevel voltage-source inverter with separate DC sources for static VAr generation. *IEEE Trans. Ind. Appl.* **1996**, *32*, 1130–1138. [CrossRef]

13. Zare, F. *Power Electronics Education Electronic-Book*; School of Engineering Systems, Queensland University of Technology: Brisbane, Australia, 2008.

14. Gupta, K.K.; Ranjan, A.; Bhatnagar, P.; Sahu, L.K.; Jain, S. Multilevel Inverter Topologies with Reduced Device Count: A Review. *IEEE Trans. Power Electron.* **2016**, *31*, 135–151. [CrossRef]

15. Debnath, S.; Qin, J.; Bahrani, B.; Saeedifard, M.; Barbosa, P. Operation, Control, and Applications of the Modular Multilevel Converter: A Review. *IEEE Trans. Power Electron.* **2015**, *30*, 37–53. [CrossRef]

16. Venkataramanaiah, J.; Suresh, Y.; Panda, A.K. A review on symmetric, asymmetric, hybrid and single DC sources based multilevel inverter topologies. *Renew. Sustain. Energy Rev.* **2017**, *76*, 788–812. [CrossRef]

17. Babaei, E.; Hosseini, S.H. New cascaded multilevel inverter topology with minimum number of switches. *Energy Convers. Manag.* **2009**, *50*, 2761–2767. [CrossRef]

18. Babaei, E.; Kangarlu, M.F.; Sabahi, M. Extended multilevel converters: An attempt to reduce the number of independent DC voltage sources in cascaded multilevel converters. *IET Power Electron.* **2014**, *7*, 157–166. [CrossRef]

19. Gupta, K.K.; Jain, S. Topology for multilevel inverters to attain maximum number of levels from given DC sources. *IET Power Electron.* **2012**, *5*, 435–446. [CrossRef]

20. Farhadi Kangarlu, M.; Babaei, E. Cross-switched multilevel inverter: An innovative topology. *IET Power Electron.* **2013**, *6*, 642–651. [CrossRef]

21. Babaei, E.; Laali, S.; Alilu, S. Cascaded multilevel inverter with series connection of novel H-bridge basic units. *IEEE Trans. Ind. Electron.* **2014**, *61*, 6664–6671. [CrossRef]

22. Babaei, E.; Alilu, S.; Laali, S. A new general topology for cascaded multilevel inverters with reduced number of components based on developed H-bridge. *IEEE Trans. Ind. Electron.* **2014**, *61*, 3932–3939. [CrossRef]

23. Shalchi Alishah, R.; Nazarpour, D.; Hosseini, S.H.; Sabahi, M. New hybrid structure for multilevel inverter with fewer number of components for high-voltage levels. *IET Power Electron.* **2014**, *7*, 96–104. [CrossRef]

24. Samadaei, E.; Gholamian, S.A.; Sheikholeslami, A.; Adabi, J. An Envelope Type (E-Type) Module: Asymmetric Multilevel Inverters with Reduced Components. *IEEE Trans. Ind. Electron.* **2016**, *63*, 7148–7156. [CrossRef]

25. Samadaei, E.; Sheikholeslami, A.; Gholamian, S.A.; Adabi, J. A Square T-Type (ST-Type) Module for Asymmetrical Multilevel Inverters. *IEEE Trans. Power Electron.* **2018**, *33*, 987–996. [CrossRef]

26. Samadaei, E.; Kaviani, M.; Bertilsson, K. A 13-levels Module (K-Type) with two DC sources for Multilevel Inverters. *IEEE Trans. Ind. Electron.* **2018**, *66*, 5186–5196. [CrossRef]

27. Vahedi, H.; Labbé, P.A.; Al-Haddad, K. Sensor-Less Five-Level Packed U-Cell (PUC5) Inverter Operating in Stand-Alone and Grid-Connected Modes. *IEEE Trans. Ind. Inform.* **2016**, *12*, 361–370. [CrossRef]

28. Metri, J.I.; Vahedi, H.; Kanaan, H.Y.; Al-Haddad, K. Real-Time Implementation of Model-Predictive Control on Seven-Level Packed U-Cell Inverter. *IEEE Trans. Ind. Electron.* **2016**, *63*, 4180–4186. [CrossRef]

29. Vahedi, H.; Al-Haddad, K. Real-Time Implementation of a Seven-Level Packed U-Cell Inverter with a Low-Switching-Frequency Voltage Regulator. *IEEE Trans. Power Electron.* **2016**, *31*, 5967–5973. [CrossRef]

30. Sun, X.; Wang, B.; Zhou, Y.; Wang, W.; Du, H.; Lu, Z. A Single DC Source Cascaded Seven-Level Inverter Integrating Switched-Capacitor Techniques. *IEEE Trans. Ind. Electron.* **2016**, *63*, 7184–7194. [CrossRef]

31. Taghvaie, A.; Adabi, J.; Rezanejad, M. A Multilevel Inverter Structure Based on a Combination of Switched-Capacitors and DC Sources. *IEEE Trans. Ind. Inform.* **2017**, *13*, 2162–2171. [CrossRef]

32. Taghvaie, A.; Adabi, J.; Rezanejad, M. Circuit Topology and Operation of a Step-Up Multilevel Inverter with a Single DC Source. *IEEE Trans. Ind. Electron.* **2016**, *63*, 6643–6652. [CrossRef]

33. Tang, Y.; Ran, L.; Alatise, O.; Mawby, P. Capacitor Selection for Modular Multilevel Converter. *IEEE Trans. Ind. Appl.* **2016**, *52*, 3279–3293. [CrossRef]

34. Zamiri, E.; Vosoughi, N.; Hosseini, S.H.; Barzegarkhoo, R.; Sabahi, M. A New Cascaded Switched-Capacitor Multilevel Inverter Based on Improved Series–Parallel Conversion with Less Number of Components. *IEEE Trans. Ind. Electron.* **2016**, *63*, 3582–3594. [CrossRef]

35. Liu, J.; Wu, J.; Zeng, J.; Guo, H. A Novel Nine-Level Inverter Employing One Voltage Source and Reduced Components as High-Frequency AC Power Source. *IEEE Trans. Power Electron.* **2017**, *32*, 2939–2947. [CrossRef]

36. Ye, Y.; Cheng, K.W.E.; Liu, J.; Ding, K. A Step-Up Switched-Capacitor Multilevel Inverter with Self-Voltage Balancing. *IEEE Trans. Ind. Electron.* **2014**, *61*, 6672–6680. [CrossRef]

37. Vijeh, M.; Rezanejad, M.; Samadaei, E.; Bertilsson, K. A General Review of Multilevel Inverters Based on Main Submodules: Structural Point of View. *IEEE Trans. Power Electron.* **2019**, *34*, 9479–9502. [CrossRef]

38. Meshram, P.M.; Borghate, V.B. A simplified nearest level control (NLC) voltage balancing method for modular multilevel converter (MMC). *IEEE Trans. Power Electron.* **2015**, *30*, 450–462. [CrossRef]

 *electronics*

Article

A Multi-Inductor H Bridge Fault Current Limiter

Amir Heidary [1], Hamid Radmanesh [2], Ali Moghim [1], Kamran Ghorbanyan [1], Kumars Rouzbehi [3], Eduardo M. G. Rodrigues [4],* and Edris Pouresmaeil [5],*

[1]   Electrical Engineering Department, Amirkabir University of Technology, 15875-4413 Tehran, Iran
[2]   Electrical Engineering Department, Aeronautical University of Science and Technology, 15914 Tehran, Iran
[3]   Department of System Engineering and Automatic Control, University of Seville, 41004 Seville, Spain
[4]   Management and Production Technologies of Northern Aveiro—ESAN, Estrada do Cercal 449, Santiago de Riba-Ul, 3720-509 Oliveira de Azeméis, Portugal
[5]   Department of Electrical Engineering and Automation, Aalto University, 02150 Espoo, Finland
*   Correspondence: emgrodrigues@ua.pt (E.M.G.R.); edris.pouresmaeil@aalto.fi (E.P.);
     Tel.: +358-505-984-479 (E.P.)

Received: 18 May 2019; Accepted: 12 July 2019; Published: 16 July 2019

Abstract: Current power systems will suffer from increasing pressure as a result of an upsurge in demand and will experience an ever-growing penetration of distributed power generation, which are factors that will contribute to a higher of incidence fault current levels. Fault current limiters (FCLs) are key power electronic devices. They are able to limit the prospective fault current without completely disconnecting in cases in which a fault occurs, for instance, in a power transmission grid. This paper proposes a new type of FCL capable of fault current limiting in two steps. In this way, the FCLs' power electronic switches experience significantly less stress and their overall performance will significantly increase. The proposed device is essentially a controllable H bridge type fault current limiter (HBFCL) that is comprised of two variable inductances, which operate to reduce current of main switch in the first stage of current limiting. In the next step, the main switch can limit the fault current while it becomes open. Simulation studies are carried out using MATLAB and its prototype setup is built and tested. The comparison of experimental and simulation results indicates that the proposed HBFCL is a promising solution to address protection issues.

Keywords: fault current limiter; microgrid protection; power quality; fault current; H bridge

1. Introduction

The immense global growth in energy demand will require additional power generation as well as an efficient, reliable complex meshed power distribution. The existing power grids will experience, in the near future, a growing burden due to an upsurge in electricity demand and will experience an ever-growing penetration of distributed power generation, which are factors that will contribute to a higher incidence of fault current levels. The massive growth of gird interconnection and integration of distributed generators (DGs) increase the network fault current level [1–4]. The solid-state fault current limiter (FCL) is a fast protection device that includes a DC reactor and solid-state switches [1–4]. The voltage source converters (VSCs) of HVDC systems are sensitive to the fault current. Recently, they have been combined with appropriate FCLs to protect them [5,6]. There are other types of FCLs that have been introduced in the literature. A resistive superconductor FCL based on variable resistance, which is very complex and costly, has been presented in the works of [7,8]. The bridge type FCLs based on DC reactor have been studied in the literature [9–12]. The AC/DC reactor based FCL has been presented in the work of [13]. In this FCL, two-stage operation decreases the voltage stress on the solid-state switches. The other well-known FCL type are the resonance type FCLs, which have high transient voltage, and this is their most important challenge [14,15]. A series two-stage FCL that behaves by operation of the solid-state switch in the secondary winding is introduced in the works of [16,17]. Saturated core FCL based on DC bias saturation and the series coil is studied in the literature [18–24]. In this type, the electronic switch connects to DC saturation current and does not have any conflict with the line current. Superconductive FCLs have been investigated for limiting the fault current in the microgrid [25,26]. FCLs can preserve microgrid from AC grid fault currents because the AC/DC microgrid should be protected in both the AC and DC sides [27]. Novel types of magnetic based FCLs are analyzed in the works of [28,29] to improve the performance of FCL for a power grid. Flux coupled FCLs and bridge type solid-state FCLs [30,31] are used to design a novel H bridge type fault current limiter (HBFCL).

The rest of this paper is organized as follows. In Section 2, the HBFCL structure is presented. In Section 3, the analytical studies are given and, in the next section, the simulation results of the proposed HBFCL are presented. In Section 5, the experimental test results are presented and, finally, the conclusion is drawn.

2. Proposed HBFCL Configuration

The proposed HBFCL is connected in series with the line to protect the point of common coupling (PCC) of the microgrid against the fault current. The HBFCL includes four inductors, L_1–L_4, as shown in Figure 1. An antiparallel power electronic IGBTs, that is, G_1 and G_2, are connected as main switches to the middle branch of the H bridge. L_3 and L_4 are coupled with L_5 and L_6, respectively. The power electronic switch, G_3 and G_4, and rectifier diodes, D_1–D_4, are connected to these coupled inductances. After switching of IGBT switches (G_3 and G_4), L_5 and L_6 are bypassed and two levels for L_3 and L_4 in the different modes are configured.

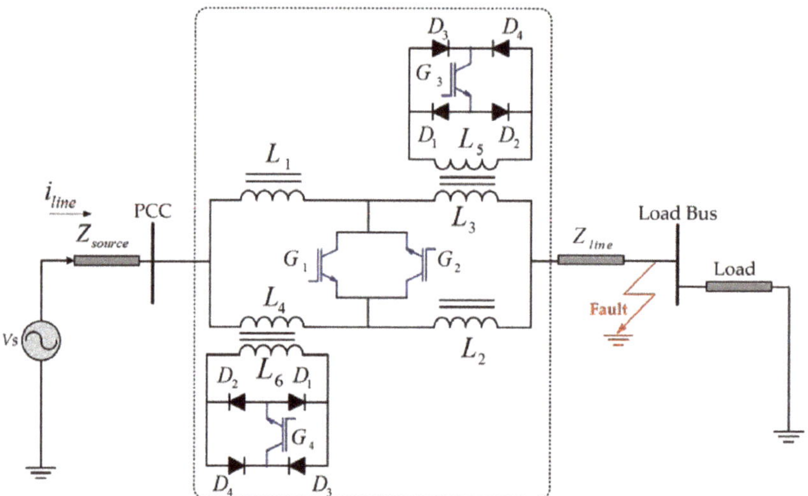

Figure 1. Proposed H bridge type fault current limiter (HBFCL) topology.

The operation of the proposed FCL is divided into three modes, as shown in Figure 2. Figure 2a–c show the HBFCL equivalent circuit during the normal operation mode after fault occurrences and during the fault limiting mode, respectively.

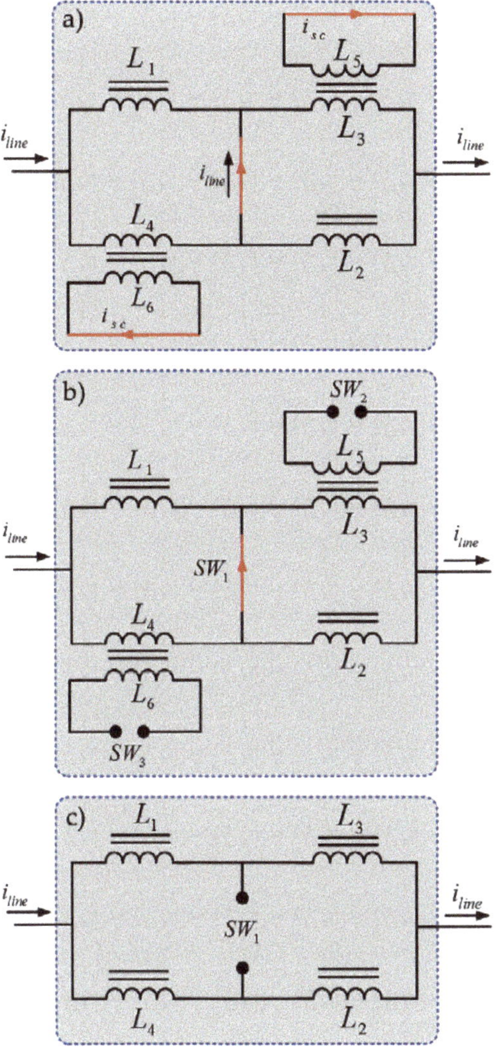

Figure 2. HBFCL equivalent circuit. (**a**) Normal operation mode, (**b**) fault operation mode in first state, and (**c**) fault operation mode in second state.

2.1. Normal Operation Mode

In this mode, as shown in Figure 2a, the secondary sides of L_3 and L_4 are short-circuited via IGBTs and the inductors are modeled by their leakage inductance and a small resistance. Considering L_1 and L_2 values, high inductive current is carried by L_3, L_4, G_1, and G_2. During the normal operation mode, all of the IGBT switches are in ON state and the maximum power flow is passed by the HBFCL.

2.2. Pre-Limiting Mode

After fault occurrence, the IGBTs G_3 and G_4 become turned-off and the main breaker SW_1, which includes series antiparallel switches that is shown as G_1, and G_2 change to turned-off state. In the off state of G_3 and G_4, the inductance of L_3 and L_4 increases and the current of the main switch, that is, SW_1, decreases to a low value. Figure 2b shows the equivalent circuit of the pre-limiting mode.

2.3. Fault Current Limiting Mode

In this mode, the current of the SW_1 decreases and it can safely be opened. In this case, the limited fault current is divided between two parallel branches, which include series connection of L_1, L_3 and L_2, L_4.

3. Analytical Studies

3.1. Steady-State Mode

Analytical studies are presented based on the three operation states of the proposed HBFCL. In the first state, there is no fault in the system. In this case, the microgrid equivalent circuit is shown in Figure 2a and the analytical study is done according to this circuit. In this case, the current and voltage is sinusoidal and we have the following:

$$i_{line} = \frac{V_S}{Z_S + Z_{HBFCL} + Z_{line} + R_{fault}}, \tag{1}$$

where

$$Z_{HBFCL} = (r_3 + r_4) + j(X_{L3} + X_{L4}), \tag{2}$$

$$V_{HBFCL} = i_{line}((r_3 + r_4) + j(X_{L3} + X_{L4})), \tag{3}$$

and

$$V_{PCC} = V_S - i_{line}((r_3 + r_4) + j(X_{L3} + X_{L4}) + Z_S), \tag{4}$$

where V_s, V_{HBFCL}, V_{PCC} are source voltage, HBFCL voltage drop, and voltage of point of common coupling, respectively. i_{line} is line current. L_{L3}, L_{L4}, r_3, and r_4 are leakage inductances and resistances of L_3 and L_4, respectively. Z_s, Z_{HBFCL}, and Z_{line} are impedances of the source, HBFCL, and line, respectively. R_{fault} is resistance of the fault.

During normal operation, the power loss is calculated with Equation (5).

$$P_{loss} = P_{Cu(L_3)} + P_{Cu(L_4)} + P_{Cu(L_5)} + P_{Cu(L_6)} + P_{SW1} + P_{SW2} + P_{SW3}, \tag{5}$$

where

$$P_{Cu} = i_{line}^2 \times r_3 + i_{line}^2 \times r_4 + i_{sc}^2 \times r_5 + i_{sc}^2 \times r_6, \tag{6}$$

$$P_{SW} = i_{line} \times V_{SW1} + 2(i_{sc} \times V_{SW2}). \tag{7}$$

The power loss depends directly on the line current, inductor secondary current, switching voltage, and coil resistance.

According to Equations (5)–(7), the HBFCL power loss is negligible by decreasing coil resistance and using the series power IGBT switch.

3.2. Pre-Fault Limiting Mode

In fault occurrence, G_3 and G_4 change the H bridge topology and limit the fault current, and we have the following equation.

$$X_{L1} \times X_{L2} = X_{L3} \times X_{L4} = (2\pi f)^2 L_1 \times L_2 = (2\pi f)^2 L_3 \times L_4, \tag{8}$$

where X_{L1} to X_{L4} are reactor impedances while the secondary side is open-circuited and f is the network frequency.

3.3. Fault Current Limiting Dynamic Mode

Considering Figure 2c, we have the following equations:

$$2L_1 = L_2, \ 2L_4 = L_3, \ L = L_1 = L_4, \tag{9}$$

$$L_{HBFCL} = \frac{(L_1 + L_3)(L_2 + L_4)}{(L_1 + L_3) + (L_2 + L_4)} = \frac{3}{2}L, \tag{10}$$

and

$$-V_S(t) + i_{line}r_{eq} + (L_{eq})\frac{di_{line}}{dt} = 0, \tag{11}$$

and in which

$$i_{line}(t) = Ae^{-\frac{r_{eq}}{L_{eq}}t} + BVm\sin(\omega t - \theta), \tag{12}$$

where A and B are determined based on initial condition.

$$r_{eq} = r_S + r_{line} + r_{HBFCL} + R_{fault} \tag{13}$$

$$L_{eq} = L_S + L_{line} + \frac{3}{2}L \tag{14}$$

4. Control Strategy

Figure 3 shows the control system block diagram based on the proposed HBFCL.

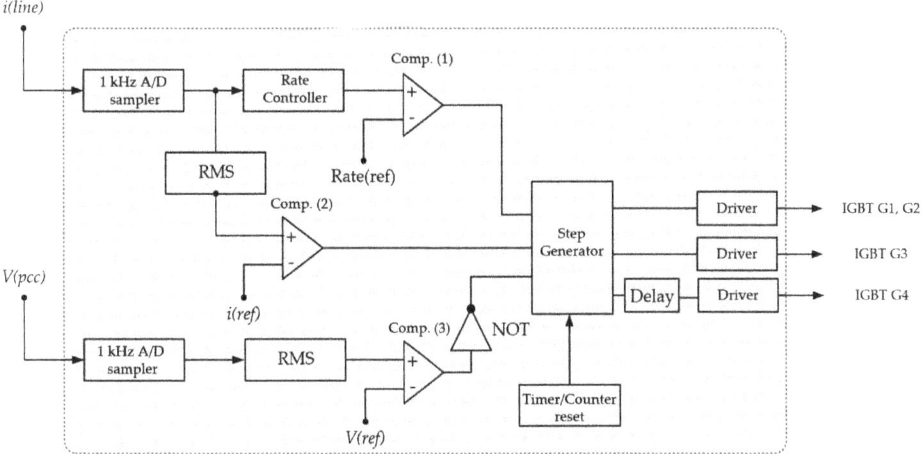

Figure 3. HBFCL control block diagram.

In this system, represented by the HBFCL control block diagram from Figure 3, the current and voltage signals are monitored via current and voltage transformers are measured and send to a digital (A/D) sampler to make the digital data. In fault cases, the current rate is raised and the *rms* value of the current is compared with the reference value, that is, 1.2 p.u. The voltage signal is sampled by the A/D block and its *rms* value is compared with the reference voltage. A step generator drives IGBT switches. The main switches are driven after a very small delay to meet the HBFCL self-protection and limit the fault current in two steps. After fault current limitation, a timer resets the step generator to turn-on G_1–G_4 for checking the fault clearance.

5. Simulation Results

In this section, simulation results are carried out considering the system configuration shown in Figure 1. The electrical network parameters are listed in Table 1.

Table 1. The values of the H bridge type fault current limiter (HBFCL) parameters.

Symbol	Description	Value
V_S	Source voltage	20 kV
r_s	Source resistance	0.1 Ω
r_{line}	Line resistance	0.1 Ω
r_f	Fault resistance	0.01 Ω
L_S	Source inductance	10 mH
L_{line}	Line inductance	10 mH
L_1	HBFCL first inductance	0.1 H
L_2	HBFCL second inductance	0.2 H
L_3	HBFCL third inductance	0.2 H
L_4	HBFCL fourth inductance	0.1 H

In order to be able to monitor the fault cases, the line to ground fault is applied to the network and the proposed HBFCL is connected in series in the line. To verify the proposed HBFCL effectiveness, two cases are considered to obtain the simulation results, that is, fault current without HBFCL effect and limited fault current with HBFCL effect, as shown in the following subsections.

5.1. Fault Condition without HBFCL Effect

In this section, the proposed electrical system shown in Figure 1 is simulated without the HBFCL effect. Figure 4 shows the line current provided by the main feeder during the normal and fault operation modes. During the normal operation mode, the line current amplitude is 200 A till t_1. After fault occurrences in t_1, the fault current is increased and its first peak amplitude reaches 6300 A. Accordingly, if bus bar base current assumes 1000 A fault first peak is 6.1 p.u, which shows studied bus-bar high strength and high possible fault current.

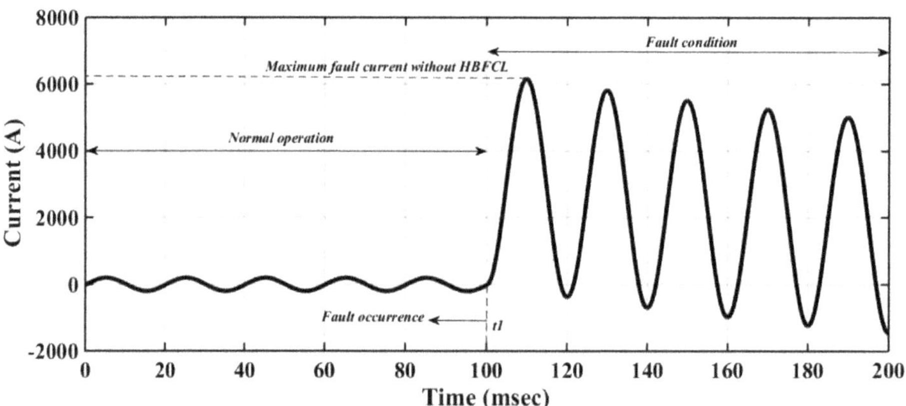

Figure 4. System voltage and current without HBFCL effect—the line current during normal and fault conditions.

As shown in Figure 5, the PCC voltage has 20 kV amplitude during the normal operation mode, and after fault occurrences, its amplitude experiences deep voltage sag and decreases to 10 kV.

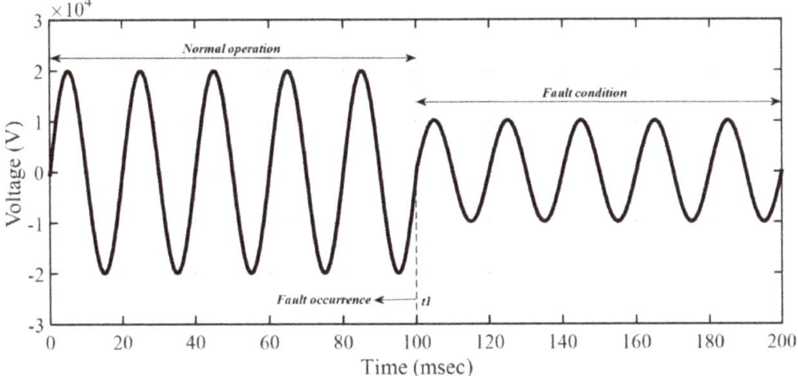

Figure 5. System voltage and current without HBFCL effect—the point of common coupling (PCC) voltage during normal and fault conditions.

5.2. Fault Condition with HBFCL Effect

Connecting the proposed HBFCL as a protection device to the line, the fault current is decreased to an acceptable level, as shown in Figure 6.

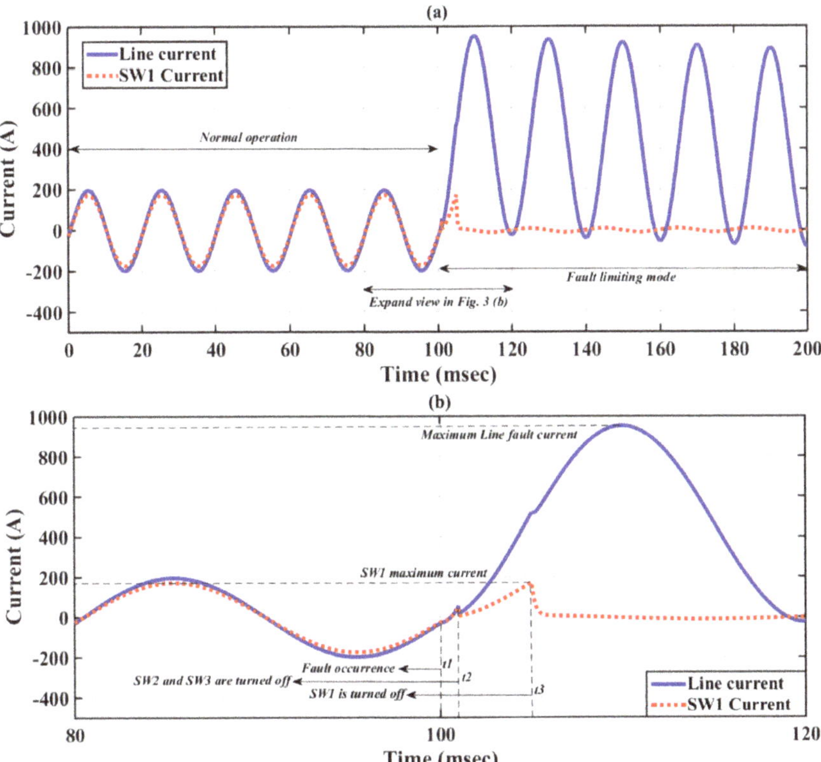

Figure 6. (a) Line and SW1 switch currents during normal and fault conditions affected by HBFCL and (b) PCC voltage during normal and fault conditions affected by HBFCL.

In order to control the fault current, IGBTs change the HBFCL topology in two steps. In t_1, 100 ms fault is occurred while between t_1 and t_2, 102 ms HBFCL control system recognizes the fault but HBFCL is not operated. In t_2, SW_2 and SW_3 are turned off and current is limited by increasing L_1 and L_2 impedance, as shown in Figure 6a. After a small delay, the main switch SW_1 is turned off and current is decreased to nominal current. Considering the HBFCL limiting strategy, the first peak of the fault current is limited to 1 kA. Figure 6b shows the PCC voltage during normal, transient, and fault states. Considering the switching transient recovery voltage (TRV) between t_2 and t_3, the TRV peak has an acceptable rate in the first switching and second switching; it is damped very well for safe switching action.

In Figure 7, it is possible to observe the SW_2 and SW_3 effect on the PCC transient recovery voltage and the SW_1 transient recovery voltage after the 100 ms instant, in which a transition from the normal operating mode to the fault limiting mode can be observed. This effect can also be observed in more detail in the expanded view of Figure 7.

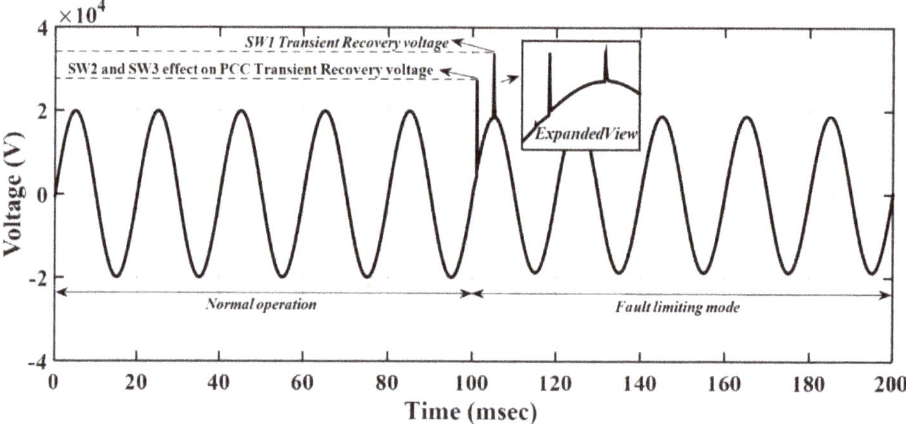

Figure 7. The effect of SW_2 and SW_3 on the PCC transient recovery voltage and the SW_1 transient recovery voltage.

Figure 8 shows the limited fault current by HBFCL where the first peak of the fault current is decreased considerably.

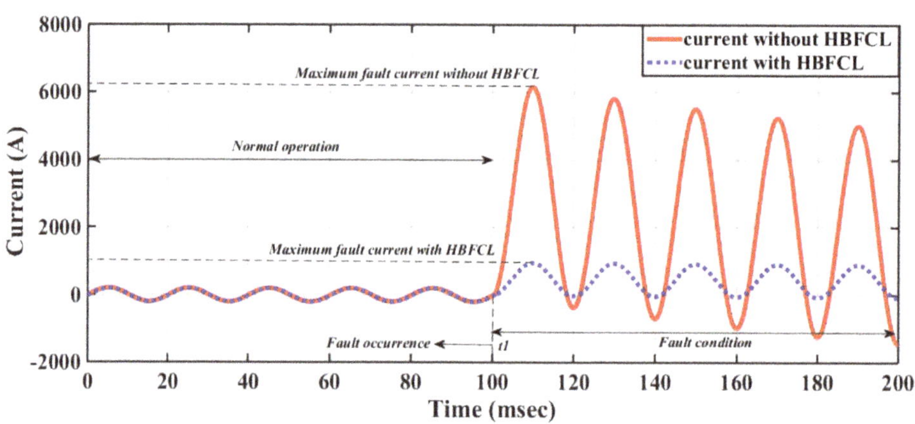

Figure 8. The line current with and without HBFCL protection.

6. Simulation Results

In this section, the laboratory test prototype is built and tested to verify the simulation results. The parameters values are listed in Table 2 and the proposed prototype is shown in Figure 9.

Table 2. The values of prototype parameters.

Symbol	Description	Value
V_S	Source voltage	20 kV
r_s	Source resistance	0.1 Ω
r_{line}	Resistance	0.1 Ω
r_f	Resistance	0.01 Ω
L_S	Source inductance	10 mH
L_{line}	Open core 30 turns inductor	10 mH
L_1	E-I core inductor	0.1 H
L_2	E-I core inductor	50 mH
L_3	E-I core inductor	0.1 H
L_4	E-I core inductor	0.2 H
R_{load}	Variable 100 W resistor	0–100 Ω

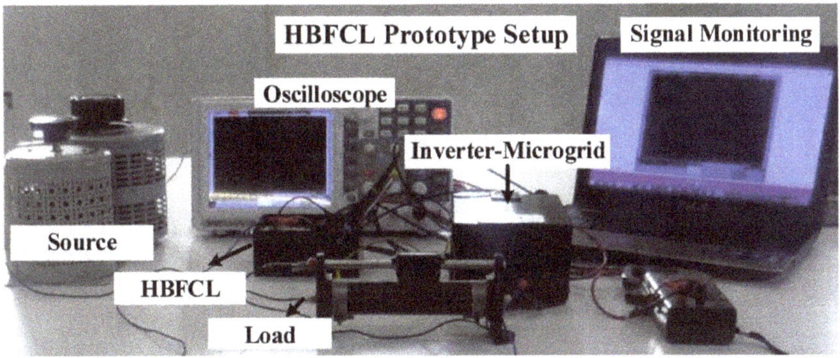

Figure 9. The proposed experimental setup.

The prototype shown in Figure 9 includes four inductors created by E-I 56 core and 0.5 mm^2 wire. Core saturation has occurred in approximately 6 A, which is out of the test range. The IGBTs with part number (STGP10NC60H) as an SW_1 and SW_3 are used in the prototype structure. The control circuit is made by NODE MCU hardware and it has independent current and voltage sensors. This hardware sends the proper pulses to IGBTs via drivers. An autotransformer is used as an electrical source and a variable resistance is used as an electrical load. The line to ground fault is applied by 25 A, 500 V solid-state relay.

The voltage and current signals during the normal and fault operation modes are presented in Figure 10a–c.

In Figure 10a, current waveform is shown during the normal and fault operation where the current amplitude in normal condition is 1 A. In t_1, fault is applied to the setup and the line current raises and reaches 3 A. This result is in fair agreement with the simulation result shown in Figure 6a. Moreover, the main switch current is measured and considered in three states, that is, normal condition, fault pre-limiting mode, and turning off the main switch. Pre-limiting operation is carried out by SW_2 and SW_3 operation, which decreases the line current. The main switch is SW_1 and its operation causes safe and easy current interruption. Figure 10b shows the PCC voltage profile during the normal and fault conditions. In the normal operation, PCC voltage is 24 V; after fault occurrences, the peak voltage reaches 40 V. By operating the main switch, transient voltage peak value decreases to 32 V and, after HBFCL operation, the PCC voltage is fixed to 23 V. This signal closely agreed with the simulation result shown in Figure 7. Figure 10c shows the SW_1 current in fair agreement with the simulation results shown in Figure 6b.

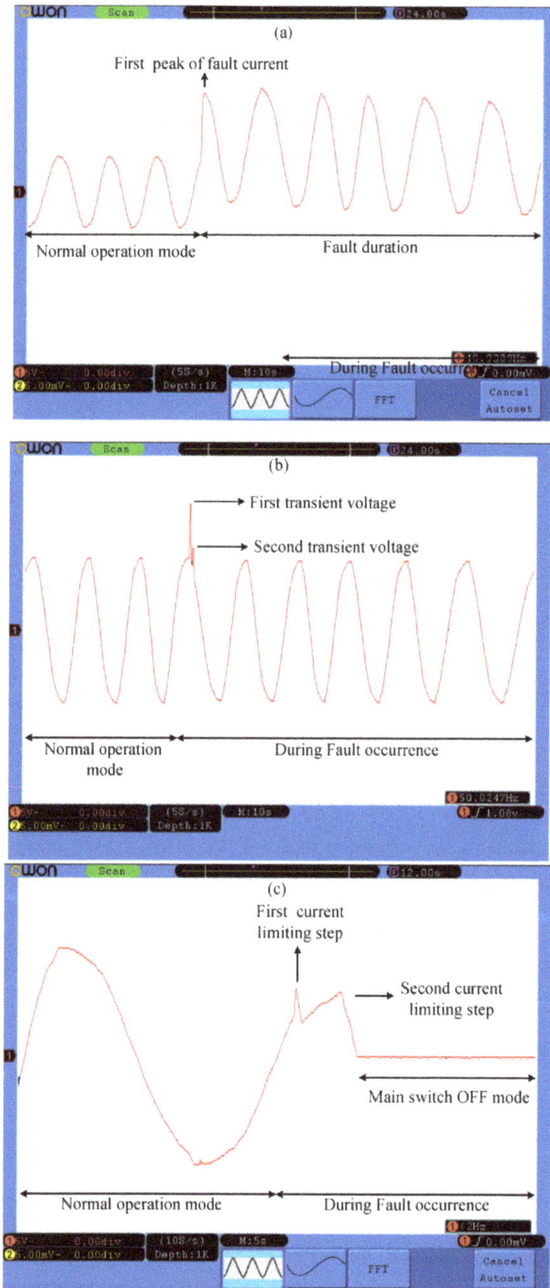

Figure 10. (**a**) Line current during normal and fault condition, (**b**) PCC voltage during normal and fault condition, and (**c**) main switch current during normal and fault condition.

7. Conclusions

Power systems will suffer a growing pressure as a result of an upsurge in electricity demand and an increasing penetration of distributed power generation, which will cause, in turn, a higher incidence of fault current levels. Therefore, in order to mitigate such potential problems, in this paper, a new type of FCL named H bridge fault current limiter (HBFCL) is proposed. The simulation and experimental results show the appropriate operation of the proposed HBFCL during the normal, transient, and fault conditions. Dissipation of fault energy in the four inductors and fault current limiting by three solid-state switches are a successful method that improves performance of the HBFCL. Experimental tests validate the performed simulations in this paper. They demonstrate that the PCC voltage can be successfully protected against the TRV.

Author Contributions: Conceptualization, A.M.; Methodology, K.G.; Software, H.R.; Validation, E.P.; Writing—original draft, A.H.; Writing—review and editing, K.R.; Supervision, E.M.G.R.

Funding: This research received no external funding.

Conflicts of Interest: The authors declare no conflict of interest.

References

1. Ueda, T.; Morita, M.; Arita, H.; Kida, Y.; Kurosawa, Y.; Yamagiwa, T. Solid-sate current limiter for power distribution system. *IEEE Trans. Power Deliv.* **1993**, *8*, 1796–1801. [CrossRef]
2. Rouzbehi, K.; Zhu, J.; Zhang, W.; Gharehpetian, G.B.; Luna, A.; Rodriguez, P. Generalized voltage droop control with inertia mimicry capability—Step towards automation of multi-terminal HVDC grids. In Proceedings of the 2015 International Conference on Renewable Energy Research and Applications (ICRERA), Palermo, Italy, 22–25 November 2015; pp. 1556–1561.
3. Hoshino, T.; Salim, K.M.; Nishikawa, M.; Muta, I.; Nakamura, T. DC reactor effect on bridge type superconducting fault current limiter during load increasing. *IEEE Trans. Appl. Supercond.* **2001**, *11*, 1944–1947. [CrossRef]
4. Salim, K.M.; Hoshino, T.; Nishikawa, M.; Muta, I.; Nakamura, T. Preliminary experiments on saturated DC reactor type fault current limiter. *IEEE Trans. Appl. Supercond.* **2002**, *12*, 872–875. [CrossRef]
5. Rouzbehi, K.; Candela, J.I.; Luna, A.; Gharehpetian, G.B.; Rodriguez, P. Flexible Control of Power Flow in Multiterminal DC Grids Using DC–DC Converter. *IEEE J. Emerg. Sel. Top. Power Electron.* **2016**, *4*, 1135–1144. [CrossRef]
6. Rouzbehi, K.; Miranian, A.; Candela, J.; Luna, A.; Rodriguez, P. Instelligent voltage control in a dc micro-grid containing PV generation and energy storage. In Proceedings of the 2014 IEEE PES T&D Conference and Exposition, Chicago, IL, USA, 14–17 April 2014; pp. 1–5.
7. Gromoll, B.; Ries, G.; Schmidt, W.; Kraemer, H.-P.; Seebacher, B.; Utz, B.; Nies, R.; Neumueller, H.-W.; Baltzer, E.; Fischer, S.; et al. Resistive fault current limiters with YBCO films 100 kVA functional model. *IEEE Trans. Appl. Supercond.* **1999**, *9*, 656–659. [CrossRef]
8. Ye, L.; Lin, L.; Juengst, K.-P. Application Studies of Superconducting Fault Current Limiters in Electric Power Systems. *IEEE Trans. Appl. Supercond.* **2002**, *12*, 900–903.
9. Hoshino, T.; Salim, K.M.; Nishikawa, M.; Muta, I.; Nakamura, T. Proposal of saturated DC reactor type superconducting fault current limiter (SFCL). *Cryogenics* **2001**, *41*, 469–474. [CrossRef]
10. Shfaghatian, N.; Heidary, A.; Radmanesh, H.; Rouzbehi, K. Microgrids Interconnection to Upstream AC grid Using a Dual-function Fault Current Limiter and Power Flow Controller: principle and test results. *IET Energy Syst. Integr.* **2019**. [CrossRef]
11. Heidary, A.; Radmanesh, H.; Bakhshi, A.; Rouzbehi, K.; Pouresmaeil, E. A Compound Current Limiter and Circuit Breaker. *Electronics* **2019**, *5*, 551. [CrossRef]
12. Radmanesh, H.; Heidary, A.; Fathi, S.H.; Gharehpetian, G.B. Dual Function Ferroresonance and Fault Current Limiter Based on DC Reactor. *IET Gener. Transm. Distrib.* **2016**, *10*, 2058–2065. [CrossRef]
13. Radmanesh, H.; Fathi, S.H.; Gharehpetian, G.B.; Heidary, A. Bridge-Type Solid-State Fault Current Limiter Based on AC/DC Reactor. *IEEE Trans. Power Deliv.* **2016**, *31*, 200–209. [CrossRef]

14. Heidary, A.; Radmanesh, H.; Fathi, S.H.; Khamse, H.R.R. Improving Transient Recovery voltage of circuit breaker using Fault Current Limiter. *Res. J. Appl. Sci. Eng. Technol.* **2012**, *4*, 5123–5128.

15. Naderi, S.B.; Jafari, M.; Hagh, M.T. Parallel-Resonance-Type Fault Current Limiter. *IEEE Trans. Ind. Electron.* **2013**, *60*, 2538–2546. [CrossRef]

16. Heidary, A.; Radmanesh, H.; Fathi, H.; Gharehpetian, G.B. Series transformer based diode-bridge-type solid state fault current limiter. *Front. Inf. Technol. Electron. Eng.* **2015**, *16*, 769–784. [CrossRef]

17. Yamaguchi, H.; Kataoka, T. An Experimental Investigation of Magnetic Saturation of a Transformer Type Superconducting Fault Current Limiter. *IEEE Trans. Appl. Supercond.* **2009**, *19*, 1876–1879. [CrossRef]

18. Kcilin, V.; Kovalcv, I.; Kmglov, S.; Stepanov, V.; Slmgaev, I.; Shchcrbzkov, V. Model of HTS Three-phase Saturated Core Fault Current Limiter. *IEEE Trans. Appl. Supercond.* **2000**, *10*, 836–839.

19. Moscrop, J.W. Experimental Analysis of the Magnetic Flux Characteristics of Saturated Core Fault Current Limiters. *IEEE Trans. Magn.* **2013**, *49*, 874–882. [CrossRef]

20. Commins, P.A.; Moscrop, J.W. Three Phase Saturated Core Fault Current Limiter Performance with a Floating Neutral. In Proceedings of the IEEE Electrical Power and Energy Conference, London, ON, Canada, 10–12 October 2012; pp. 249–254.

21. Chen, X.; Chen, B.; Tian, C.; Yuan, J.; Liu, Y. Modeling and Harmonic Optimization of a Two-Stage Saturable Magnetically Controlled Reactor for an Arc Suppression Coil. *IEEE Trans. Ind. Electron.* **2012**, *59*, 2824–2831. [CrossRef]

22. Moriconi, F.; de la Rosa, F.; Darmann, F.; Nelson, A.; Masur, L. Deployment of Saturated-Core Fault Current Limiters in Distribution and Transmission Substations. *IEEE Trans. Appl. Supercond.* **2011**, *21*, 1288–1293. [CrossRef]

23. Xin, Y.; Gong, W.Z.; Niu, X.Y.; Gao, Y.Q.; Guo, Q.Q.; Xiao, L.X.; Cao, Z.J.; Hong, H.; Wu, A.G.; Li, Z.H.; et al. Manufacturing and Test of a 35 kV/90 MVA Saturated Iron-Core Type Superconductive Fault Current Limiter for Live-Grid Operation. *IEEE Trans. Appl. Supercond.* **2009**, *19*, 1934–1937. [CrossRef]

24. Xin, Y.; Hong, H.; Wang, J.Z.; Gong, W.Z.; Zhang, J.Y.; Ren, A.L.; Zi, M.R.; Xiong, Z.Q.; Si, D.J.; Ye, F. Performance of the 35 kV/90 MVA SFCL in Live-Grid Fault Current Limiting Tests. *IEEE Trans. Appl. Supercond.* **2011**, *21*, 1294–1297. [CrossRef]

25. Zheng, F.; Deng, C.; Chen, L.; Li, S.; Liu, Y.; Liao, Y. Transient Performance Improvement of Micro-grid by a Resistive Superconducting Fault Current Limiter. *IEEE Trans. Appl. Supercond.* **2015**, *25*. [CrossRef]

26. Hwang, J.-S.; Khan, U.A.; Shin, W.-J.; Seong, J.-K.; Lee, J.-G.; Kim, Y.; Lee, B.-W. Validity Analysis on the Positioning of Superconducting Fault Current Limiter in Neighboring AC and DC Microgrid. *IEEE Trans. Appl. Supercond.* **2013**, *23*. [CrossRef]

27. Abdolkarimzadeh, M.; Nazari-Heris, M.; Abapour, M.; Sabahi, M. A Bridge Type Fault Current Limiter for Energy Management of AC/DC Microgrids. *IEEE Trans. Power Electron.* **2017**, *32*, 9043–9050. [CrossRef]

28. Heidary, A.; Radmanesh, H.; Rouzbehi, K.; Pou, J. A DC-Reactor Based Solid-State Fault Current Limiter for HVDC Applications. *IEEE Trans. Power Deliv.* **2019**, *34*, 720–728. [CrossRef]

29. Eladawy, M.; Metwally, I.A. A Novel Five-Leg Design for Performance Improvement of Three-Phase Presaturated Core Fault-Current Limiter. *IEEE Trans. Magn.* **2018**, *54*. [CrossRef]

30. Yan, S.; Tang, Y.; Ren, L.; Xu, Y.; Wang, Z.; Liang, S.; Zhang, Z. Design and Verification Test of a Flux-Coupling-Type Superconducting Fault Current Limiter. *IEEE Trans. Magn.* **2018**, *54*. [CrossRef]

31. Tseng, H.-T.; Jiang, We.; Lai, J. A Modified Bridge Switch-Type Flux-Coupling Non-superconducting Fault Current Limiter for Suppression of Fault Transients. *IEEE Trans. Power Deliv.* **2018**, *33*, 2624–2633. [CrossRef]

Article

An Alternative Carrier-Based Implementation of Space Vector Modulation to Eliminate Common Mode Voltage in a Multilevel Matrix Converter

Janina Rząsa

Department of Power Electronics and Power Engineering, Rzeszow University of Technology, 35-959 Rzeszow, Poland; jrzasa@prz.edu.pl; Tel.: +48-017-865-1976

Received: 13 December 2018; Accepted: 30 January 2019; Published: 6 February 2019

Abstract: The main aim of the paper is to find a control method for a multilevel matrix converter (MMC) that enables the elimination of common mode voltage (CMV). The method discussed in the paper is based on a selection of converter configurations and the instantaneous output voltages of MMC represented by rotating space vectors. The choice of appropriate configurations is realized by the use of space vector modulation (SVM), with the application of Venturini modulation functions. A multilevel matrix converter, which utilizes a multilevel structure in a traditional matrix converter (MC), can achieve an improved output voltage waveform quality, compared with the output voltage of MC. The carrier-based implementation of SVM is presented in this paper. The carrier-based implementation of SVM avoids any trigonometric and division operations, which could be required in a general space vector approach to the SVM method. With use of the proposed control method, a part of the high-frequency output voltage distortion components is eliminated. The application of the presented modulation method eliminates the CMV in MMC what is presented in the paper. Additionally, the possibility to control the phase shift between the appropriate input and output phase voltages is obtained by the presented control strategy. The results of the simulation and experiment confirm the utility of the proposed modulation method.

Keywords: multilevel matrix converter; rotating voltage space vector; common move voltage; space vector pulse width modulation; venturini control method

---

1. Introduction

A multilevel matrix converter (MMC) is a frequency converter, whose topology [1–8] was proposed by analogy to multilevel inverters and its aim is the reduction of the voltage rating of the switches with respect to the supply voltages and the further improvement of the synthesized current and voltage waveforms. Two scientific centres paid attention to the analysis of MMC operations and two different control methods were developed there. The authors of papers [5–8] concentrate on the use of the space vector modulation (SVM) method, whereas in papers [2–4], the implementation of the Venturini control method is presented. The main goal in using either of the modulation methods in controlling MMC is the synthesis of the referenced sinusoidal output voltage and sinusoidal input current by controlling the input displacement angle. The application of these methods is involved with appearing CMV on the output terminals. The problem with appearing CMV is concerned with all the converters being controlled by the use of the pulse width modulation (PWM) method, both indirect frequency converters with a DC link, as well as a direct matrix converter (MC) and MMC. Because the topology of MMC is the modification of conventional MC, the analysis of the cancelation methods of CMV used in MC would be valuable in finding the control method to eliminate the CMV in MMC.

As for MC, many different methods have been reported to mitigate the detrimental influences of the CMV. The majority of the methods are based on a modification of SVM by the elimination of the zero-space vector, as a complement of the switching cycle and replacement of the zero vectors by rotating space vectors or by active vectors, with minimum absolute values [9–12]. The authors of References [9,10] found that, in MC controlled by the use of SVM using rotating vectors, instead of zero vectors, had 42% lower CMV. The voltage transfer ratio (VTR) was found to be higher than 0.5. Next, in Reference [12], the controlling method of CMV reduction is achieved by using the switch configurations that connect each input phase to a different output phase, which means that rotating space vectors are used or the configurations connect all the output phases to the input phase, with the minimum absolute voltage. The authors found that the result of the CMV peak value reduction was 45.4%, while the VTR was 0.5% and 42.3%, when the VTR was higher. The next is the method that targeted the operation of the drive for a higher modulation index range ($0.577 \leq m \leq 0.866$) [12]. This method eliminates the zero vectors but continues to use active voltage vectors, with normalized duty ratios. The elimination of zero vectors reduces the peak value of CMV by 42%. Even though these methods, with their own modulation strategies, can produce a sound output performance within the specified operating range, they are applicable only to a limited VTR range. To achieve a sound output performance for the whole VTR range, these different modulation strategies should be properly combined. However, it is inconvenient to combine each method with the different switching patterns, because each modulation method, with its own formulae, uses different vectors to calculate the duty cycles. The main goal of the article is to present the control method that results in the elimination of CMV in MMC. The entire elimination of CMV in MC and in MMC is possible only by the use of such configurations, resulting from on-off states of bidirectional switches, which realize the rotating voltage space vectors.

In the application to MC, the use of the only rotating voltage space vectors to obtain the entire elimination of CMV is presented in References [13–19]. The author of Reference [13] compares the CMV in MC, controlled by the use of the Venturini method, which solely applies rotating space vectors, using the scalar control method and SVM control method, applying active and zero-space vectors. To obtain the cancelation of CMV in MC, the authors of Reference [14] introduce the new SVM and develop the modification of four-step commutation. The modification of the four-step commutation is dictated by the fact that, during the four-step commutation, such switch configurations arise, which are represented by active space vectors, what results in the high value of CMV. The SVM technique, solely using rotating space vectors, is also applied by the authors of References [15,19] in the modulation of dual MCs. Next, in Reference [20], the authors present a carrier-based implementation of SVM for dual MCs using only rotating space vectors. The advantage of the proposed strategy is an alternative way to achieve SVM, which does not involve the knowledge of space vectors, when it is derived. Additionally, it avoids any trigonometric and division operations that could be needed to implement the SVM using the general space vector approach.

In reference to MMC, an elimination of CMV is discussed in References [21,22] but the method presented there does not rely on the use solely of rotating space vectors in the synthesis of the output voltage. The authors of References [21,22] use space vector modulation, applying active and zero space voltage vectors. This method demands many trigonometric and division operations. The method analysed in References [21,22] results in only a 50% reduction in the peak value of CMV.

To solve the problem of the elimination of CMV in MMC, the author of the paper proposes the application of the modulation method, using solely rotating space vectors. To determine the switch duty cycles of MMC, the carrier-based implementation of SVM was used, which makes the application of the modulation method easy. The advantage of the proposed strategy is an alternative way to achieve SVM, which does not involve the knowledge of space vectors, when it is implemented. Additionally, it avoids any trigonometric and division operations that could be needed to implement the SVM, using the general space vector approach. The elaboration of the proposed method required the analysis of admissible switch configurations and the characteristics of their corresponding space vectors. The switch configurations and space vectors, providing the synthesis of the required output voltage and the elimination of CMV, were selected and presented in the paper.

In the MMC controlled by the implementation of the proposed method, the output voltage containing the fundamental component and high-frequency distortion components, with considerably less amplitudes, compared with the distortion component in the output voltage of MC, is synthesized. The next advantage of the method is that it obtains, in simulation tests, the entire cancelation of CMV and the reduction of the peak value of CMV to 12% of the peak value of the output voltage. This paper also presents the possibility to control the phase shift between the output voltage and appropriate input voltage. Controlling the phase shift between the output and input voltage may be important when working with the same output and input frequency of MMC and it could be applied as a converter in a Flexible AC Transmission System (FACTS) [23] to control the power flow or compensate the voltage dips.

The paper is organized in a total of five sections. The second section describes the topology of MMC, the admissible configuration of the analysed converter and the output voltage space vectors. The proposed modulation method is described in the third section. The fourth section contains the simulation and experimental results to validate the proposed method and the last section has the discussion.

2. Multilevel Matrix Converter

2.1. Topology of MMC

A multilevel matrix converter (MMC) (Figure 1) consists of 18 bidirectional switches, $S_{Aa1} - S_{Cc2}$ and nine clamp capacitors, $C_1 - C_9$. The bidirectional switches constitute two integral semiconductor modules. The manufacturers, Yaskawa, ABB, Alstom, Siemens and so forth, have shown their interest in the production of these semiconductor power modules, consisting of bidirectional switches. In fact, Yaskawa has introduced many standard units of matrix converters of medium voltages and several megawatts.

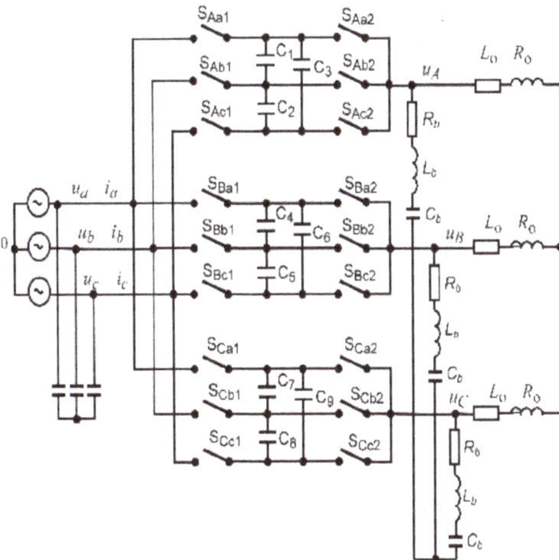

Figure 1. Scheme of multilevel matrix converter (MMC).

The clamp capacitors in MMC are connected by two switch matrixes and play a role analogous to flying capacitors in multicell converters, principally providing an additional intermediate voltage level in the process of synthesizing the output voltage. The capacitance of clamp capacitors depends on the value of the load current, switching the frequency and assumed value of the voltage ripple on the clamp capacitors [1–3]. Based on studies [1] and [24–26], concerned with the multicell converter, a capacitance value (1) of the capacitor should be determined using the admissible value of the voltage ripple on that capacitor ΔU_C, number p of multilevel converter cells, load current I_o and switching frequency f_s of the converter switches.

$$C = \frac{I_o}{\Delta U_C p f_s} \tag{1}$$

To eliminate improper clamp capacitor voltages, shaping additional balancing circuits are used in MMC. The balancing circuit, providing an automatic maintenance of the clamp capacitor voltages, consists of R_b resistance, L_b inductance and C_b capacitance, connected in a series. The parameters of the balancing circuits are selected so that the resonance frequency f_o of those circuits is equal to the switching frequency f_s of the switches (2).

$$f_o = \frac{1}{2\pi\sqrt{L_b C_b}} = f_s \tag{2}$$

The resonance frequency is the most important parameter of the balancing circuit but the effectiveness of the balancing process, with a passive RLC circuit, depends also on the balancing circuit characteristic impedance, the converter parameters and the converter operating conditions [26,27].

2.2. Admissible Switch Configurations of MMC

Taking into account the voltage characteristics of the MMC supplying source, the switch configurations that do not make short circuit of the input phases and simultaneously assure the current path in the resistive-inductive load, are admissible. Therefore, each of the three output phases could be connected into one of the input phases directly through two series-connected bidirectional switches or through two switches and a clamp capacitor. Clamp capacitors C1 – C9 perform additional intermediate voltage levels in waveforms of the output voltage. In the ideal conditions of charging, voltages on the clamp capacitors should be equal to half of the appropriate line-line voltages (3).

$$\begin{aligned}
u_{C1} = u_{C4} = u_{C7} = (u_a - u_b)/2 \\
u_{C2} = u_{C5} = u_{C8} = (u_b - u_c)/2 \\
u_{C3} = u_{C6} = u_{C9} = (u_c - u_a)/2
\end{aligned} \tag{3}$$

As a result of the use of clamp capacitors, there exist additional current flow paths between the input and output phases. Taking into account only the admissible states of the bidirectional switches of MMC, each of the output phases may be connected with the three supply phases in nine different ways, corresponding to the converter configurations. These configurations arise from the 'on' or 'off' states of the bidirectional switches, which, for the output phase A, are shown in Table 1.

Table 1. Switch configuration of MMC in the output phase A.

Configuration	Switch State ('on': 1, 'off': 0)						Instantaneous Value of Voltage in Output Phase A
	S_{Aa1}	S_{Aa2}	S_{Ab1}	S_{Ab2}	S_{Ac1}	S_{Ac2}	
1	1	1	0	0	0	0	u_a
2	0	0	1	1	0	0	u_b
3	0	0	0	0	1	1	u_c
4	1	0	0	1	0	0	$(u_a + u_b)/2$
5	0	1	1	0	0	0	$(u_a + u_b)/2$
6	0	0	0	1	1	0	$(u_b + u_c)/2$
7	0	0	1	0	0	1	$(u_b + u_c)/2$
8	0	1	0	0	1	0	$(u_a + u_c)/2$
9	1	0	0	0	0	1	$(u_a + u_c)/2$

In the case of the first three configurations (1, 2, 3), shown in Table 1, every output phase is directly connected with one of the input phases, which is characteristic of a conventional MC. This means that the phase output voltage is equal the appropriate phase input voltage. The next six configurations (4, 5, 6, 7, 8, 9) implement the connections across clamp capacitors. For instance, while the switches S_{Aa1} and S_{Ab2} (configuration 4) or switches S_{Ab1} and S_{Aa2} (configuration 5) are 'on,' the current of the output phase A is running across the capacitor C_1 and connected in parallel capacitors C_2 and C_3. The instantaneous voltage of the output phase A is the same in both cases and equals $u_{C1} = (u_a + u_b)/2$, because of the voltage of the capacitor C_1 and because it is connected in parallel capacitors C_2 and C_3, equalling $(u_a - u_b)/2$. The choice between configurations 4 and 5 is the choice between different capacitor current directions, allowing for the possibility to control the capacitor voltage. The same regularity is fulfilled in the next two couples of configurations, that is, configurations 6 and 7 or 8 and 9. Recapitulating, we can conclude that each output phase can be connected to u_a, u_b, u_c, as the voltage supply (henceforth termed "full-amplitude voltage supply") or to $(u_a + u_b)/2$, $(u_b + u_c)/2$, $(u_a + u_c)/2$ (henceforth, "half-amplitude voltage supply").

Considering a three-phase to three-phase MMC, one has to take into account $9^3 = 729$ possible switch configurations, which can be used practically in the process of the synthesis of output voltages and the synthesis of the voltages of clamp capacitors, determining the intermediate levels of the supply voltages.

2.3. Output Voltage Space Vectors in MMC

The output voltages in three-input to three-output circuit of MMC could be presented using space vectors, as defined by (4). Instantaneous output voltages u_A, u_B and u_C in the relation (4) are appropriately equal to one of values defined in the output phase A in Table 1. The label xxx in the name $\overrightarrow{V}_{xxx}$ of the space vector means the type of configuration appropriately chosen in the output phase A, B and C.

$$\overrightarrow{V}_{xxx} = \frac{2}{3}\left[u_A + au_B + a^2u_C \right] \tag{4}$$

Analysis of the output voltages, corresponding to 729 switch configurations, allows for it to be noticed that the instantaneous output voltage space vectors could be split into the following groups:

- 27 zero space vectors, where the output voltages in each phase are the same;
- 360 active space vectors, where two of the output voltages have the same values; and
- 342 rotating space vectors.

The active voltage space vectors correspond to the connection of two output phases to the same voltage supply. The zero vectors arise when the output phases are connected with the same "full-amplitude voltage supply" or the same "half-amplitude voltage supply."

2.4. Output Voltage Space Vectors Reducing CMV in MMC

Among 342 configurations of MMC, with instantaneous output voltages represented by rotating space vectors, only 54 could be considered to reduce CMV. This conclusion is drawn from the analysis of rotating space vectors, determined with the assumption that the MMC is supplied by a balanced input voltage (5).

$$\begin{bmatrix} u_a \\ u_b \\ u_c \end{bmatrix} = \begin{bmatrix} U_{im}\cos\omega_i t \\ U_{im}\cos(\omega_i t - 120°) \\ U_{im}\cos(\omega_i t + 120°) \end{bmatrix} \tag{5}$$

All 342 rotating voltage space vectors representing instantaneous output voltages could be split into the following five groups:

- Rotating voltage space vectors, with a constant module equal to the amplitude of the input voltage U_{im}. These vectors correspond to switch configurations, where three output voltages are synthesized by the use of three full-amplitude supplying voltages. Six vectors belonging to this group create two sets, consisting of three rotating space vectors shifted by 120°. One set rotates in a positive direction (CCW vectors) along a complex plane and the next set rotates in a negative one (CW vectors). Two of these vectors are represented by Equations (6) and (7), as well as the relation (6)—for CCW vectors and (7)—for CW vectors. The digits in the label of the voltage vector name should be interpreted as follows: the first digit defines the configuration of the switches in the output phase A, the second and third digits, the output phase B and C, appropriately. The application of the rotating space vectors belonging to this group in the modulation of switch duty cycles results in a zero value of CMV (8).

$$\vec{V}_{123} = \frac{2}{3}\left[u_a + au_b + a^2 u_c\right] = U_{im}e^{j\omega_i t} \tag{6}$$

$$\vec{V}_{132} = \frac{2}{3}\left[u_a + au_c + a^2 u_b\right] = U_{im}e^{-j\omega_i t} \tag{7}$$

where: $a = e^{j120°}$, $\quad a^2 = e^{j240°}$

$$u_{CMV(123)} = \frac{u_a + u_b + u_c}{3} = 0 \tag{8}$$

- Rotating voltage space vectors, with a constant module equal to half of the input phase voltage amplitude that corresponds to 48 configurations with a connection of three output phases to three different half-amplitude voltage supplies. Half of these vectors complete 8 sets of three vectors rotating in a positive direction (CCW vectors) and half of them form 8 sets of vectors rotating in a negative one (CW vectors). Two of these vectors are assigned as (9) (CCW vectors) and (10) (CW vectors). The application of the rotating space vectors belonging to this group in the modulation of switch duty cycles also results in a zero value of CMV (11).

$$\vec{V}_{468} = \frac{2}{3}\left[\frac{1}{2}(u_a + u_b) + a\frac{1}{2}(u_b + u_c) + a^2\frac{1}{2}(u_a + u_c)\right] = -\frac{1}{2}U_{im}e^{j(\omega_i t - 120°)} \tag{9}$$

$$\vec{V}_{486} = \frac{2}{3}\left[\frac{1}{2}(u_a + u_b) + a\frac{1}{2}(u_a + u_c) + a^2\frac{1}{2}(u_b + u_c)\right] = -\frac{1}{2}U_{im}e^{-j(\omega_i t + 120°)} \tag{10}$$

$$u_{CMV(468)} = \frac{\frac{1}{2}(u_a + u_b) + \frac{1}{2}(u_b + u_c) + \frac{1}{2}(u_a + u_c)}{3} = 0 \tag{11}$$

- Rotating voltage space vectors, with a constant module equal to half of the input phase voltage amplitude, which corresponds to 72 configurations, with a connection of two output phases to two different half-amplitude voltage supplies and a third output phase connected to the full-amplitude voltage supply. Two examples are shown: CCW vector, as Equation (12) and CW vector, as Equation (13). The application in the modulation of switch duty cycles the rotating space vectors, belonging to the group being discussed, results in a value of CMV (14) that is not zero.

$$\vec{V}_{148} = \frac{2}{3}\left[u_a + a\frac{1}{2}(u_a + u_b) + a^2\frac{1}{2}(u_a + u_c)\right] = \frac{1}{2}U_{im}e^{j\omega_i t} \tag{12}$$

$$\vec{V}_{184} = \frac{2}{3}\left[u_a + a\frac{1}{2}(u_a + u_c) + a^2\frac{1}{2}(u_a + u_b)\right] = \frac{1}{2}U_{im}e^{-j\omega_i t} \tag{13}$$

$$u_{CMV(148)} = \frac{u_a + \frac{1}{2}(u_a + u_b) + \frac{1}{2}(u_a + u_c)}{3} = \frac{1}{2}u_a \tag{14}$$

- Rotating voltage space vectors, with a changeable module, that correspond to 72 configurations, with a connection of two output phases to two different full-amplitude voltage supplies and a third output phase connected to a half-amplitude voltage supply. The Equation (15) is an example of the vectors belonging to this group. The application of the rotating space vectors, belonging to this group, in the modulation of switch duty cycles, results in a value of CMV (16) that is not zero.

$$\vec{V}_{126} = \tfrac{2}{3}\left[u_a + au_b + a^2\tfrac{1}{2}(u_b + u_c)\right] = U_{im}e^{j\omega_i t} + \tfrac{1}{3}a^2 u_{bc}$$
$$= U_{im}e^{j\omega_i t} + \tfrac{\sqrt{3}}{3}U_{im}\sin\omega_i t\, e^{j240°} \tag{15}$$

$$u_{CMV(126)} = \frac{u_a + u_b + \tfrac{1}{2}(u_b + u_c)}{3} = \frac{1}{6}u_{bc} \tag{16}$$

- Rotating voltage space vectors, with a changeable module, that correspond to 144 configurations, with a connection of two output phases to two different half-amplitude voltage supplies and a third output phase connected to a full-amplitude voltage supply. The Equation (17) is an example of the vectors belonging to this group. The application of the rotating space vectors, belonging to this group, in the modulation of switch duty cycles, results in a value of CMV (18) that is not zero.

$$\vec{V}_{146} = \frac{2}{3}\left[u_a + a\frac{1}{2}(u_a + u_b) + a^2\frac{1}{2}(u_b + u_c)\right] =$$
$$= U_{im}e^{j\omega_i t} - \frac{\sqrt{3}}{3}U_{im}\sin(\omega_i t - 60°)\,e^{j120°} + \frac{\sqrt{3}}{3}\sin\omega_i t\, e^{j240°} \tag{17}$$

$$u_{CMV(146)} = \frac{u_a + \tfrac{1}{2}(u_a + u_b) + \tfrac{1}{2}(u_b + u_c)}{3} = \frac{1}{6}u_{ac} \tag{18}$$

Finally, only 54 of the 729 switch configurations of MMC could be chosen, while the CMV elimination is required in the proposed control method. The lay-out of a complex plane of the rotating space vectors, belonging to the mentioned groups, is shown in Figure 2. In Figure 2, the initial position of the rotating space vectors is shown. The digits used in the label of vectors, instead of the full names of vectors, are shown in the figure.

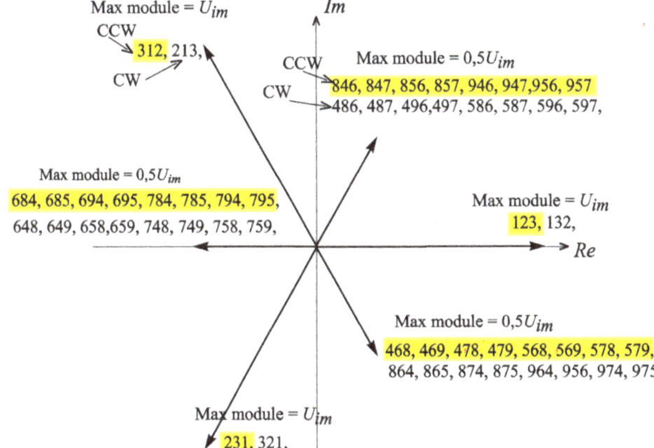

Figure 2. Lay-out of a complex plane of the voltage rotating space vectors, whose implementation assures the elimination of CMV.

The performed analysis of all of the allowable configuration of MMC and the corresponding voltage space vectors allows the vectors synthesizing the sinusoidal output voltage, sinusoidal input current and elimination of CMV to be chosen. As a result, the proposed modulation method is based on the application solely of the switch configurations that correspond to the rotating space vectors, with a constant module. The rotating space vectors are not usually used in modulation strategies, because they lay in different positions, so it is difficult to create a repetitive pattern. However, in Reference [13], one can find that the implementation of the Venturini modulation function in the determination of switch duty cycles in conventional MC could provide the control using only rotating space vectors, while simultaneously eliminating CMV. By analogy, the Venturini modulation functions are used in the proposed strategy for modulation duty cycles in MMC.

3. Proposed modulation method

Method of the Output Voltage Synthesis

The synthesis of the output voltage by the use of rotating space vectors could be realized by applying a carrier-based implementation of SVM. The Venturini modulation functions are used in the proposed method to set out the switch duty cycles. The determination of the duty cycles for bidirectional switches of conventional MC and MMC by the use of Venturini modulation functions has been presented in References [13] and [2,3]. Here, the Venturini modulation function, in consideration of angle ψ, in the form of (19) or (20), is used. A value of angle ψ defines the phase shift between that defined by (5) input phase voltages and the appropriate output phase voltages. The application of the modulation function (19) results in CCW output voltage rotating space vectors and a lagging input displacement angle, whereas the modulation function (20) gives CW output voltage rotating space vectors and leading input displacement angle [2]. Both of them, that is, the modulation function (19), depending on the difference $\omega_o - \omega_i$ of the output and input frequency and modulation function (20), depending on the sum $\omega_o + \omega_i$ result in the output voltage amplitude, equal to half of the input voltage amplitude, at most.

$$
\begin{aligned}
d_1^- &= m_{Aa}^- = m_{Bb}^- = m_{Cc}^- = \tfrac{1}{3}\left(1 + 2k_U\cos(\omega_o - \omega_i)\,t + \psi\right) \\
d_2^- &= m_{Ab}^- = m_{Bc}^- = m_{Ca}^- = \tfrac{1}{3}\left(1 + 2k_U\cos\left((\omega_o - \omega_i)\,t - \tfrac{2\pi}{3} + \psi\right)\right) \\
d_3^- &= m_{Ac}^- = m_{Ba}^- = m_{Cb}^- = \tfrac{1}{3}\left(1 + 2k_U\cos\left((\omega_o - \omega_i)\,t + \tfrac{2\pi}{3} + \psi\right)\right)
\end{aligned}
\tag{19}
$$

$$
\begin{aligned}
d_1^+ &= m_{Aa}^+ = m_{Bc}^+ = m_{Cb}^+ = \tfrac{1}{3}\left(1 + 2k_U\cos(\omega_o + \omega_i)\,t + \psi\right) \\
d_2^+ &= m_{Ab}^+ = m_{Ba}^+ = m_{Cc}^+ = \tfrac{1}{3}\left(1 + 2k_U\cos\left((\omega_o + \omega_i)\,t - \tfrac{2\pi}{3} + \psi\right)\right) \\
d_3^+ &= m_{Ac}^+ = m_{Bb}^+ = m_{Ca}^+ = \tfrac{1}{3}\left(1 + 2k_U\cos\left((\omega_o + \omega_i)\,t + \tfrac{2\pi}{3} + \psi\right)\right)
\end{aligned}
\tag{20}
$$

As carrier signals, two-phase shifted carrier signals are adopted. The displacement of carrier signals, involved in the control of switches S_{ij1} and switches S_{ij2}, is $T_s/2$, where T_s is the carrier signal period. Duty cycles, in which switches S_{ij1} are switched-on, arise from the comparison of the corresponding modulation functions with one of the carrier signals. A carrier signal shifted by half of the switching cycle T_s determines the duty cycles of S_{ij2} switches (Figure 3). The digits in Figure 3, placed below the duty cycles, mean the numbers of configurations defined in Table 1. At the bottom in Figure 3, the names of the appropriate space vectors synthesizing the load voltage are placed. The mentioned vectors are the same as the rotating space vectors shown in Figure 2.

The application of modulation functions, with phase shift angle ψ, provides the phase shift between the output and the appropriate input voltage. This feature is important when the MMC works with the same input and output frequency and it allows the MMC to be used as a converter in Flexible AC Transmission System (FACTS) devices.

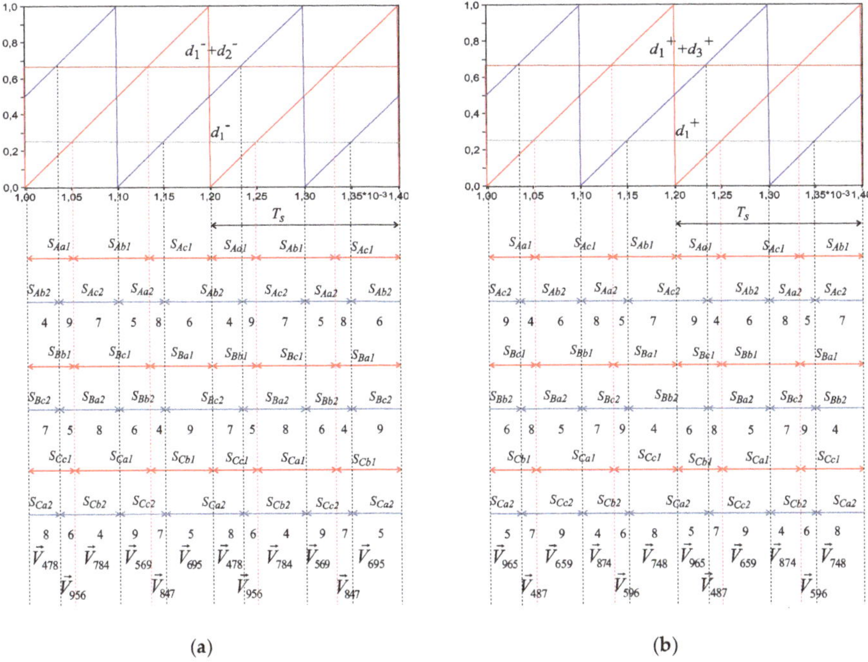

(a) (b)

Figure 3. Switch duty cycles and rotating space vectors of the output voltage in MMC for CCW-type space vectors (**a**) and CW-type space vectors (**b**).

4. Simulation and Experiment

4.1. Simulation

Simulation tests were realized in an EMTP-ATP program. The matrix of the bidirectional switches was modelled as a matrix of the ideal bidirectional switches, controlled using signals generated in TACS subroutine. The supply grid is represented by ideal sinusoidal voltage sources, with an RMS value of 220 V and a frequency of 50 Hz, while the load consists in star-connected resistance and inductance elements, with values of 2 Ω and 10 mH. The carrier frequency was f_{carr} = 5 kHz. The capacitance of the clamp capacitors is equal 10μF. The performances obtained for MMC, controlled by the use of the carrier-based implementation of SVM combined with the Venturini modulation functions, are presented in Figures 4–7. The waveforms of the output voltage and CMV, as depicted in Figure 4 and in Figure 5, illustrate the control, with a lagging input displacement angle (CCW rotating space vectors). Waveforms in Figures 6 and 7 correspond to the leading input displacement angle, when the CW rotating space vectors are used. One can see that in both cases, the CMV is equal to zero. All these waveforms illustrate the control, with a value of the shift angle of ψ = 0. The Fourier analysis was performed, with an accuracy of 10 Hz. It could be observed that, besides fundamental harmonics, the output voltage consists of high frequency components, concentrated as sidebands around each multiple of the carrier frequency. However, it is also seen that, in comparison with conventional MC (Figure 8b), the amplitudes of the first group of these harmonics are significantly decreased. The comparison of the distortion components of the first groups in MC and MMC output voltages (Figures 4b and 8b), obtained with the same controlling parameters, indicates that the amplitude of these components decreased from 180.9 V in MC to 8.7 V in MMC. The waveforms chosen for presentation and FFT analysis prove that the applied modulation method results in a significant reduction of the distortion components of the MMC output voltage, compared with the MC output voltage, which is a basic demand of a proper controlling method used for MMC. The second important feature of the proposed modulation method is the entire cancelation of CMV.

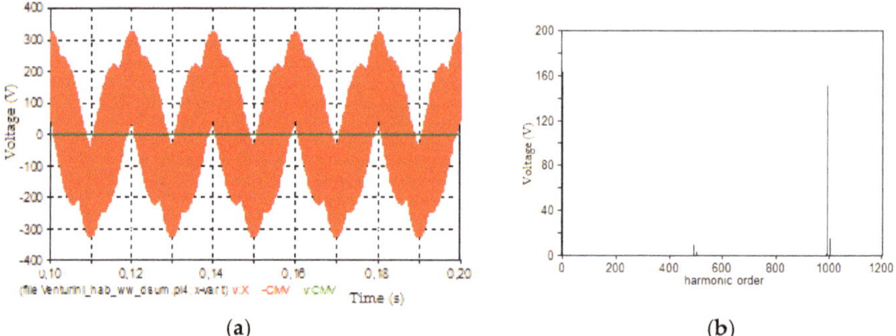

Figure 4. Waveforms of the output voltages (red lines) and CMV (green lines) (**a**) and Fourier analysis of the output voltage (**b**), synthesized in MMC and controlled by the use of CCW rotating space vectors, for the angle $\psi = 0°$ and output frequency of 50 Hz.

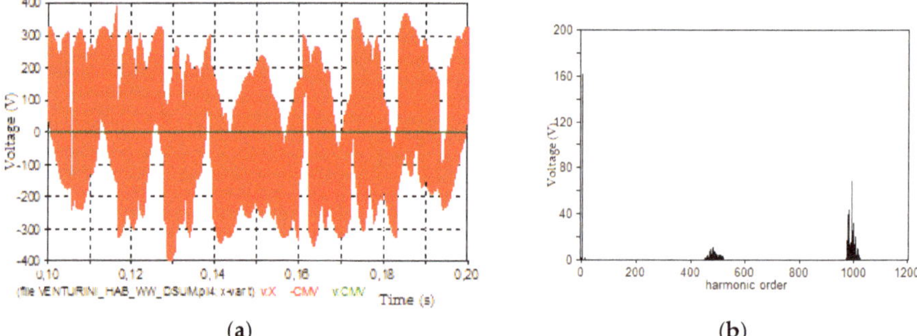

Figure 5. Waveforms of the output voltages (red lines) and CMV (green lines) (**a**) and Fourier analysis of the output voltage (**b**), synthesized in MMC and controlled by the use of CCW rotating space vectors, for the angle $\psi = 0°$ and output frequency of 80 Hz.

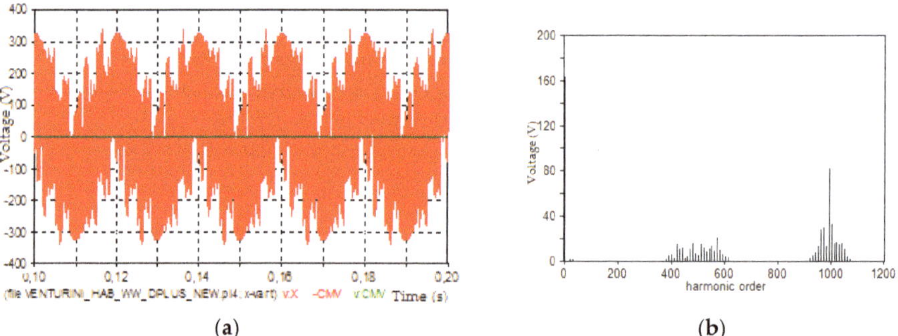

Figure 6. Waveforms of the output voltages (red lines) and CMV (green line) (**a**) and Fourier analysis of the output voltage (**b**), synthesized in MMC and controlled by the use of CW rotating space vectors, for the angle $\psi = 0°$ and output frequency of 50 Hz.

The next analysis (Figures 9 and 10) deals with the control of the phase shift between the output and input phase voltages. MMC works with the output frequency the same as it does with an input one. In Figure 9, the waveform of the output voltages, together with the appropriate input voltages, for different shift angle values ψ between the input and output voltages, is shown. Performed simulation analysis proves that the control of the phase shift between the output and input voltage in MMC, controlled by the use of the carrier-based implementation of SVM, with Venturini modulation functions, is possible and is characterized by a linear relation (Figure 10).

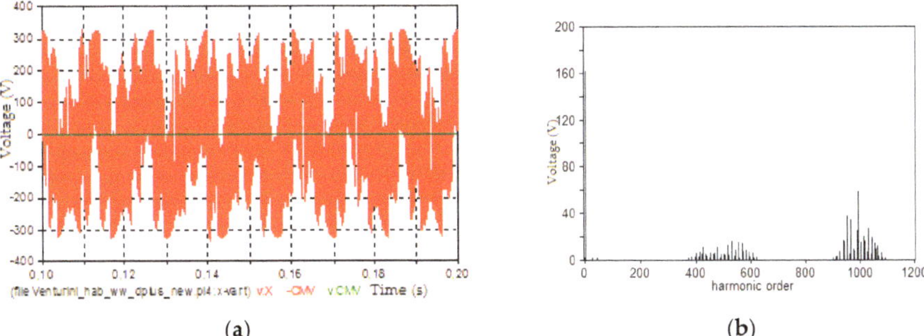

(a)   (b)

Figure 7. Waveforms of the output voltages (red lines) and CMV (green line) (**a**) and Fourier analysis of the output voltage (**b**), synthesized in MMC and controlled by the use of CW rotating space vectors, for the angle $\psi = 0°$ and output frequency of 80 Hz.

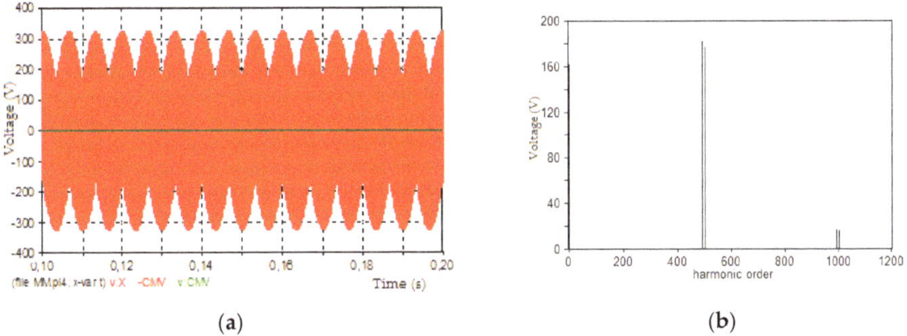

(a)   (b)

Figure 8. Waveforms of the output voltage (red lines) and CMV (green line) (**a**) and Fourier analysis (**b**) of the output voltage, synthesized in conventional MC by the use of CCW rotating space vectors, for the output frequency of 50 Hz.

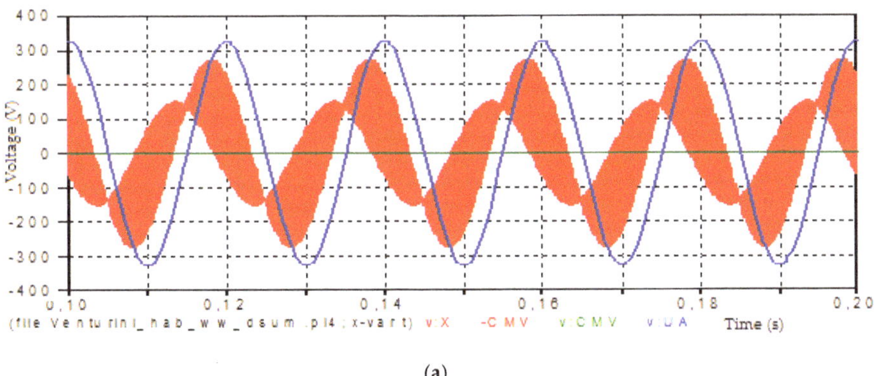

(a)

Figure 9. *Cont.*

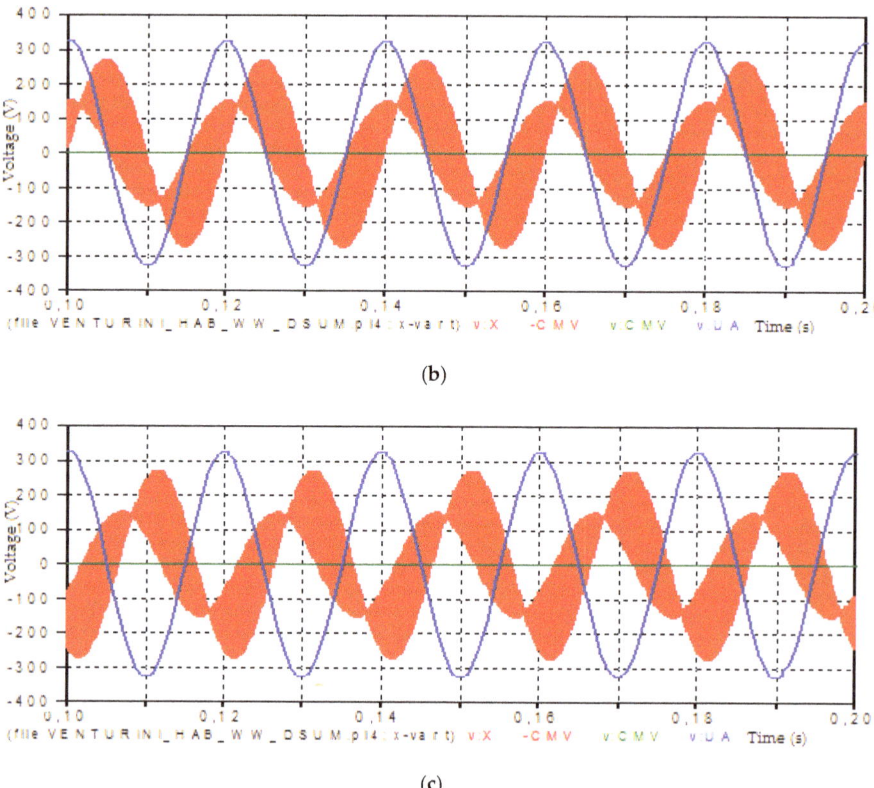

(b)

(c)

Figure 9. Waveform of the output voltages (red lines), input voltages (blue line) and CMV (green line) in MMC, controlled by the use of CCW rotating space vectors, for the output frequency of 50 Hz and angle $\psi = 60°$ (**a**); angle $\psi = 6-0°$ (**b**) and angle $\psi = 180°$ (**c**).

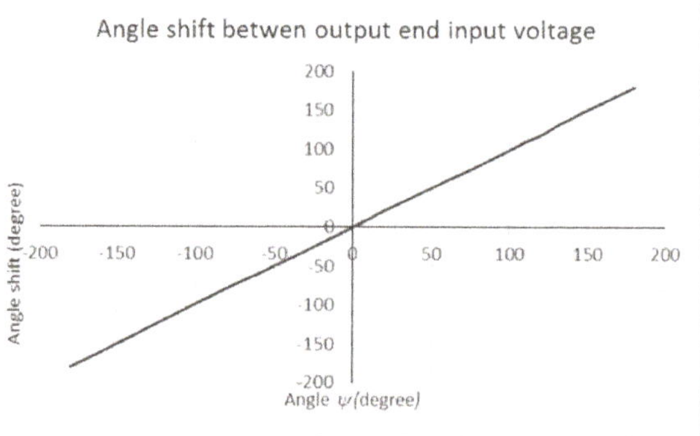

Figure 10. Angle shift between the output and input voltage versus the reference angle ψ.

4.2. Experiment

To further verify the proposed control method, measurement tests were performed. The experimental parameters are shown in Table 2.

Table 2. Experimental Parameters.

Parameter Name	Parameter Value
clamp capacitors C_1–C_9	1.1 μF
balancing circuit C_b; L_b; R_b	1 μF; 2.6 mH; 65 Ω
input frequency f	50 Hz
RMS value of input phase voltage	220 V
carrier frequency	5 kHz

Presented in Figures 11–13, waveforms represent the chosen results of measurements. They were registered in MMC, controlled by the use of modulation function (19) and were therefore the CCW-type rotating voltage space vectors. In Figure 11, the output voltage, together with the appropriate input voltage, is shown. The referenced angle shift is equal to 60°, −60° or 180° and the measurements confirm the accomplishment of these values. In Figure 12, the synthesized output voltage and CMV are shown. The maximum instantaneous values of CMV appear when the angle shift ψ is equal to −60° (Figure 12b). These maximum values of CMV do not exceed 40 V (Figure 13b), which is less than 12% of the peak values of the output voltage and amplitude of the input voltage.

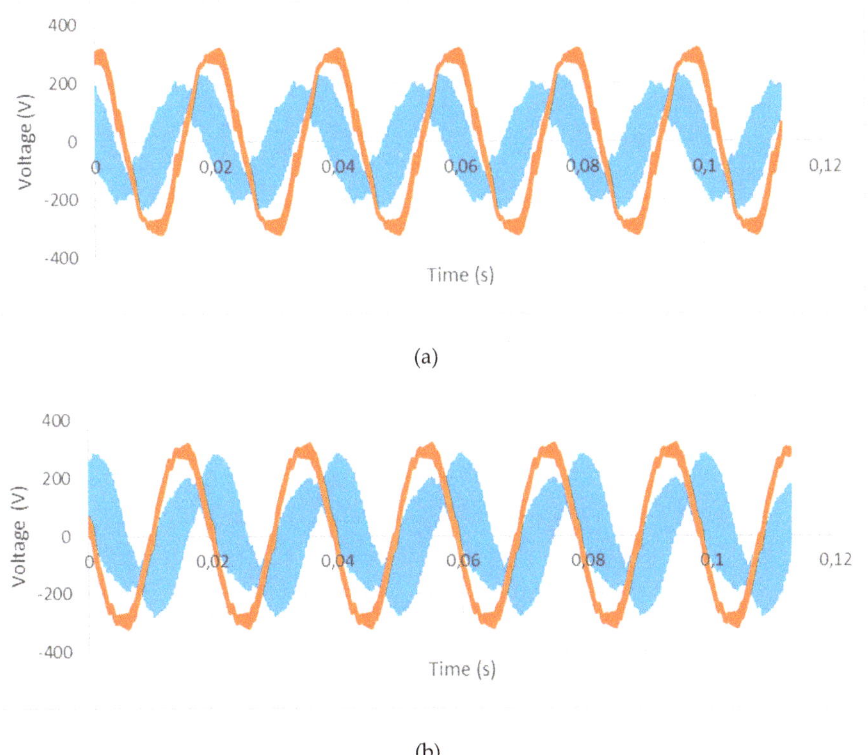

(a)

(b)

Figure 11. *Cont.*

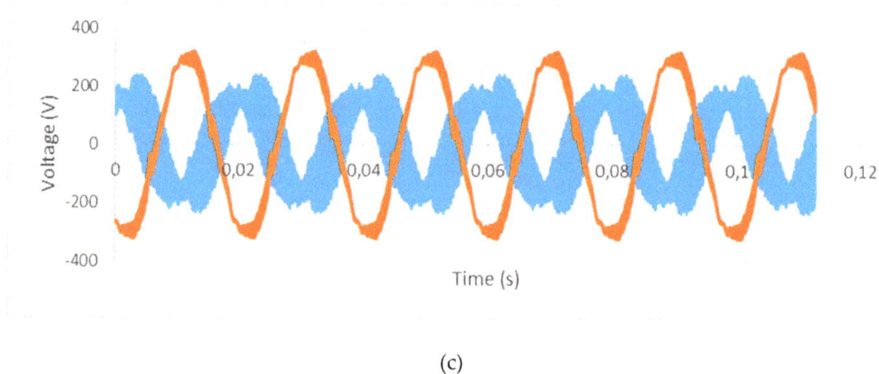

(c)

Figure 11. Waveform of the input phase voltage (orange) and output phase voltage (blue) for the referenced shift angle $\psi = 60°$ (**a**); $\psi = 6-0°$ (**b**) and $\psi = 180°$ (**c**).

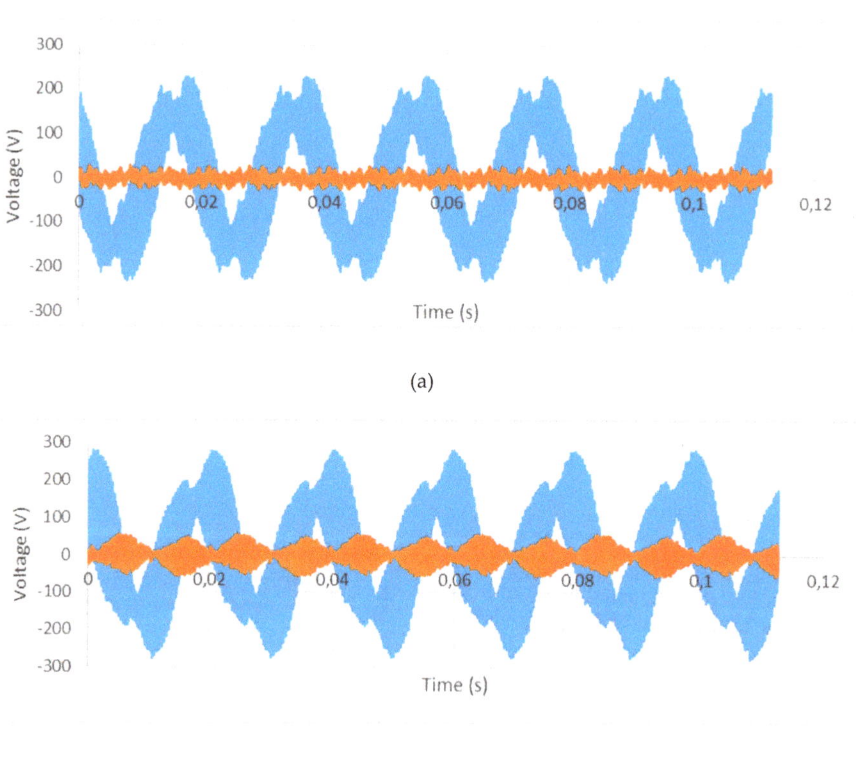

(a)

(b)

Figure 12. *Cont.*

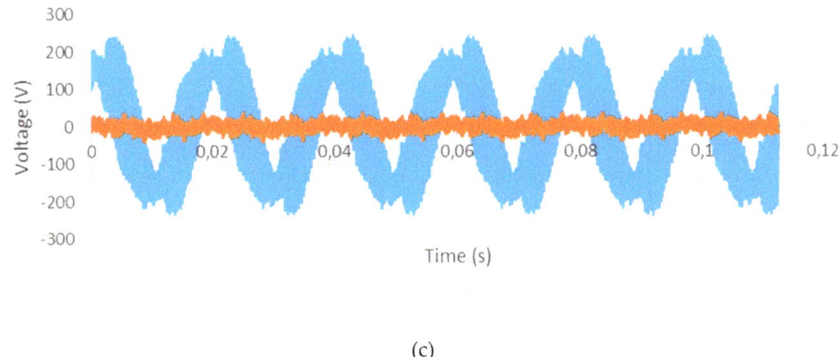

(c)

Figure 12. Waveform of the output phase voltage (blue) and CMV (orange), for the shift angle $\psi = 60°$ (**a**); $\psi = 6-0°$ (**b**) and $\psi = 180°$ (**c**).

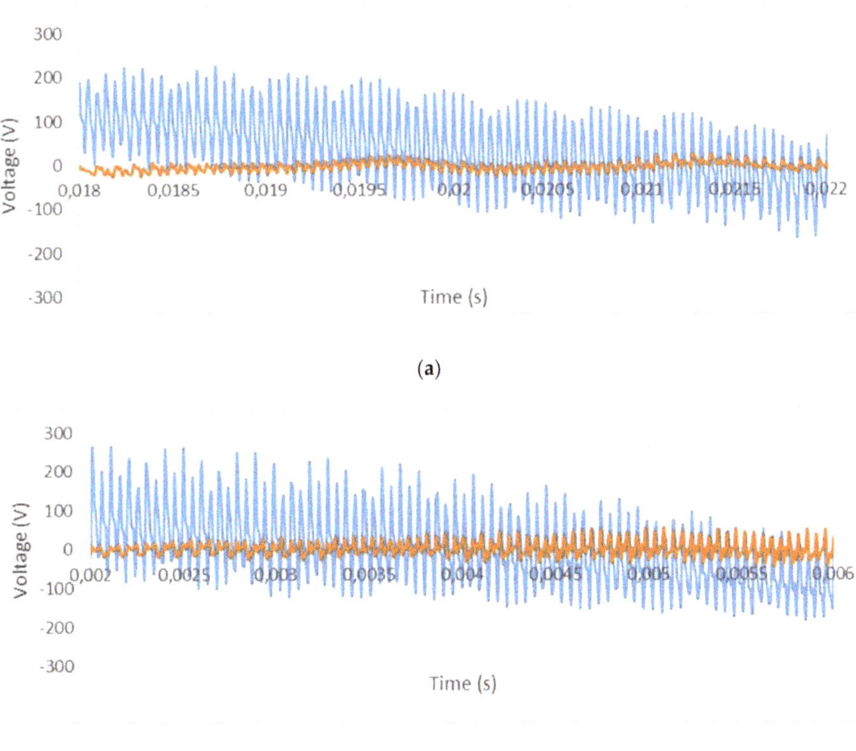

(**a**)

(**b**)

Figure 13. *Cont.*

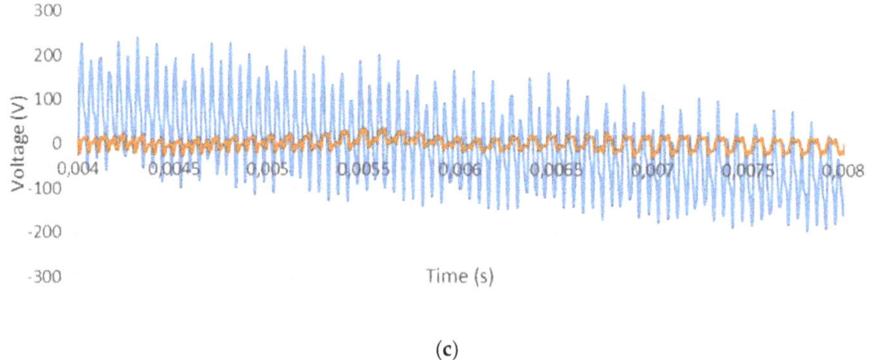

(c)

Figure 13. The enlarged waveform of the output phase voltage (blue) and CMV (orange), for the shift angle $\psi = 60°$ (**a**); $\psi = 6-0°$ (**b**) and $\psi = 180°$ (**c**).

5. Discussion

The proposed carrier-based SVM, using Venturini modulation functions, is valuable because it realizes control whilst improving the waveform of the MMC output voltage compared with the MC output voltage. In MMC, controlled using the proposed method, the amplitudes of the first group of output voltage distortion components, concentrated near the first multiple of the carrier frequency, are only near 5% of the appropriate amplitude of output voltage distortion components in MC.

An important achievement, obtained by implementing the proposed modulation method, is the entire elimination of CMV, which was confirmed by the results of simulation tests. In the experiment, the peak value of CMV is less than 12% of the amplitude of the supplying voltage and the peak value of the output voltage. The occurrence of the CMV, measured higher than zero, may be explained by the noise activated by the four-step commutation process of the bidirectional switches. During the short period of the commutation steps, the active space vectors can appear and cause a higher than zero CMV. To avoid the problem with commutation noise, the four-step commutation in the experimental model of MMC should be modified in future research.

An additional advantage of the presented modulation method, not presented until now in the papers concerned with analysis of MMC, is the possibility to control the phase shift between the output voltage and the appropriate input voltage.

The drawback of the proposed controlling method is the fact that, using only rotating space vectors, the elimination of CMV is possible; this is concerned with the application of modulation functions, which determine the constant input displacement angle between the input voltage and the phase current. On the other hand, this feature may be utilized in FACTS devices that realize series compensation when the input terminals of MMC are connected with the supply network in a shunt manner. MMC, controlled using CW rotating space vectors (modulation function (20)) at the input terminals, draws the current which precedes the appropriate supplying voltage, so MMC works as a source of reactive power for the AC system and also realizes shunt compensation.

Funding: This research received no external funding.

Conflicts of Interest: The author declares no conflict of interest.

Abbreviations

AC	Alternating Current
CCW rotating space vector	Counter Clockwise rotating space vector
CMV	Common mode Voltage
CW	Clockwise rotating space vector
DC	Direct Current
EMTP-ATP	Electromagnetic Transients Program—with version ATP
FACTS	Flexible AC Transmission System
FFT	Fast Fourier Transformation
MC	Matrix Converter
MMC	Multilevel Matrix Converter
PWM	Pulse Width Modulation
SVM	Space Vector Modulation
VTR	Voltage Transfer Ratio

References

1. Shi, Y.; Yang, X.; He, Q.; Wang, Z. Research on Novel Capacitor Clamped Multilevel Matrix Converter. *IEEE Trans. Power Electron.* **2005**, *20*, 1055–1065. [CrossRef]
2. Rząsa, J. Multilevel Matrix Converter Controlled with use of Venturini Method. *Przegląd Elektrotechniczny Rok* **2007**, *Nr 2*, 57–64. (In Polish)
3. Rząsa, J. Capacitor Clamped Multilevel Matrix Converter Controlled with Venturini Method. In Proceedings of the EPE-PEMC, Poznań, Poland, 1–3 September 2008.
4. Rząsa, J. Switch Current Harmonic Distortion in Classic and Capacitor-Clamped Multilevel Matrix Converters. In Proceedings of the 35th Annual Conference of the IEEE Industrial Electronics Society IECON'09, Porto, Portugal, 3–5 November 2009.
5. Lie, X.; Clare, J.C.; Wheeler, P.W.; Empringham, L. Space Vector Modulation for a Capacitor Clamped Multi-level Matrix Converter. In Proceedings of the EPE-PEMC 2008 13th International Power Electronics and Motion Conference, Poznań, Poland, 1–3 September 2008.
6. Lie, X.; Clare, J.C.; Wheeler, P.W.; Empringham, L. Capacitor Clamped Multi-Level Matrix Converter: Space Vector Modulation and Capacitor Balance. In Proceedings of the 2008 34th Annual Conference of IEEE Industrial Electronics, Orlando, FL, USA, 10–13 November 2008.
7. Lie, X.; Yongdong, L.; Kui, W.; Clare, J.C.; Wheeler, P.W. Research on the Amplitude Coefficient for Multilevel Matrix Converter Space Vector Modulation. *IEEE Trans. Power Electron.* **2012**, *27*, 3544–3556. [CrossRef]
8. Lie, X.; Clare, J.C.; Wheeler, P.W.; Empringham, L.; Yongdong, L. Capacitor Clamped Multilevel Matrix Converter Space Vector Modulation. *IEEE Trans. Ind. Electron.* **2012**, *59*, 105–115. [CrossRef]
9. Espina, J.; Ortega, C.; de Lillo, L.; Empringham, L.; Balcells, J.; Arias, A. Reduction of Output Common Mode Voltage Using a Novel SVM Implementation in Matrix Converters for Improved Motor Lifetime. *IEEE Trans. Ind. Electron.* **2014**, *61*, 5903–5911. [CrossRef]
10. Espina, J.; Arias, A.; Balcells, J.; Ortega, C. Common mode output waveforms reduction for Matrix Converters drives. In Proceedings of the 2009 35th Annual Conference of IEEE Industrial Electronics, Porto, Portugal, 3–5 November 2009; pp. 4499–4504. [CrossRef]
11. Guan, Q.; Wheeler, P.; Guan, Q.; Yang, P. Common-mode voltage reduction for matrix converters using all valid switch states. *IEEE Trans. Power Electron.* **2016**, *31*, 8247–8259. [CrossRef]
12. Padhee, V.; Sahoo, A.K.; Mohan, N. Modulation technique for common mode voltage reduction in a matrix converter drive operating with high voltage transfer ratio. In Proceedings of the 2016 IEEE Applied Power Electronics Conference and Exposition (APEC), Long Beach, CA, USA, 20–24 March 2016; pp. 1982–1988. [CrossRef]
13. Rzasa, J. Control of a matrix converter with reduction of a common mode voltage. In Proceedings of the IEEE Compatibility in Power Electronics, Gdynia, Poland, 6 January 2005.
14. Nguyen, H.; Lee, H. A Modulation Scheme for Matrix Converters with Perfect Zero Common-Mode Voltage. *IEEE Trans. Power Electron.* **2016**, *31*, 5411–5422. [CrossRef]

15. Mohapatra, K.K.; Ned, M. Open-End Winding Induction Motor Driven with Matrix Converter for Common-Mode Elimination. In Proceedings of the Power Electronics, Drives and Energy Systems, New Delhi, India, 12–15 December 2006.

16. Gupta, R.K.; Mohapatra, K.K.; Somani, A.; Mohan, N. Direct-Matrix-Converter-Based Drive for a Three-Phase Open-End-Winding AC Machine with Advanced Features. *IEEE Trans. Ind. Electron.* **2010**, *57*, 4032–4042. [CrossRef]

17. Tewari, S.; Mohan, N. Matrix Converter Based Open-End Winding Drives with Common-Mode Elimination: Topologies, Analysis, and Comparison. *IEEE Trans. Power Electron.* **2018**, *33*, 8578–8595. [CrossRef]

18. Rząsa, J. Research on Dual Matrix Cnverter Feeding an Open-End-Winding Load Controlled with the Use of Rotating Space Vectors. Part I. In Proceedings of the 39th Annual Conference of the IEEE Industrial Electronics Society IECON 2013, Vienna, Austria, 10–13 November 2013.

19. Rząsa, J.; Garus, G. Research on Dual Matrix Converter Feeding an Open-End-Winding Load Controlled with the Use of Rotating Space Vectors. Part II. In Proceedings of the 39th Annual Conference of the IEEE Industrial Electronics Society IECON 2013, Vienna, Austria, 10–13 November 2013.

20. Baranwal, R.; Basu, K.; Mohan, N. An alternative carrier based implementation of Space Vector PWM for dual matrix converter drive with common mode voltage elimination. In Proceedings of the IECON 2014—40th Annual Conference of the IEEE Industrial Electronics Society, Dallas, TX, USA, 29 October–1 November 2014; pp. 1208–1213. [CrossRef]

21. Jianglei, Q.; Lie, X.; Wang, L.; Lin, Q.; Yongdong, L. The modulation of common mode voltage suppression for a three-level matrix converter. In Proceedings of the 2016 IEEE International Conference on Aircraft Utility Systems (AUS), Beijing, China, 10–12 October 2016; pp. 533–538. [CrossRef]

22. Jianglei, Q.; Lie, X.; Lina, W.; Yannian, H. Research on the modulation and control of multilevel matrix converter. *J. Eng.* **2018**, *2018*, 614–621. [CrossRef]

23. Monteiro, J.; Silva, J.F.; Pinto, S.G.; Palma, J. Matrix Converter-Based Unified Power-Flow Controllers: Advanced Direct Power Control Method. *IEEE Trans. Power Deliv.* **2011**, *26*, 420–430. [CrossRef]

24. Meynard, H.; Foch, T.A.; Thomas, P.; Courault, J.; Jakob, R.; Nahrstaedt, M. Multicell Converters: Basic Concepts and Industry Applications. *IEEE Trans. Ind. Electron.* **2002**, *49*, 978–987. [CrossRef]

25. Meynard, T.A.; Fadel, M.; Aouda, N. Modeling of Multilevel Converters. *IEEE Trans. Power Electron.* **1997**, *44*, 356–364. [CrossRef]

26. Pirog, S.; Stala, R. Selection of parameters for a balancing circuit of dc–dc and ac–ac multicell converters. In Proceedings of the 2005 European Conference on Power Electronics and Applications, Dresden, Germany, 11–14 September 2005; p. 910.

27. Stala, R.; Mondzik, A. A study of the balancing process in multicell ac/ac converter. *Przegląd Elektrotechniczny* **2009**, *nr 7*, 168–172.

Article

A Compound Current Limiter and Circuit Breaker

Amir Heidary [1], Hamid Radmanesh [2], Ali Bakhshi [1], Kumars Rouzbehi [3] and Edris Pouresmaeil [4],*

[1]   Electrical Engineering Department, Amirkabir University of Technology, 15914 Tehran, Iran; amir.powersys@gmail.com (A.H.); bakhshiialii@gmail.com (A.B.)
[2]   Electrical Engineering Department, Shahid Sattari Aeronautical University of Science and Technology, 15914 Tehran, Iran; radmanesh@ssau.ac.ir
[3]   Department of System Engineering and Automatic Control, University of Seville, 41004 Seville, Spain; krouzbehi@us.es
[4]   Department of Electrical Engineering and Automation, Aalto University, 02150 Espoo, Finland
*   Correspondence: edris.pouresmaeil@aalto.fi

Received: 18 April 2019; Accepted: 7 May 2019; Published: 16 May 2019

Abstract: The protection of sensitive loads against voltage drop is a concern for the power system. A fast fault current limiter and circuit breaker can be a solution for rapid voltage recovery of sensitive loads. This paper proposes a compound type of current limiter and circuit breaker (CLCB) which can limit fault current and fast break to adjust voltage sags at the protected buses. In addition, it can act as a circuit breaker to open the faulty line. The proposed CLCB is based on a series *L-C* resonance, which contains a resonant transformer and a series capacitor bank. Moreover, the CLCB includes two anti-parallel power electronic switches (a diode and an IGBT) connected in series with bus couplers. In order to perform an analysis of CLCB performance, the proposed structure was simulated using MATLAB. In addition, an experimental prototype was built, tested, and the experimental results were reported. Comparisons show that experimental results were in fair agreement with the simulation results and confirm CLCB's ability to act as a fault current limiter and a circuit breaker.

Keywords: circuit breaker; fault current limiter

1. Introduction

Faults in electrical power systems are inevitable. They can lead to high transients and thermal stresses on power system equipment such as overhead lines, cables, transformers, and switchgears. Therefore, the fault current protection schemes are important. The simplest solution to limit the short-circuit current would be the application of a source with high impedance. The main drawback of this solution is that it also influences the system during normal operation conditions, and it results in a considerable voltage drop for high current loads [1,2]. Therefore, electric networks require efficient and reliable equipment to limit the short-circuit current. Another solution to this problem is the use of technologies such as fault current limiters (FCLs). The FCL is one of the protection devices, which is used to limit the fault current. The FCL should limit the fault current passing through it within the first half-cycle and the best FCL should limit the fault current before the first peak [3]. However, high price, power losses, continuous current after fault current flow limitation, and harmonic distortion are some of the main problems of typical FCLs. Since the 1970s, several types of FCLs have been investigated such as fuses with fault-current limitation, series current limiting reactors [4], series transformers [5], superconducting fault current limiters (SCFCL) [6–8], solid-state FCLs (SSFCL) [9–13], and fault current limiting circuit breakers (FCLCB). In the recent years, researchers have focused on the SSFCLs and FCLCBs, such as: Purely resistive FCL [14], hybrid-resistive FCL [15], saturable core FCL [16], IGBTs controlled series reactor FCL [17], solid-state FCLCB (SSFCL-CB) [18], and bridge type

FCL [19]. These new protection devices usually use inductors to decrease the fault current. In these structures, the reactor is ignorable during the normal operation mode and has a fixed impedance during the fault episode, which decreases the system fault current and in some cases can improve the system stability [19]. The influence of the FCL on the short-circuit level of the substation bus bar splitter circuit breaker has been investigated in [20,21]. A rectifier-type SFCL with non-inductive reactor has been reported in [22]. In [23], the power electronic switches selection for 20 kV distribution network application are discussed. A DC circuit breaker for voltage source converter (VSC) has been proposed in [24]. The fast-closing switch application in solid-state circuit breaker and its optimization process has been studied in [25]. Application of current-limiting circuit breakers to control the arc-flash energy has been presented in [26]. Classification of solid-state circuit breakers and application of solid-state circuit breaker, to improve grid voltage quality during the fault is reported in [27]. In [28], the comparison of two control methods of power swing reduction in a power system with unidirectional power flow controller (UPFC) is discussed. Analysis and control of fault current by firing angle control of solid-state fault current limiter is an important issue which depends on the strategy of power electronic switch control [29].

This paper presents a new type of current limiter circuit breaker (CLCB) with series compensation. This protection device is invisible during the normal operation mode. During the fault period, it disconnects the loads from the source. The operational effectiveness of this device is verified by MATLAB simulations and confirmed by the developed experimental tests. The results show the fast-closing switch based CLCB has more advantages than the former FCLs with low cost and can improve the system protection against fault by fast current limiting and breaking.

Expected advantages of the proposed CLCB over other FCLs are as the following:

- Ability to remain invisible to the grid under normal operation mode, introducing negligible impedance in the network;
- Short recovery time and ability to limit the fault current before initiation of the first peak;
- By connecting the proposed CLCB to the grid, the mechanical circuit breaker can be replaced;
- Using the proposed CLCB in the network decreases the grid short-circuit levels;
- Fast recovery after fault removal.

This paper has been organized as follows:

In Section 2, the system topology including proposed CLCB is discussed. In Section 3, the analytical analysis of the CLCB operation during normal and fault operation modes, voltage sag at sensitive bus, and power losses are studied. Then, in Section 4, the control system is studied. In the next section, the MATLAB software was used to simulate the operational behavior of the CLCB. In Section 6, experimental results are presented and finally a conclusion is drawn.

2. Electrical Network Modeling

Figure 1 shows a single line diagram of the power grid, in which CLCB connects bus 3 and bus 4 as bus coupler.

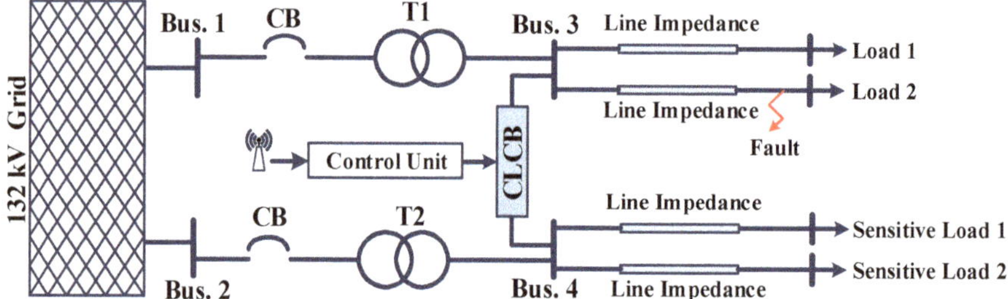

Figure 1. Single line diagram of the distribution network.

Bus 3 is assumed to be faulty and bus 4 is connected to the sensitive loads by feeders. The CLCB topology is shown in Figure 2.

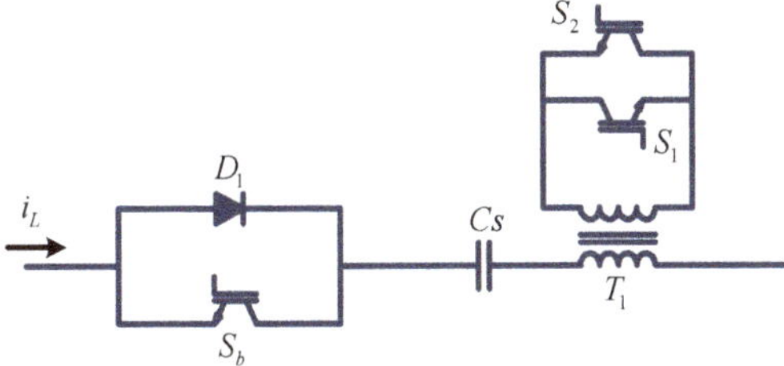

Figure 2. Current limiting circuit breaker topology.

In this circuit, D_1 and S_b are power electronics diode and IGBT switch, respectively, and C_s is a series capacitor bank. In addition, the primary side of transformer T_1 is connected in series to a line and its secondary is connected to two anti-parallel IGBTs. During normal operation mode, the resonance transformer and series capacitor form a series resonance *L-C* tank with resonance frequency equal to electrical network frequency. In this case, D_1 for positive half-cycles and S_b for negative half-cycles, are in on-state and voltage drop on the CLCB components is negligible. The CLCB configuration during normal operation mode is shown in Figure 3a.

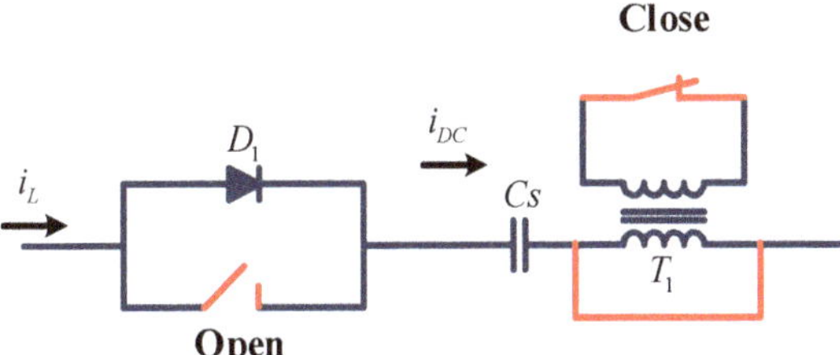

Figure 3. Proposed CLCB topology (**a**) normal operation, (**b**) fault current limiting, and (**c**) fault current breaking.

During a fault, the fault current increases and passes the threshold current level (I_L). In this case, the control circuit detects the fault and turns on the antiparallel IGBTs. Therefore, the secondary side of the resonance transformer is short-circuited and the resonance transformer shows negligible impedance. The series capacitor impedance then limits the fault current. Figure 3b shows the CLCB topology in the fault current limiting mode. To open the faulty line, the control circuit turns off S_b after one cycle delay. In this case, D_1 passes a positive half-cycle and the induced DC voltage on the series capacitor charges it. Then, the series capacitor opens the faulty line successfully. The CLCB topology in circuit breaker mode is shown in Figure 3c.

3. Analytical Studies

3.1. CLCB Operation in Normal Mode

In this mode, the secondary side of the transformer is open and the series resonance *LC* tank is in resonance condition. Therefore, the electrical network equivalent circuit in steady state condition is an *R-L* circuit where *R* and *L* equal to source, line, and load resistances and inductances, respectively. In addition, the source voltage is denoted with $V_s(t)$ and is equal to $V_m \, sin(\omega t)$. By applying Kirchhoff law to the network, the line current for the steady-state condition,

$$V_m \sin(\omega t) = L\frac{di_L(t)}{dt} + Ri_L(t) \tag{1}$$

then

$$i_L(t) = \frac{V_m}{\sqrt{R^2 + \omega^2 L^2}} \sin\left(\omega t - \tan^{-1}\frac{\omega L}{R}\right) \tag{2}$$

Equation (2) shows the sinusoidal nature of the line current during the normal operation mode.

3.2. CLCB Operation in Fault Current Limiting Mode

During a fault, the resonance transformer is by-passed via IGBTs and the equivalent circuit of the network is an *R-L-C* circuit, where *R* and *L* include the source, line, and CLCB (transformer leakage and magnetization) resistances and inductances, respectively, and *C* is the series capacitor bank. In this case, the *RLC* circuit current can be obtained using the Equation (3)

$$LC\frac{d^2 V_C(t)}{dt^2} + RC\frac{dV_C(t)}{dt} + V_C(t) = V_s(t) - (V_D + V_{IGBT}) \tag{3}$$

where initial conditions for *L* is $i_L(0^-) = i_L(0^+) = I_0$, for *C* is $V_C(0^-) = V_C(0^+) = V_0$, and V_D and V_{IGBT} is IGBT voltage drop, respectively

$$\frac{dV_C(0^-)}{dt} = \frac{i_L(0^-)}{C} = \frac{I_0}{C} \tag{4}$$

Solving this equation results in the following equation:

$$V_C(t) = e^{-\alpha t}(A_1 \cos\beta t + A_2 \sin\beta t) + \frac{V_m}{\sqrt{(1-LC\omega^2)^2+(RC\omega)^2}} \sin\left(\omega t + \frac{\pi}{2} + \tan^{-1}\left(\frac{RC\omega}{(1-LC\omega^2)}\right)\right) - (V_D + V_{IGBT}) \tag{5}$$

where $\alpha = \frac{R}{2L}$, $\omega_0 = \frac{1}{\sqrt{LC}}$, $\beta = \sqrt{\alpha^2 - \omega_0^2}$, and the value of A_1 and A_2 can be obtained using initial conditions. Then,

$$i_L(t) = C\left(e^{-\alpha t}(A_1 \cos\beta t + A_2 \sin\beta t)\right)' + \frac{\omega V_m}{\sqrt{(1 - LC\omega^2)^2 + (RC\omega)^2}} \cos\left(\omega t - \tan^{-1}\left(\frac{RC\omega}{(1 - LC\omega^2)}\right)\right) \tag{6}$$

The obtained value for $i_L(t)$ includes two-term responses and one steady-state term. The transient responses are dampened after some milliseconds. The steady-state response includes the phase angle shift as shown in the simulation results.

3.3. CLCB Operation in Circuit Breaking Mode

In this case, the electrical network is in faulty condition and the suggested CLCB should open the faulty line. Therefore, the control system turns off S_b and induces the DC voltage on the series capacitor. The charged capacitor then opens the faulty line and the transmission line current reaches zero. In this case, we have:

$$i_L(t) = C\left(e^{-\alpha t}(A_3 \cos\beta t + A_4 \sin\beta t)\right)' \tag{7}$$

The Equation (7) includes two exponential parts and, the line current reaches zero.

4. Control Strategy

The control block diagram of the proposed CLCB is shown in Figure 4. In the normal mode, the S_b was in on-state for negative half-cycles and IGBTs were in off-state. Therefore, the line current (i_L) passed through the series resonance *LC* tank and the CLCB showed negligible impedance.

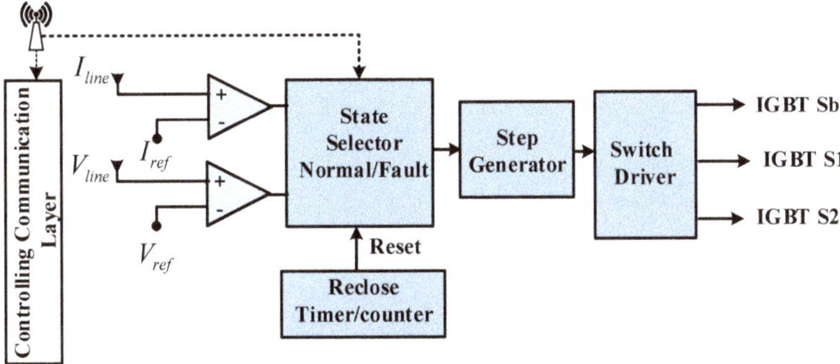

Figure 4. Bock diagram presentation of CLCB control logic.

At fault inception, the I_L becomes greater than the maximum permissible current (I_{ref}) and the control circuit turns on the anti-parallel IGBTs and turns off the S_b after the one cycle delay. Therefore, the resonance transformer is bypassed and the impedance of the series capacitor limits the fault current. By turning off the T_{s1}, the faulty line is opened and the CLCB acts as a circuit breaker. After fault removal, the step generator resets the gates pulses of the power electronics switches and returns the network to the pre-fault condition.

5. Simulation Results

The single line diagram of the electrical network including CLCB and shown in Figure 1 is simulated. The parameters of the suggested CLCB and electrical network are listed in Table 1. The results are obtained considering a single-phase to the ground short-circuit fault at bus A. The simulation results are studied for the system with and without using the CLCB.

Table 1. Parameters of electrical network and CLCB.

Parameters	Value	Description
$V_S(t)$	20 kV	Nominal voltage
ω	314 rad/s	Nominal frequency
R_s	0.5 Ω	Source resistance
L_s	9 mH	Source inductance
C_s	56 uF	Series capacitor
L_p	20 mH	Primary inductance of the transformer
L_m	0.18 H	Magnetization inductance of the transformer
L_t	50 mH	Secondary inductance of the transformer
R_p	2 Ω	Primary resistance of the transformer
R_t	2 Ω	Secondary resistance of the transformer
Z_L	0.27 + j0.35 Ω/km	Line impedance
Z_T	0.07 + j2.16 Ω	Transformer impedance

In normal operation mode, both buses delivered power to the loads at half capacity (12.5 MVA). In this case, there was no voltage drop on the CLCB devices and because of the system symmetry; no current was circulated through the interconnected CLCB. In addition, it is assumed that there was no CLCB connected to the feeder and line current was in normal condition as shown in Figure 5. A fault at bus (A) could cause severe voltage sag, which would affect the sensitive load. In this case, the fault current increased and its amplitude reached 6.8 kA as shown in Figure 5.

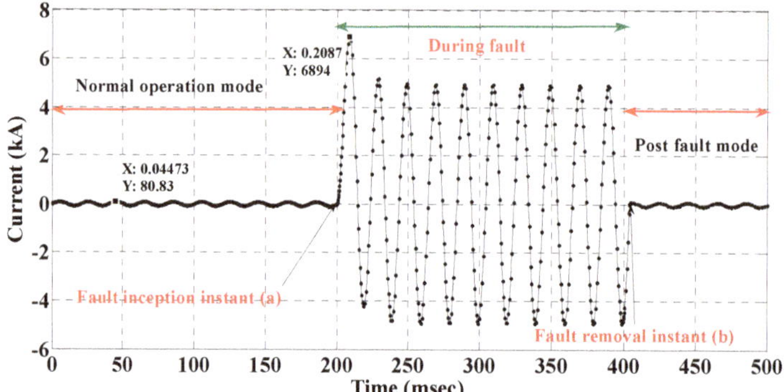

Figure 5. Fault current at bus A without using CLCB.

It is assumed that the fault occurred at bus (A) which produced an increase of current in the interconnection CLCB and bus (A) experienced a transient voltage. To prevent the service interruption at a sensitive load, the CLCB was connected in series with the feeder and interconnection bus as shown in Figure 1. In fault case, the CLCB impedance increased and its series *LC* tank was in series with the interconnection bus during the increase of the current. Therefore, its impedance decreased the faulty line current to an acceptable level and compensated the voltage sag at bus (A). Figures 6 and 7 show the fault current and bus (A) voltage for both cases with and without using CLCB.

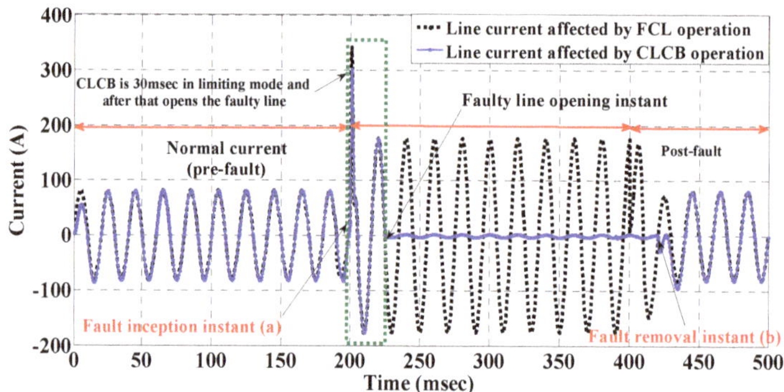

Figure 6. Fault current during the normal operation and fault with connected CLCB.

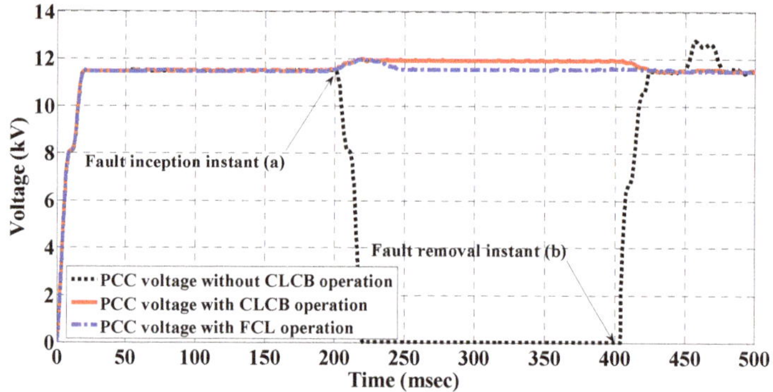

Figure 7. RMS value of bus A voltage with and without using CLCB.

In the case 1 (without CLCB), the fault current increased to the peak value of 6.8 kA but by using the CLCB, the fault current was limited to the peak value of 200 A. It is shown that in case 1, the voltage of the bus (A) decreased approximately to zero. However, CLCB not only reduced the voltage sag to 0.9 pu, but also it opened the faulty line and fixed the bus (A) voltage to 1 pu. During the normal operation mode, the impedance of the series resonance *LC* tank was close to zero and there was no voltage drop on it. During the fault, the resonance transformer was bypassed and a considerable voltage drop was seen on the series capacitor. In circuit breaking mode, the induced DC voltage on the series capacitor charged it higher than the peak voltage of the network and caused it to open the faulty line. Figure 8 shows the series capacitor voltage during normal operation and fault for AC and DC operation cases.

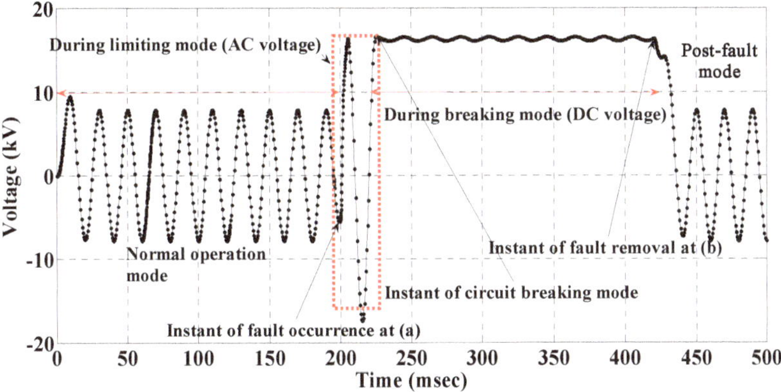

Figure 8. Series capacitor voltage during the normal operation and fault including fault current limiting mode (AC operation) and circuit breaking mode (DC capacitor charging).

As shown in Figure 8, after fault inception, the fault current increased but the impedance of the series capacitor in AC mode decreased the fault current. After one cycle delay, the controller turned off T_{s1} and induced DC voltage on the series capacitor charged it with DC voltage. In this case, the faulty line was opened via a series capacitor and the fault current reached zero.

The load voltage during normal and fault operation modes is shown in Figure 9. The fault occurred at instant (a) and the voltage of the load decreased to zero. At instant (b), the fault was cleared and the load voltage returned to the pre-fault value.

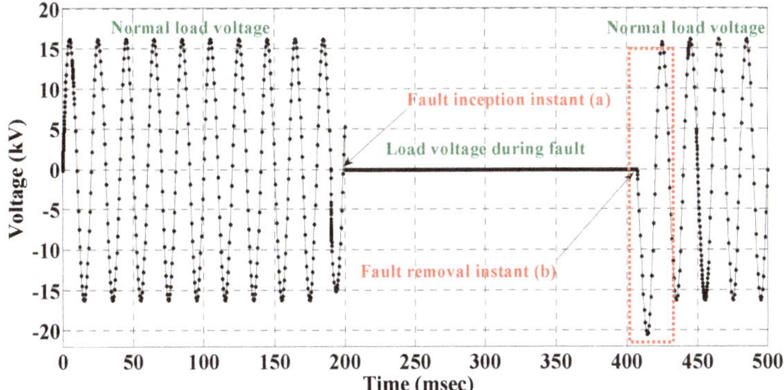

Figure 9. Load voltage during the normal and fault operation modes.

After the fault removal, the voltage of the electrical load was distorted for a first half-cycle. The stored energy on the series capacitor during the fault period caused this voltage fluctuation.

The CLCB operation and its effect on faulty line current are shown in Figure 10. These comparative plots show the CLCB influence on both decreasing the fault current and opening the faulty line. The dotted plot shows the fault current when there is no connected FCL in series with the feeder. By FCL utilization, the fault current was decreased as shown with dash line in Figure 10.

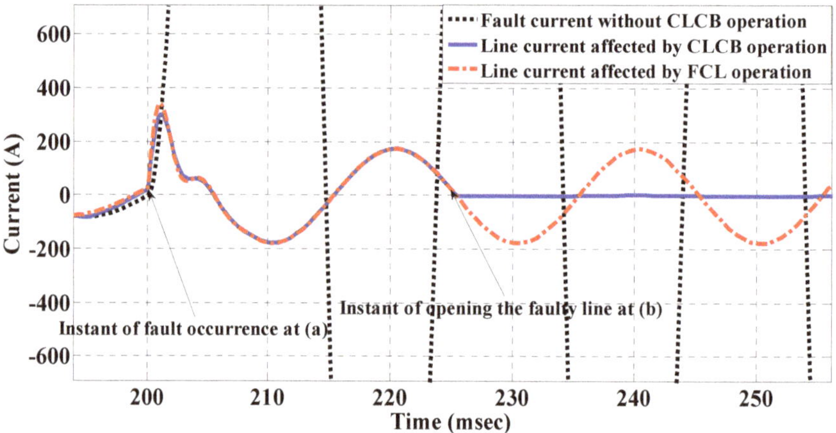

Figure 10. Comparison of line current during the normal and fault operation modes; with and without using CLCB.

In the instant of fault inception, the first peak of the fault current decreased and after that the limited fault current reached an acceptable level. The blue solid plot shows the line current during normal and fault operation modes affected by the proposed CLCB. At the first cycle of the line fault current, the proposed CLCB acted as a fault current limiter. Then it opened up the faulty line, and current decayed to zero.

6. Experimental Results

To verify the simulation results, a CLCB prototype was built as shown in Figure 11. The CLCB prototype was tested in normal and fault operation modes. Table 2 lists the experimental setup parameters.

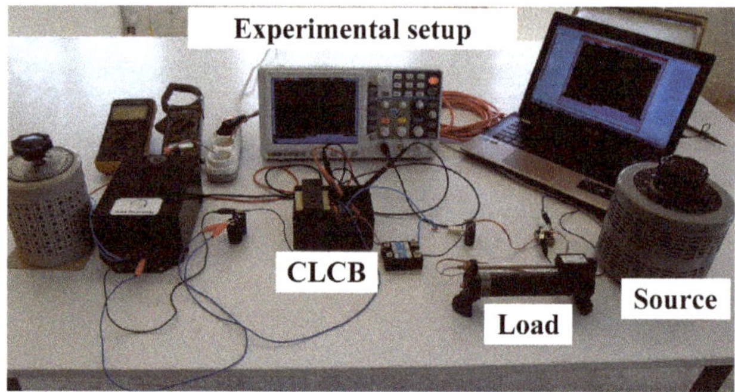

Figure 11. Prototype of the proposed CLCB.

Table 2. Experimental setup characteristics.

Parameters	Value	Description
$V_S(t)$	110 V	Nominal voltage (rms)
ω	314 rad/sec	Nominal frequency
r_s	0.5 Ω	Source resistance
L_s	10 mH	Source inductance
C_s	56 uF	Series capacitor
L_p	20 mH	Primary inductance of transformer
L_m	0.18 H	Magnetization inductance of transformer
r_{eL}	0.016 Ω	Linkage resistance
X_{eL}	0.65 Ω	Linkage inductance
R_{cL}	29.62 Ω	Transformer core resistance
R_L	600 Ω	Transformer impedance
LTS25-NP	25 A	Current sensor
Atmega32		Pulse generator
TLP-250		IGBTS gate drivers
IGBT(NGTB25N120IHL)	1200 V, 25 A	Fast-closing switches
Power Diode(SEMIKRON)	1200 V, 25 A	Transmission line switches

Using a mechanical switch, a single line to ground fault was implemented. The controlling circuit included a voltage transformer, a current transducer (*LTS 25-NP*), IGBT gate drivers (TLP250), RC filter, and an Atmel XMEGA microcontroller. Measurements of line voltage and current in faulty condition were processed and detected by microcontroller and operation command was generated in two stages. In the first stage, by operating a switch of the transformer, secondary fault current magnitude was limited. In the second stage, by operating series IGBT, fault current was broken.

Figure 12 shows the line current during the normal and fault operation modes. In this plot, the phase to ground fault occurred at instant (a) via a mechanical switch and was cleared up at instant (b) by the opening of the mechanical switch. As shown here, after the fault occurrence, the CLCB limited the fault current, opened the faulty line, and decreased the fault current to zero. After fault clearance, CLCB recovered the faulty line in less than 20 ms. This measured curve is in fair agreement with Figure 6.

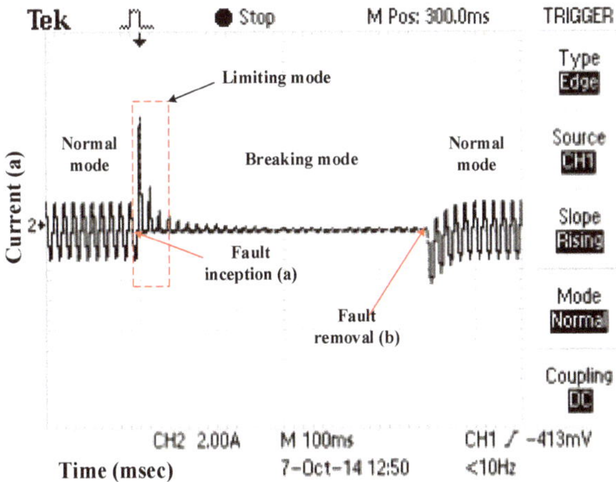

Figure 12. Line current during the normal and fault operation modes.

Figure 13 shows the protected bus voltage during the normal and fault operation modes. As shown in this figure, CLCB can successfully fix protected bus voltage to an acceptable level during the fault. This figure is in agreement with Figure 7. In this figure, the duration of the normal, fault operation modes, and its effect on the line current can be seen in the upper curve.

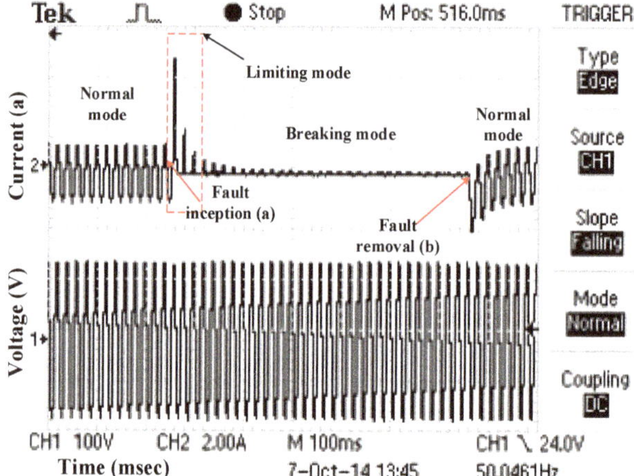

Figure 13. Protected bus voltage during the normal and fault operation modes.

The voltage of the series capacitor is shown in Figure 14. The series capacitor voltage during the normal operation mode was sinusoidal and this capacitor was in resonance with the series transformer primary. After the fault, by operating series IGBT operation, voltage changed to the DC voltage, which opened the faulty line. This figure is in agreement with Figure 8.

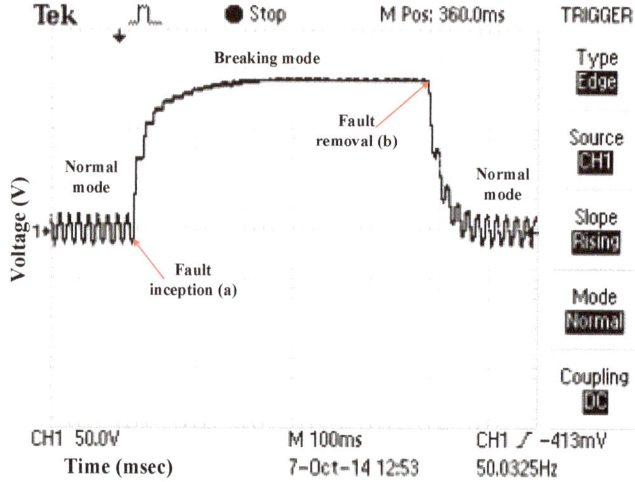

Figure 14. The voltage of series capacitor during the normal and fault operation modes.

Comparing the proposed CLCB with traditional CB and power electronic based CB, superiority of the proposed structure can be listed as follows:

- A low number of series power electronic switches (two switches);
- Series switch low voltage stress;
- Low current magnitude in breaking state;
- Combination of fault current limiting structure with solid-state breaker;
- Very fast operation in comparison with mechanical breakers.

7. Conclusions

In this paper, a new type of CLCB is proposed. This device acts by dual-function protection, not only limiting the fault current but also open the faulty line similar to a circuit breaker. In practice, its fast response to faults can successfully limit the first peak of the fault current. In addition, the proposed CLCB assists to recover the protected buses voltage to an acceptable level. Therefore, the sensitive loads do not experience a significant voltage sag. The CLCB can be placed as a solid-state circuit breaker (instead of the traditional circuit breakers) and behaves as a fault current limiter. Performance of proposed CLCB is proved by simulation and experimental test results.

Author Contributions: All authors contributed equally to this work and all authors have read and approved the final manuscript.

Funding: This research received no external funding.

Conflicts of Interest: The authors declare no conflict of interest.

References

1. Elmitwally, A. Proposed hybrid superconducting fault current limiter for distribution systems. *Int. J. Electr. Power Energy Syst.* **2009**, *31*, 619–625. [CrossRef]
2. Didier, G.; Lévêque, J. Influence of fault type on the optimal location of superconducting fault current limiter in electrical power grid. *Int. J. Electr. Power Energy Syst.* **2014**, *56*, 279–285. [CrossRef]
3. Prigmore, J.R.; Mendoza, J.A.; Karady, G.G. Comparison of four different types of ferromagnetic materials for fault current limiter applications. *IEEE Trans. Power Deliv.* **2013**, *28*, 1491–1498. [CrossRef]
4. Chen, S.; Yang, P.; Xu, W.; Bian, X.; Wang, W. Side effects of Current-Limiting Reactors on power system. *Int. J. Electr. Power Energy Syst.* **2013**, *45*, 340–345. [CrossRef]
5. Tian, M.; Li, Q.; Li, Q. A controllable reactor of transformer type. *IEEE Trans. Power Deliv.* **2004**, *19*, 1718–1726. [CrossRef]
6. Didier, G.; Leveque, J.; Rezzoug, A. A novel approach to determine the optimal location of SFCL in electric power grid to improve power system stability. *IEEE Trans. Power Deliv.* **2013**, *28*, 978–984. [CrossRef]
7. Aracil, J.C.; Lopez-Roldan, J.; Coetzee, J.C.; Darmann, F.; Tang, T. Analysis of electromagnetic forces in high voltage superconducting fault current limiters with saturated core. *Int. J. Electr. Power Energy Syst.* **2012**, *43*, 1087–1093. [CrossRef]
8. Rouzbehi, K.; Candela, J.I.; Luna, A.; Gharehpetian, G.B.; Rodriguez, P. Flexible Control of Power Flow in Multiterminal DC Grids Using DC–DC Converter. *IEEE J. Emerg. Sel. Top. Power Electron.* **2016**, *4*, 1135–1144. [CrossRef]
9. Heidary, A.; Radmanesh, H.; Fathi, S.H.; Gharehpetian, G.B. Series transformer based diode-bridge-type solid state fault current limiter. *Front. Inf. Technol. Electron. Eng.* **2015**, *16*, 769–784. [CrossRef]
10. Ghanbari, T.; Farjah, E. Development of an efficient solid-state fault current limiter for microgrid. *IEEE Trans. Power Deliv.* **2012**, *27*, 1829–1834. [CrossRef]
11. Fereidouni, A.R.; Vahidi, B.; Mehr, T.H. The impact of solid state fault current limiter on power network with wind-turbine power generation. *IEEE Trans. Smart Grid* **2013**, *4*, 188–1196. [CrossRef]
12. Heidary, M.A.; Radmanesh, H.; Rouzbehi, K.; Pou, J. A dc-reactor based solid-state fault current limiter for hvdc applications. *IEEE Trans. Power Deliv.* **2019**, *34*, 720–728. [CrossRef]
13. Javadi, H.; Ali Mousavi, S.M.; Khederzadeh, M. A novel approach to increase FCL application in preservation of over-current relays coordination in presence of asynchronous DGs. *Int. J. Electr. Power Energy Syst.* **2013**, *44*, 810–815. [CrossRef]
14. Naderi, S.B.; Jafari, M.; Hagh, M.T. Controllable resistive type fault current limiter (CR-FCL) with frequency and pulse duty-cycle. *Int. J. Electr. Power Energy Syst.* **2014**, *61*, 11–19. [CrossRef]
15. Rouzbehi, K.; Miranian, A.; Candela, J.; Luna, A.; Rodriguez, P. Intelligent voltage control in a dc micro-grid containing PV generation and energy storage. In Proceedings of the 2014 IEEE PES T&D Conference and Exposition, Chicago, IL, USA, 14–17 April 2014; pp. 1–5.

16. Rouzbehi, K.; Miranian, A.; Luna, A.; Rodriguez, P. Towards fully controllable multi-terminal DC grids using flexible DC transmission systems. In Proceedings of the 2014 IEEE Energy Conversion Congress and Exposition (ECCE), Pittsburgh, PA, USA, 14–18 September 2014; pp. 5312–5316.

17. Kang, B.; Park, J.-D. Application of thyristor-controlled series reactor for fault current limitation and power system stability enhancement. *Int. J. Electr. Power Energy Syst.* **2014**, *63*, 236–245. [CrossRef]

18. Lanes, M.; Matusalem, B.; Barbosa, P.G. Fault current limiter based on resonant circuit controlled by power semiconductor devices. *IEEE Lat. Am. Trans.* **2007**, *5*, 311–320. [CrossRef]

19. Radmanesh, H.; Heidary, A.; Fathi, S.H.; Gharehpetian, G.B. Dual function ferroresonance and fault current limiter based on DC reactor. *IET Gener. Trans. Distrib.* **2016**, *10*, 2058–2065. [CrossRef]

20. Heidary, A.; Radmanesh, H.; Fathi, S.H.; Khamse, H.R.R. Improving transient recovery voltage of circuit breaker using fault current limiter. *Res. J. Appl. Sci. Eng. Technol.* **2012**, *4*, 5123–5128.

21. Javadi, H. Fault current limiter using series impedance combined with bus sectionalizing circuit limiter. *Int. J. Electr. Power Energy Syst.* **2011**, *33*, 731–736. [CrossRef]

22. Hoshino, T.; Itsuya, M.; Nakamura, T.; Salim, K.M.; Yamada, M. Non-inductive variable reactor design and computer simulation of rectifier type superconducting fault current limiter. *IEEE Trans. Appl. Supercond.* **2005**, *15*, 2063–2066. [CrossRef]

23. Radmanesh, H.; Fathi, S.H.; Gharehpetian, G.B.; Heidary, A. A novel solid-state fault current-limiting circuit breaker for medium-voltage network applications. *IEEE Trans. Power Deliv.* **2016**, *31*, 236–244. [CrossRef]

24. Sano, K.; Takasaki, M. A surge-less solid-state DC circuit breaker for voltage source converter based HVDC systems. *IEEE Trans. Ind. Appl.* **2014**, *50*, 2690–2699. [CrossRef]

25. Zou, J.; Chen, J.; Dong, E. Study of fast-closing switch based fault current limiter with series compensation. *Int. J. Electr. Power Energy Syst.* **2002**, *24*, 719–722. [CrossRef]

26. Gregory, G.; Lippert, K.J. Applying low-voltage circuit breakers to limit arc-flash energy. *IEEE Trans. Ind. Appl.* **2012**, *48*, 1225–1229. [CrossRef]

27. Meyer, C.; De Doncker, R.W. Solid-state circuit breaker based on active thyristor topologies. *IEEE Trans. Power Electron.* **2006**, *21*, 450–458. [CrossRef]

28. Gharibpour, H.; Monsef, H.; Ghanaatian, M. The comparison of two control methods of power swing reduction in power system with UPFC compensator. In Proceedings of the 20th Iranian Conference on Electrical Engineering (ICEE2012), Tehran, Iran, 15–17 May 2012; pp. 386–391.

29. Sharma, J.P.; Chauhan, V. Analysis and control of fault current by firing angle control of solid state fault current limiter. In Proceedings of the 2015 International Conference on Energy Systems and Applications, Pune, India, 30 October 2015; pp. 527–530.

 *electronics*

Article

Nonlinear Effects of Three-Level Neutral-Point Clamped Inverter on Speed Sensorless Control of Induction Motor

Peifei Li, Lei Zhang, Bin Ouyang and Yong Liu *

Department of Electrical Engineering, Navy Engineering University, Jiangxia District,
Wuhan 430200, Hubei, China; peifeilee@163.com (P.L.); leizhangtian@163.com (L.Z.); yangbinou@126.com (B.O.)
* Correspondence: kouqing20@163.com

Received: 12 March 2019; Accepted: 30 March 2019; Published: 4 April 2019

Abstract: In the model reference adaptive speed observer, the induction motor supply voltage is used as the input of the reference model. However, measuring the supply voltage complicates the system and increases the cost, so the command voltage calculated by the controller is generally used instead of the actual supply voltage in the drive system. However, due to the nonlinear effects of the inverter, the voltage calculated by the controller is different from the actual supply voltage, resulting in a speed observation deviation. This paper analyzes the multiple effects that cause the three-level neutral-point clamped (TL-NPC) inverter output voltage and command voltage deviation. A voltage deviation compensation measure based on the volt-second balance principle is proposed. In this context, the expression of the rotational speed deviation caused by the voltage deviation is derived rigorously and in detail. Finally, the effectiveness of the voltage compensation measure is verified by experiments. The experimental results are basically consistent with the theoretical derivation expressions. The method and analysis in this paper is applicable to induction motor speed sensorless control systems driven by two-level and other multilevel inverters.

Keywords: nonlinear effects; three-level neutral-point clamped inverter; induction motor; speed observation; compensation

---

1. Introduction

In recent years, speed sensorless vector control of induction motors driven by three-level neutral-point clamped (TL-NPC) inverters have become a research hotspot in the field of high-power motor drive [1–3]. Induction motors are mechanically robust, low cost, simple to manufacture, highly reliable, and are suitable for high-voltage and high-power drive applications. At present, the high-voltage high-power motor drive device often adopts a three-level neutral-point clamped structure, which has the advantages of small loss, high efficiency, and small harmonics compared with a conventional two-level inverter [4,5]. Among all the control strategies of induction motors, vector control or field-oriented control is the most popular. It can realize the independent control of torque and flux linkage to achieve fast torque response [6]. The magnetic field orientation control can be realized by measuring the magnitude and direction of magnetic flux directly by the magnetic flux sensor or Hall effect sensor in the machine (direct vector control). The magnetic field orientation (indirect vector) vector control can also be applied indirectly by the slip frequency component in the rotor dynamics. The latter is more feasible because it does not require additional flux sensors, which will take up additional space and cost. By decoupling the excitation component and the torque component of the stator current in the synchronous rotating reference frame, the indirect field-oriented control strategy realizes the independent control of the torque and flux, thus making the control of the induction motor simple. However, in order to achieve control in a synchronous coordinate system, it is necessary to use a speed encoder to measure the speed of the rotor. The use of speed encoders means additional electronic equipment, cost, and installation space. Therefore, the speed estimation technique is used to eliminate the shaft speed encoder [7].

Rotor speed observation technology based on an adaptive full-order observer has become a hot research question [8]. The observer has high precision for rotor speed, good robustness to motor parameters, and better performance than traditional methods. The speed sensorless vector control system composed of the adaptive flux observer scheme has the advantages of fast recognition speed and good dynamic performance [9,10]. The supply voltage of the motor is the input of the full-order observer. In order to simplify the system, the command phase voltage calculated by the controller is often used instead of the actual phase voltage. However, due to the nonlinear effects of the TL-NPC inverter, there is an error between the actual phase voltage and the command phase voltage, which affects the observation of the speed.

A dead-time compensation method based on the pulse width modulation (PWM) was proposed, which was used in a three-level inverter-fed induction motor drive system and is only applicable to a self-balancing space vector pulse width-modulated scheme [11]. The dead-time effect in a three-level inverter was analyzed, and the compensations method was proposed, however, the effect of forward voltage drop was not considered [12]. A generic compensation scheme to accommodate the effects of multilevel converters was proposed, which mainly consider the ON-state device voltage drops [13]. A comprehensive distortion compensation method by injecting offset voltage in the modulated signal was proposed to improve the line current waveform quality, which mainly considers the effect of dead-time delay and device voltage drop of the single-phase TL-NPC converters [14].

In order to reduce the influence of inverter nonlinearity on the observation of rotor speed, this paper proposes a method to compensate the error between the command phase voltage and the actual phase voltage. This paper is organized as follows. Section 2 deduces the nonlinearity of the TL-NPC inverter by mainly considering the dead-time delay and forward voltage drop of insulated gate bipolar transistor (IGBT) and diode. Section 3 deduces the influence of the voltage error caused by the nonlinearity of the TL-NPC inverter on the observation of the rotational speed and obtains the rotational speed error transfer function. Section 4 proposes a compensation scheme for the nonlinear effects of the TL-NPC inverter. Section 5 verifies the compensation scheme by experimental tests. Section 6 concludes this paper.

2. Nonlinear Effects of Three-Level Neutral-Point Clamped Inverter

In this paper, we mainly analyzed the influence of the nonlinear effect of the three-level inverter on its output voltage. We analyzed the two major types of nonlinear effect of the TL-NPC inverter, namely, the dead-time delay and device forward voltage drop. The dead-time delay includes dead time, turn-on delay, and turn-off delay. These two major types of nonlinear effect have a great influence on the output of the inverter. For this reason, we only analyzed these two major types of nonlinear effect of the TL-NPC inverter.

2.1. Dead-Time Delays

The drive system of induction motor with TL-NPC inverter is shown in Figure 1. The hardware of the three phases are generally identical in a motor drive system. The output voltage of the inverter is a phase voltage referenced to the neutral point voltage, and the three phases are generally independently controlled. The phase voltage used for the speed observation is the voltage output from the inverter, which is also referenced to the neutral point voltage, so the three phases voltages are independent. This paper only introduces the method of the voltage compensation. This method assumes that the three phases are identical, so only one of the phases is analyzed below, but the specific compensation voltages of the three phases differ according to the voltage and current of each phase.

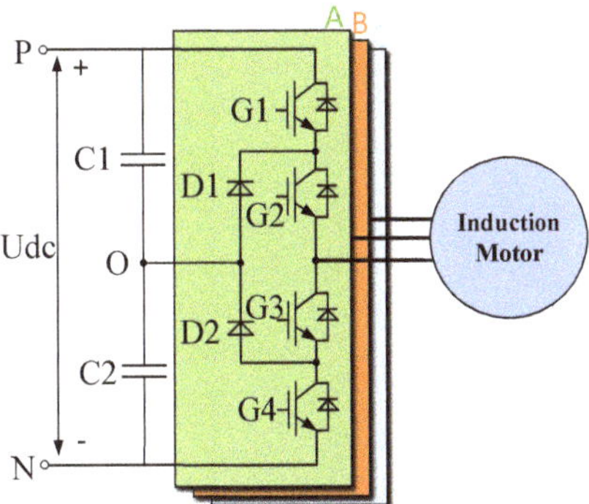

Figure 1. The drive system of induction motor with the three-level neutral-point clamped (TL-NPC) inverter.

The switching signal and the ideal supply voltage of the TL-NPC inverter in different switching states are shown in Figure 2.

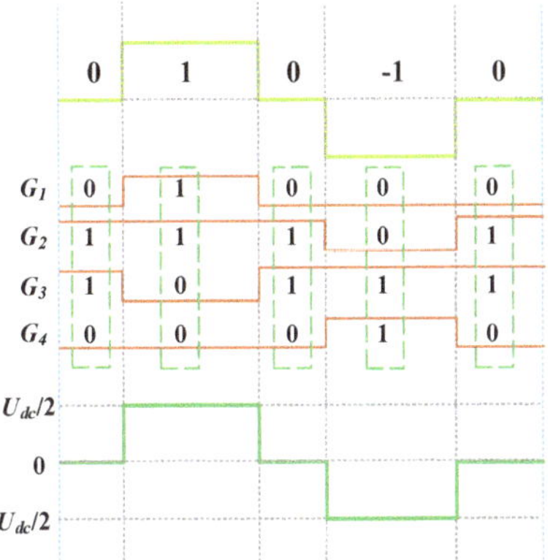

Figure 2. The switching signal and ideal supply voltage of the TL-NPC inverter.

Due to it taking a certain amount of time for the IGBT to turn off, in order to prevent a short circuit in the TL-NPC during the switching process, it is necessary to set a dead-time (T_d) delay between G_1/G_3 and G_2/G_4. The gate signal G_1', G_2', G_3', G_4' with dead-time delay was set as shown in Figure 3. In addition, after the gate drive signal was applied to the IGBT, it takes a certain amount of time to turn on and off, and the required time is T_{on} and T_{off}, respectively. T_{on} and T_{off} are not only related to the characteristics of the IGBT, but also the stray parameters of the circuit, and for this reason, the test method is generally used to obtain T_{on} and T_{off}. Due to the effects of T_d, T_{on}, and T_{off}, the actual supply voltage and the desired voltage will deviate. The T_d, T_{on}, and T_{off} are primarily determined by device characteristics and can be obtained from the device datasheet or measured by test methods.

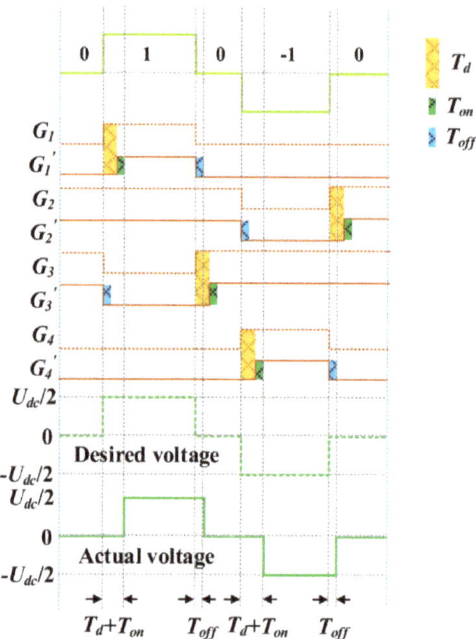

Figure 3. The gate signal with the dead-time delay and actual voltage of the TL-NPC.

As can be seen from Figure 3, the voltage deviation due to dead-time delay, turn-on time, and turn-off time during each switching cycle is

$$u_d = sign(u^*)\frac{(T_d + T_{on} - T_{off})}{T_s}\frac{U_{dc}}{2},\tag{1}$$

where u^* is the desired voltage and T_s is the switching period.

2.2. Forward Voltage Drops

In a general control system, the influence of forward voltage drop is negligible. However, in order to obtain a precise supply voltage and achieve high-performance rotor speed observation in speed sensorless control systems, it cannot be ignored.

The current path in different switching modes is shown in Figure 4, where the current flows through two devices in different paths. The IGBTs and the diodes generally have the same voltage- and current-level devices in a TL-NPC inverter, and their difference in forward voltage drop is small. For example, the forward voltage drop of Infineon's IGBT (Neubiberg, Germany) (FZ3600R17KE3_B2) and diode (DZ3600S17K3_B2) are 2.0 and 1.8 V respectively with the difference being about 10%, and they are often used together in TL-NPC inverters. For the IGBT module (Infineon's F3L300R12ME4_B23) used in the experimental platform of this paper, which contains the clamped diodes, the forward drop of the IGBT and the diode are 1.75 and 1.65 V respectively, with a difference of about 9%. Hence, if we ignore the difference in voltage drop between the IGBT, antiparallel freewheeling diode, and clamp diode, the total forward voltage drop can be expressed as

$$u_f = 2u_{ce},\tag{2}$$

where u_{ce} is the average forward voltage drop of a single device and can be obtained from the datasheet.

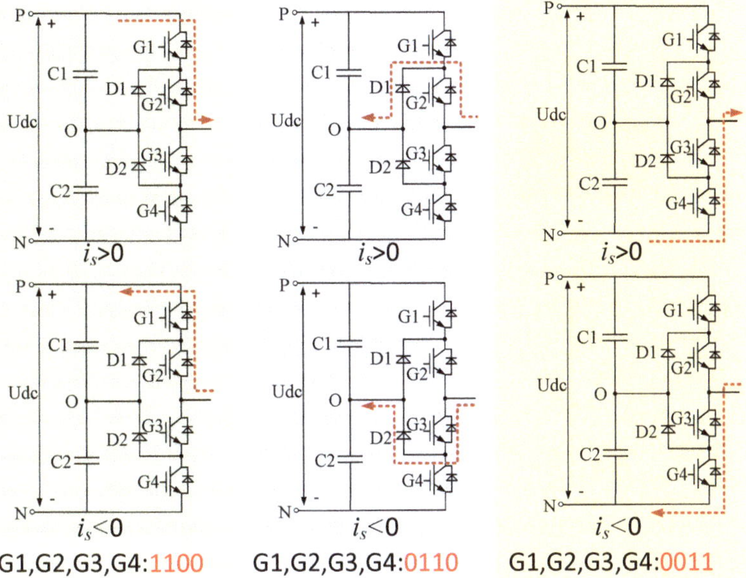

Figure 4. The current path in the different switching modes of the TL-NPC.

The forward voltage drop in different current directions is shown in Figure 5. Hence, the voltage deviation due to the forward voltage drop can be expressed as

$$u_{forward} = sign(i_s)u_f. \tag{3}$$

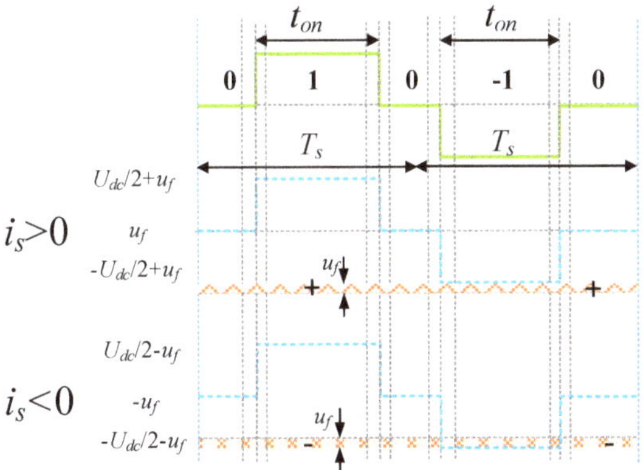

Figure 5. The forward voltage drops in different current directions.

3. Analysis of the Influence of Voltage Error on Rotor Speed Observation

Adaptive control is mainly used for parameter adaptation. The essence of an adaptive control mechanism is to adapt to the controlled system with parameters that need to be estimated. The configuration of model reference adaptive system (MRAS) is illustrated in Figure 6. There is a reference motor model and the adaptive model as a function of the parameter to be estimated.

The adaptive mechanism is used to ensure that the state of the observer converges to the state of the motor.

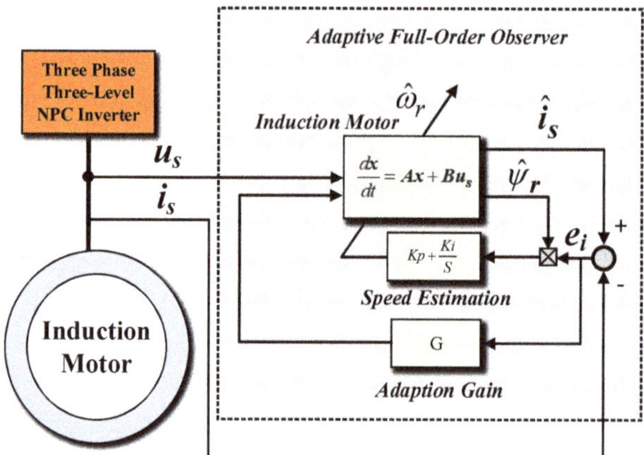

Figure 6. The general configuration of the model reference adaptive systems.

The equation of state model of a three-phase induction motor in a two-phase synchronous coordinate system can be obtained by the voltage, flux linkage, and inductance equation of the induction motor. If the induction motor stator current and rotor flux are selected as state variables, the induction motor model is as shown in Equation (4)

$$\frac{dx}{dt} = Ax + Bu_s, \tag{4}$$

where $x = [i_{sd}, i_{sq}, \psi_{rd}, \psi_{rq}]^T$, $u_s = [u_{sd}, u_{sq}]^T$, $A = \begin{bmatrix} a_{11} & \omega_e & a_{13} & a_{14}\omega_r \\ -\omega_e & a_{11} & -a_{14}\omega_r & a_{13} \\ a_{31} & 0 & a_{33} & \omega_e - \omega_r \\ 0 & a_{31} & -\omega_e + \omega_r & a_{33} \end{bmatrix}$, $B = \begin{bmatrix} b & 0 \\ 0 & b \\ 0 & 0 \\ 0 & 0 \end{bmatrix}$, $C = \begin{bmatrix} 1 & 0 & 0 & 0 \\ 0 & 1 & 0 & 0 \end{bmatrix}$, $a_{11} = -\frac{1-\sigma}{\sigma T_r} - \frac{R_s}{\sigma L_s}$, $a_{13} = \frac{L_m}{\sigma L_s L_r} \cdot \frac{1}{T_r}$, $a_{14} = \frac{L_m}{\sigma L_s L_r}$, $a_{31} = \frac{L_m}{T_r}$, $a_{33} = -\frac{1}{T_r}$,

$b = \frac{1}{\sigma L_s}$, $\sigma = 1 - \frac{L_m^2}{L_s L_r}$ and $T_r = \frac{L_r}{R_r}$. R_s, R_r, L_s, L_r, and L_m are stator resistance, rotor resistance, stator inductance, rotor inductance, and magnetizing inductance, respectively.

From the state Equation (4), we can build an adaptive observer as shown in Equation (5).

$$\frac{d\hat{x}}{dt} = \hat{A}\hat{x} + B\overline{u}_s, \tag{5}$$

where $\hat{A} = \begin{bmatrix} a_{11} & \omega_e & a_{13} & a_{14}\hat{\omega}_r \\ -\omega_e & a_{11} & -a_{14}\hat{\omega}_r & a_{13} \\ a_{31} & 0 & a_{33} & \omega_e - \hat{\omega}_r \\ 0 & a_{31} & -\omega_e + \hat{\omega}_r & a_{33} \end{bmatrix}$, $\hat{x} = [\hat{i}_{sd}, \hat{i}_{sq}, \hat{\psi}_{rd}, \hat{\psi}_{rq}]^T$, $\overline{u}_s = u_s + \Delta u_s$,

$\Delta u_s = [\Delta u_{sd}, \Delta u_{sq}]$ is the error between the command voltage and the actual voltage. The observed speed can be expressed as

$$\hat{\omega}_r = (k_{p\omega} + \frac{k_{i\omega}}{s})\hat{\psi}_{rd}(i_{sq} - \hat{i}_{sq}). \tag{6}$$

The error equation derived from the induction motor model Equation (4), together with the adaptive full-order observer Equation (5), can be expressed by the following equations:

$$\begin{bmatrix} sIe_i \\ sIe_\psi \end{bmatrix} = \begin{bmatrix} a_{11}I - \omega_e J & a_{13}I - a_{14}\omega_r J \\ a_{31}I & a_{33}I + (\omega_r - \omega_e)J \end{bmatrix} \begin{bmatrix} e_i \\ e_\psi \end{bmatrix} + \begin{bmatrix} 0 & a_{14}\Delta\omega_r J \\ 0 & -\Delta\omega_r J \end{bmatrix} \begin{bmatrix} \hat{i}_s \\ \hat{\psi}_r \end{bmatrix} + \begin{bmatrix} bI \\ 0 \end{bmatrix} \Delta u_s, \quad (7)$$

where $I = \begin{bmatrix} 1 & 0 \\ 0 & 1 \end{bmatrix}, J = \begin{bmatrix} 0 & -1 \\ 1 & 0 \end{bmatrix}, e_i = [i_{sd} - \hat{i}_{sd}, i_{sq} - \hat{i}_{sq}], e_\psi = [\psi_{rd} - \hat{\psi}_{rd}, \psi_{rq} - \hat{\psi}_{rq}], \Delta\omega_r = \omega_r - \hat{\omega}_r.$

According to Equation (7), we can get the current error equation as shown in Equations (8)–(10).

$$\begin{cases} C_d e_i &= C_\omega J\hat{\psi}_r \Delta\omega_r + C_{\Delta u}\Delta u_s \\ C_d &= [s^2 - (a_{33} + a_{11})s + a_{11}a_{33} - a_{31}a_{13} - \omega_e(\omega_r - \omega_e)]I \\ &\quad + [(2\omega_e - \omega_r)s + a_{11}(\omega_r - \omega_e) - \omega_e a_{33} + a_{31}a_{14}\omega_r]J \\ &= C_{d1}I + C_{d2}J \\ C_\omega &= a_{14}sI + a_{14}\omega_e J = C_{\omega 1}I + C_{\omega 2}J \\ C_{\Delta u} &= (bs - a_{33}b)I - (\omega_r - \omega_e)bJ = C_{\Delta u1}I + C_{\Delta u2}J \end{cases} \quad (8)$$

$$e_i = G_\omega(s)J\hat{\psi}_r\Delta\omega_r + G_{\Delta u}(s)\Delta u_s, \quad (9)$$

$$\begin{cases} G_\omega(s) = C_d^{-1}C_\omega = \dfrac{(C_{d1}C_{\omega 1}+C_{d2}C_{\omega 2})I+(C_{d1}C_{\omega 2}-C_{d2}C_{\omega 1})J}{C_{d1}^2+C_{d2}^2} = \begin{bmatrix} G_{\omega 1}(s) & -G_{\omega 2}(s) \\ G_{\omega 2}(s) & G_{\omega 1}(s) \end{bmatrix} \\ G_{\Delta u}(s) = C_d^{-1}C_{\Delta u} = \dfrac{(C_{d1}C_{\Delta u1}+C_{d2}C_{\Delta u2})I+(C_{d1}C_{\Delta u2}-C_{d2}C_{\Delta u1})J}{C_{d1}^2+C_{d2}^2} = \begin{bmatrix} G_{\Delta u1}(s) & -G_{\Delta u2}(s) \\ G_{\Delta u2}(s) & G_{\Delta u1}(s) \end{bmatrix} \end{cases} \quad (10)$$

The observed rotor speed can be expressed as Equation (11). It can be represented by a block diagram as Figure 7.

$$\hat{\omega}_r = (k_{p\omega} + \tfrac{k_{i\omega}}{s})e_i^T J\hat{\psi}_r = (k_{p\omega} + \tfrac{k_{i\omega}}{s})[G_\omega(s)J\hat{\psi}_r\Delta\omega_r + G_{\Delta u}(s)\Delta u_s]J\hat{\psi}_r$$
$$= (k_{p\omega} + \tfrac{k_{i\omega}}{s})G_{\omega 1}(s)\Delta\omega_r|\hat{\psi}_{rd}|^2 + (k_{p\omega} + \tfrac{k_{i\omega}}{s})[G_{\Delta u2}(s)\Delta u_{sd} + G_{\Delta u1}(s)\Delta u_{sq}]|\hat{\psi}_{rd}| \quad (11)$$

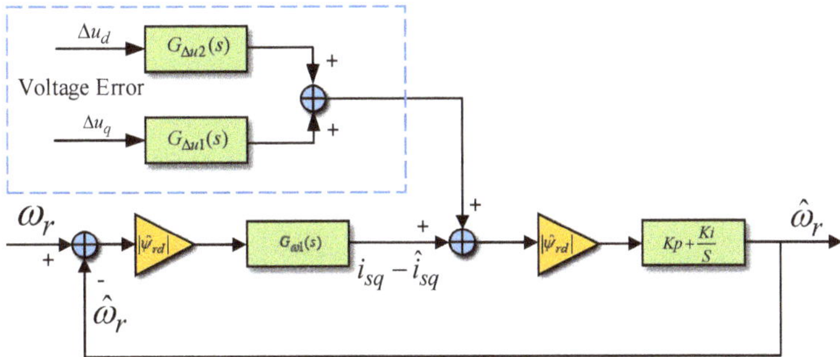

Figure 7. The block diagram of rotor speed observation with input voltage error.

We assume that $G_{\omega 1} = \dfrac{G_{\omega 1n}(s)}{G_{\omega 1d}(s)}, G_{\Delta u1} = \dfrac{G_{\Delta u1n}(s)}{G_{\omega 1d}(s)}, G_{\Delta u2} = \dfrac{G_{\Delta u2n}(s)}{G_{\omega 1d}(s)}.$

According to Equation (8), we can get $G_{\omega 1n}(s) = C_{d1}C_{\omega 1} + C_{d2}C_{\omega 2}, G_{\Delta u1n}(s) = C_{d1}C_{\Delta u1} + C_{d2}C_{\Delta u2}, G_{\Delta u2n}(s) = C_{d1}C_{\Delta u2} - C_{d2}C_{\Delta u1}.$

Using Mason's gain formula to process Equation (11), we can get Equation (12). Hence, Figure 7 can be further represented as Figure 8.

$$\begin{cases} \hat{\omega}_r = H_\omega(s)\omega_r + H_{\Delta ud}(s)\Delta u_{sd} + H_{\Delta uq}(s)\Delta u_{sq} \\ H_\omega(s) = \dfrac{|\hat{\psi}_{rd}|G_{\omega 1n}(s)(sk_{p\omega}+k_{i\omega})}{sG_{\omega 1d}(s)+|\hat{\psi}_{rd}|^2 G_{\omega 1n}(s)(sk_{p\omega}+k_{i\omega})} \\ H_{\Delta ud}(s) = \dfrac{|\hat{\psi}_{rd}|G_{\Delta u2n}(s)(sk_{p\omega}+k_{i\omega})}{sG_{\omega 1d}(s)+|\hat{\psi}_{rd}|^2 G_{\omega 1n}(s)(sk_{p\omega}+k_{i\omega})} \\ H_{\Delta uq}(s) = \dfrac{|\hat{\psi}_{rd}|G_{\Delta u1n}(s)(sk_{p\omega}+k_{i\omega})}{sG_{\omega 1d}(s)+|\hat{\psi}_{rd}|^2 G_{\omega 1n}(s)(sk_{p\omega}+k_{i\omega})} \end{cases} \tag{12}$$

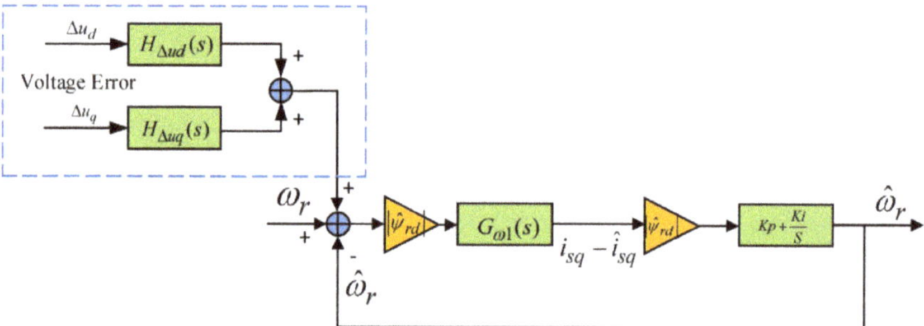

Figure 8. The general configuration of the model reference adaptive systems.

Figure 9 shows the amplitude of the transfer function under steady state conditions. From the figure, we can see that the lower the rotor speed, the more sensitive the observed speed is to the voltage error. The parameters of the induction motor are shown in Table 1.

Table 1. Parameters of the induction motor.

Rs	Stator resistance	2.33 Ω
Rr	Rotor resistance	2.12 Ω
Ls	Stator inductance	0.2994 H
Lr	Rotor inductance	0.3007 H
Lm	Magnetizing inductance	0.2866 H
P	Pole pairs	2

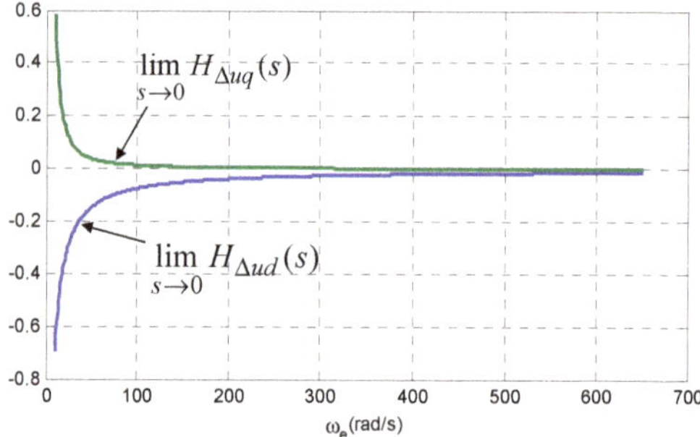

Figure 9. The amplitude of the transfer function of voltage error and observed speed.

4. Nonlinear Effects Compensation Scheme

In order to improve the performance of the speed observation, it was necessary to make the input voltage of the observer coincide with the supply voltage of the induction motor. In this paper, the output voltage of the inverter was compensated by the principle of volt-second balance. The speed observation uses the voltage command before compensation as the input voltage, thus achieving the goal that the input voltage of the observer is consistent with the supply voltage of the induction motor.

A comprehensive compensation scheme was used to consider multiple factors of dead time and forward voltage drops. Figure 10 shows the comprehensive voltage compensation block diagram of one phase of a TL-NPC inverter. Figure 11 shows the control flowchart of the whole procedure of the controller using proposed method.

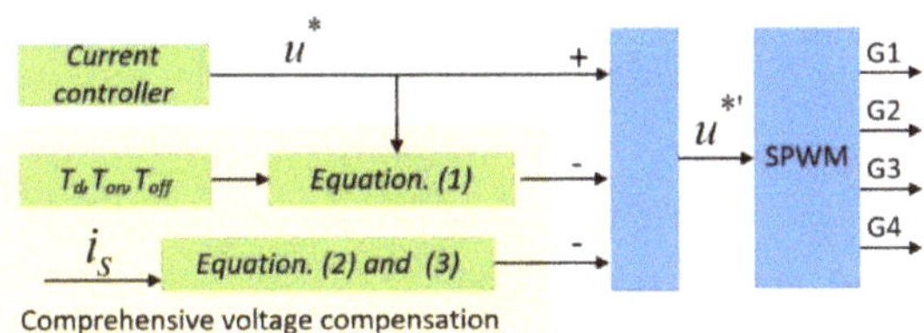

Figure 10. The comprehensive voltage compensation block diagram of one phase of a TL-NPC inverter.

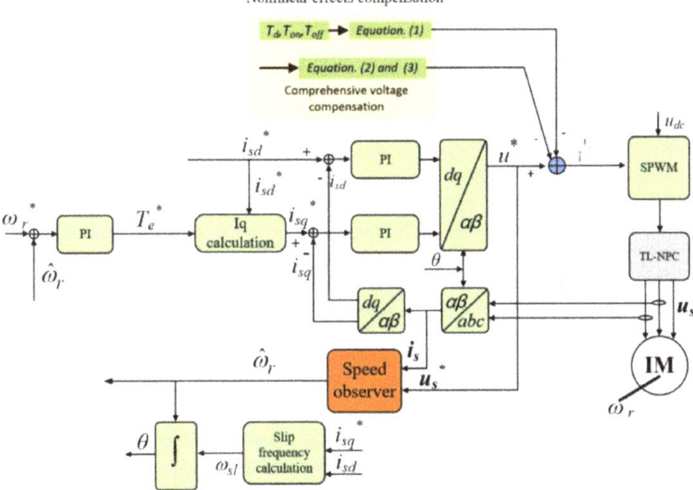

Figure 11. The control flowchart showing the whole procedure of the controller with the proposed method.

5. Experimental Results

In order to verify the previous analysis, an experimental platform was set up. A TL-NPC inverter fed from a constant DC voltage supply drives an induction motor. The TL-NPC inverter was designed using Infineon FF300R12ME4_B11 IGBT modules. The current controller was implemented on Texas Instruments TMS320F28335 and Altera Cyclone III EP3C25F324 digital signal controllers. The experimental platform control block diagram is shown in Figure 12. In order to reduce the influence of speed control on the actual speed of the motor, firstly, the speed closed-loop control strategy was used to accelerate the motor to the target speed, and then a constant torque control strategy was adopted, and the constant torque was 0 N·m by setting the i_{sq}^* to 0 A. A high-precision speed encoder was also arranged on the motor to measure the actual speed of the motor. The parameters of nonlinear effects of TL-NPC inverter of the experimental platform are shown in Table 2.

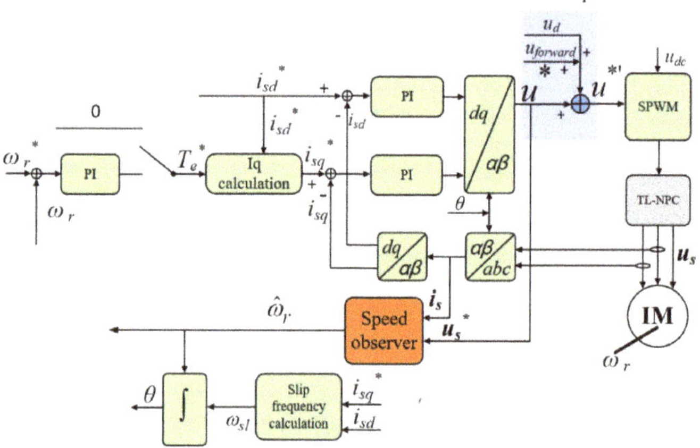

Figure 12. The amplitude of the transfer function of voltage error and observed speed.

Table 2. The parameters of the nonlinear effects of the experimental platform.

Ts	Switching cycle	250 µs
T_d	Dead time	5 µs
T_{on}	Turn on delay time	2 µs
T_{off}	Turn off delay time	2.5 µs
u_{ce}	IGBT and diode forward voltage	1.75 V

The experimental results of actual speed and observer speed at different speeds are shown in Figure 13 (7 rad/s), Figure 14 (70 rad/s), Figure 15 (140 rad/s), Figure 16 (280 rad/s), Figure 17 (420 rad/s), and Figure 18 (623 rad/s). From these figures, we can see that the lower the speed, the greater the fluctuation in the observed speed when no voltage error compensation is performed. The lower the speed, the more obvious the voltage error compensation is when compared to the accurate speed observation. The speed errors of the observer speed without and with the proposed compensation are shown in Figure 19. The experimental results were consistent with the previous analysis.

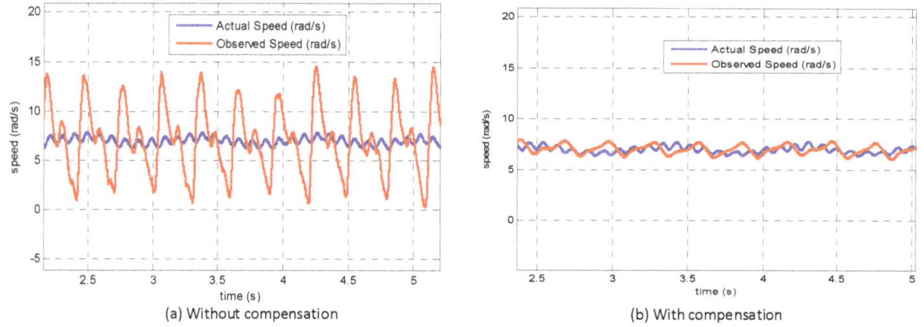

Figure 13. The experimental results of actual speed and observer speed at 7 rad/s.

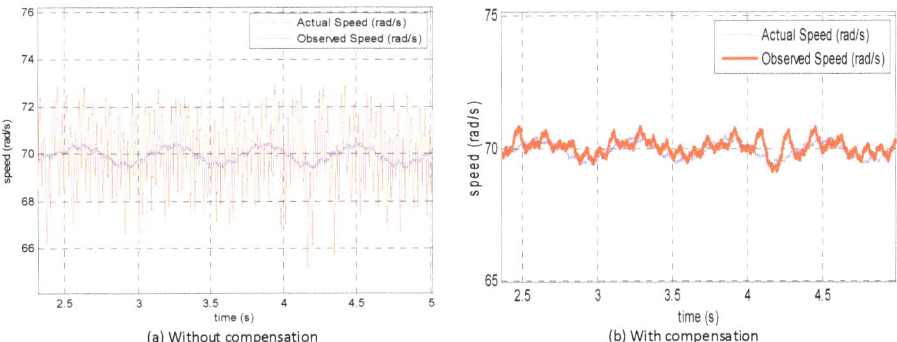

(a) Without compensation (b) With compensation

Figure 14. The experimental results of actual speed and observer speed at 70 rad/s.

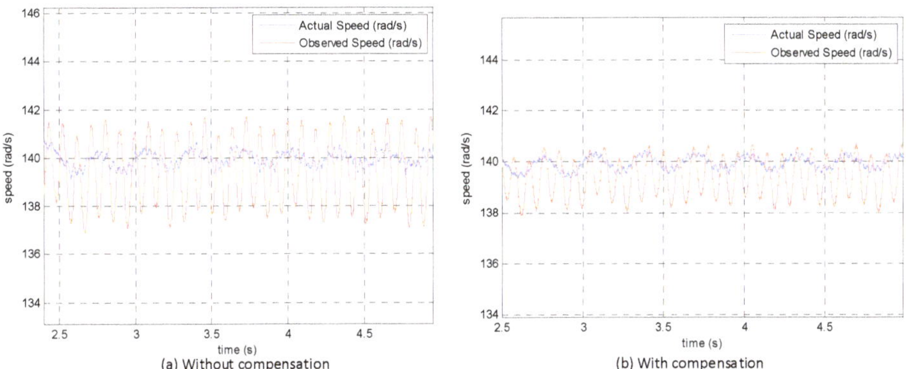

(a) Without compensation (b) With compensation

Figure 15. The experimental results of actual speed and observer speed at 140 rad/s.

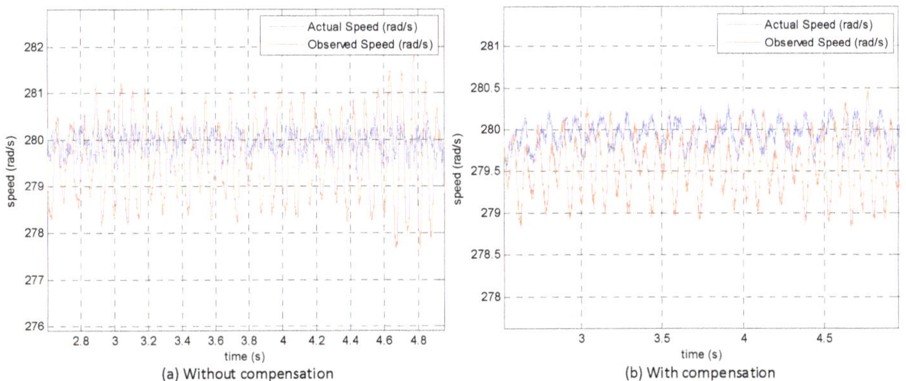

(a) Without compensation (b) With compensation

Figure 16. The experimental results of actual speed and observer speed at 280 rad/s.

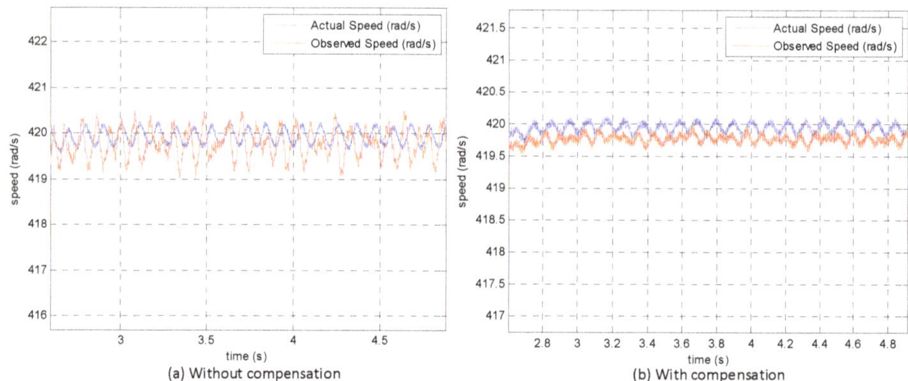

Figure 17. The experimental results of actual speed and observer speed at 420 rad/s.

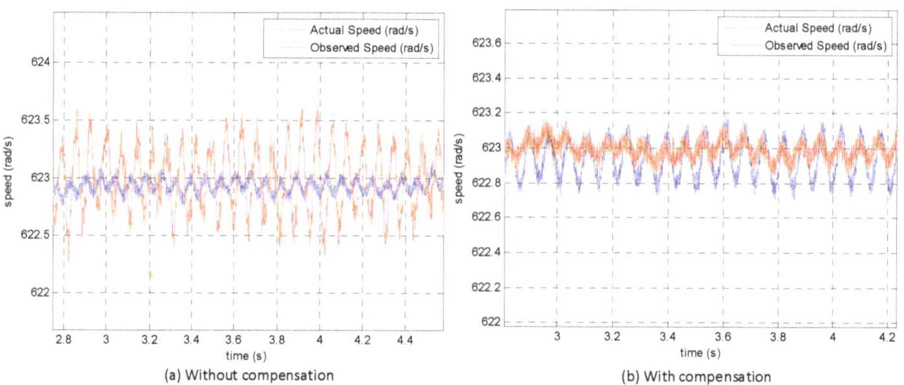

Figure 18. The experimental results of actual speed and observer speed at 623 rad/s.

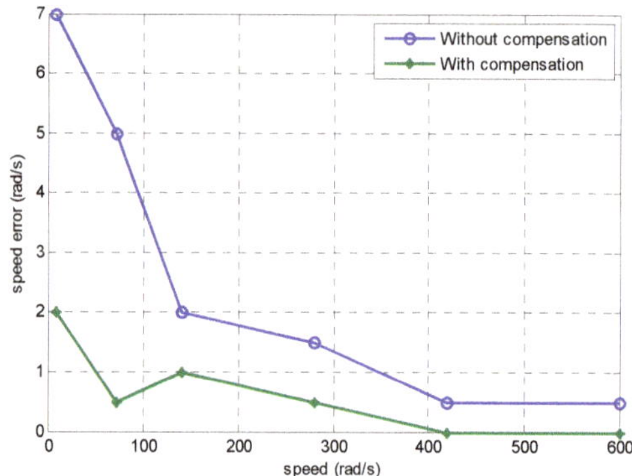

Figure 19. The speed errors of observer speed without and with the proposed compensation.

6. Conclusions

In the model reference adaptive speed observer, the induction motor supply voltage is used as the input of the reference model. The command voltage calculated by the controller is generally used instead of the actual supply voltage in the drive system. However, due to nonlinear effects of the inverter, the voltage calculated by the controller is different from the actual supply voltage, resulting in a speed observation deviation. A comprehensive compensation scheme considering the dead-time delay and device forward voltage drop was proposed to improve the accuracy of speed observation, which was verified in an experimental platform. The main conclusions of this paper are summarized as follows:

- The transfer function expression of the inverter output voltage error and the observed rotational speed error was derived. The transfer function indicated that the inverter output voltage error will result in inaccurate observation speed.
- The transfer function indicated that the lower the speed, the more obvious the influence of nonlinear effects of the TL-NPC inverter on the speed observation.
- The proposed nonlinear effects of the TL-NPC inverter compensation method were effective for speed observation. The compensation method was verified by the experiment.

In conclusion, the nonlinear effects compensation method is also suitable for five-level and other multilevel inverters for induction motor sensorless drive systems.

Author Contributions: Conceptualization, P.L. and L.Z.; Methodology, B.O.; Software, Y.L.; Validation, P.L., L.Z. and B.O.; Formal Analysis, Y.L.; Writing—Original Draft Preparation, P.L.; Writing—Review & Editing, Y.L.; Visualization, L.Z.; Supervision, B.O.

Funding: This research received no external funding.

Conflicts of Interest: The authors declare no conflict of interest.

References

1. Zhang, Y.; Zhu, J. An Improved Direct Torque Control for Three-Level Inverter-Fed Induction Motor Sensorless Drive. *IEEE Trans. Power Electron.* **2012**, *27*, 1502–1513. [CrossRef]
2. Habibullah, M.; Lu, D.; Xiao, D.; Fletcher, J.E. Muhammed Fazlur Rahman. Predictive Torque Control of Induction Motor Sensorless Drive Fed by a 3L-NPC Inverter. *IEEE Trans. Ind. Inform.* **2017**, *13*, 60–70. [CrossRef]
3. Zhang, Y.; Bai, Y.; Yang, H.; Zhang, B. Low Switching Frequency Model Predictive Control of Three-Level Inverter-Fed IM Drives with Speed-Sensorless and Field-Weakening Operations. *IEEE Trans. Ind. Electron.* **2019**, *66*, 4262–4272. [CrossRef]
4. Liu, P.; Duan, S.; Yao, C.; Chen, C. A Double Modulation Wave CBPWM Strategy Providing Neutral-Point Voltage Oscillation Elimination and CMV Reduction for Three-Level NPC Inverters. *IEEE Trans. Ind. Electron.* **2018**, *65*, 16–26. [CrossRef]
5. López, I.; Ceballos, S.; Pou, J.; Zaragoza, J.; Andreu, J.; Ibarra, E.; Konstantinou, G. Generalized PWM-Based Method for Multiphase Neutral-Point-Clamped Converters with Capacitor Voltage Balance Capability. *IEEE Trans. Power Electron.* **2017**, *32*, 4878–4890. [CrossRef]
6. Orfanoudakis, G.I.; Sharkh, S.M.; Yuratich, M.A. Analysis of Dc-Link Capacitor Current in Three-Level Neutral Point Clamped and Cascaded H-Bridge Inverters. *IET Power Electron.* **2013**, *6*, 1376–1389. [CrossRef]
7. Yang, G.; Chin, T.-H. Adaptive-speed identification scheme for a vector-controlled speed sensorless inverter-induction motor drive. *IEEE Trans. Ind. Appl.* **1993**, *29*, 820–825. [CrossRef]
8. Lee, C.-M.; Chen, C.-L. Observer-based speed estimation method for sensorless vector control of induction motors. *Proc. IEE Control Theory Appl.* **1998**, *145*, 359–363. [CrossRef]
9. Yan, Z.; Jin, C.; Utkin, V.I. Sensorless sliding-mode control of induction motors. *IEEE Trans. Ind. Electron.* **2000**, *47*, 1286–1297.
10. Rehman, H.; Derdiyok, A.; Gûven, M.K.; Xu, L. A new current model flux observer for wide speed range sensorless control of an induction machine. *IEEE Trans. Power Electron.* **2002**, *17*, 1041–1048. [CrossRef]
11. Patel, P.J.; Patel, V.P.; Tekwani, P.P.N. Pulse-based dead-time compensation method for self-balancing space vector pulse width-modulated scheme used in a three-level inverter-fed induction motor drive. *IET Power Electron.* **2011**, *4*, 624–631. [CrossRef]
12. Zhou, D.; Rouaud, D.G. Dead-time effect and compensations of three level neutral point clamp inverters for high performance drive applications. *IEEE Trans. Power Electron.* **1997**, *14*, 397–402.

13. Minshull, S.R.; Bingham, C.M.; Stone, D.A.; Foster, M.P. Compensation of Nonlinearities in Diode-Clamped Multilevel Converters. *IEEE Trans. Ind. Electron.* **2010**, *57*, 2651–2658. [CrossRef]

14. Wang, S.; Song, W.; Ma, J.; Zhao, J.; Feng, X. Study on Comprehensive Analysis and Compensation for the Line Current Distortion in Single-Phase Three-Level NPC Converters. *IEEE Trans. Ind. Electron.* **2018**, *65*, 2199–2211. [CrossRef]

 *electronics*

Article

An Improved Model Predictive Torque Control for a Two-Level Inverter Fed Interior Permanent Magnet Synchronous Motor

Guozheng Zhang [1], Chen Chen [1], Xin Gu [1,*], Zhiqiang Wang [2] and Xinmin Li [1]

[1] School of Electrical Engineering and Automation, Tianjin Polytechnic University, Tianjin 300387, China
[2] School of Artificial Intelligence, Tianjin Polytechnic University, Tianjin 300387, China
* Correspondence: guxin@tju.edu.cn; Tel.: +86-1382-113-4212

Received: 21 June 2019; Accepted: 8 July 2019; Published: 10 July 2019

Abstract: In conventional model predictive control, the dimensions of the control variables are different from each other, which makes adjusting the weighted factors in the cost function complicated. This issue can be solved by adopting the model predictive flux control. However, the performance of the electromagnetic torque is affected by the change of the cost function. A novel model predictive torque control of the interior permanent magnet synchronous motor is presented in this paper, and the cost function involving the excitation torque and reluctance torque is established. Combined with the model predictive flux control and discrete space vector modulation, the current ripple and torque ripple are reduced. The performance of torque under an overload condition is superior to model predictive flux control. The effectiveness of the proposed algorithm is verified by the simulation and experimental results.

Keywords: weighted factor; model predictive flux control; interior permanent magnet synchronous; discrete space vector modulation

---

1. Introduction

The interior permanent magnet synchronous motor (IPMSM) is widely used in the fields of industry, transportation, and aerospace, because of advantages such as high-power density, high-torque density, and high efficiency [1,2]. Because of the asymmetry rotor magnetic circuit structure, the reluctance torque can be generated by IPMSM. In the traditional control strategy, the d-axis stator current is equal to zero, such that the q-axis current is proportional to the torque required for the PMSMs. However, with the so-called $i_d = 0$ control algorithm, the reluctance torque of the IPMSM is not fully employed. Therefore, the maximum torque per ampere control is presented for IPMSM in order to increase the output torque and the efficiency of the motor [3,4].

Traditional control methods of IPMSM mainly include space vector control (SVM) and direct torque control (DTC) [5–7]. DTC has advantages of a simple structure, good dynamic performance, and strong robustness, but the torque ripple is high because of adopting the hysteresis controller. An online hysteresis loop adjustment controller is proposed in the literature [8]. It reduces the torque ripple using a proportional-integral (PI) controller to adjust the width of the torque and flux hysteresis loop. Based on duty cycle modulation, other methods combine active vectors with zero vectors to suppress the torque ripple [9,10], but the duty ratio calculation is complex in these methods.

The finite control set model predictive control (FCS-MPC) is adopted to the motor drive system with the development of a digital signal processor (DSP). This method can solve non-linear problems easily [11,12]. The model predictive torque control (MPTC) uses a mathematical model and cost function to replace the torque and flux hysteresis controller and look-up table, compared with the traditional DTC. The optimal switching state is selected from all of the possible switching states in each control period of MPTC. The optimal switching state will be applied in the next control period. Thus, the torque ripple can be reduced, and the output performance of the system will be improved.

The cost function plays a key role in the selection of the optimal switching state [13]. In traditional MPTC, the cost function includes a torque component and stator flux component, which have different units. Weighted factors need to be designed for the two components. The weighted factor is usually determined by the trial-and-error method. Thus, knowing how to avoid the adjusting of the weighted factor has attracted the attention of researchers from all over the world. A multi-objective sorting method is used to eliminate the weighted factor of cost function in the literature [14], which sorts the errors generated by the different switching states from small to large. A new cost function, which is free of a weighted factor, can be established by the sorting results. The authors of [15] use the VIKOR sorting method to eliminate the weighted factor. The authors of [16] built a cost function of the vector duty cycle without a weighted factor. The torque and flux errors in a traditional cost function are replaced by the two deadbeat duty ratios of the adopted vectors. This method solves the problem of duty cycle optimization in traditional MPTC. A model predictive flux control (MPFC) is presented in the literature [17]. Based on the online predictive control of the stator flux vector, the weighted factor is removed from the cost function, and the algorithm is simplified.

In this paper, the torque characteristics of the IPMSM under the MTPA control is analyzed, and an improved MPTC method is proposed based on an IPMSM system driven by a three-phase two-level voltage source inverter. Firstly, instead of the torque and flux components, excitation torque and reluctance torque components are adopted to consist the cost function. As the two torque components share the same unit, the weighted factor is eliminated. Furthermore, the torque ripple under an overload condition is reduced by the adoption of MPFC, and the torque ripple caused by the traditional single vector control is also suppressed by the improved discrete space vector modulation method. Finally, the effectiveness of the proposed method is verified in the whole speed range by the simulation and experimental results.

2. The Two-Level Voltage Source Inverters

The topology of a three-phase two-level voltage source inverter (VSI) is shown in Figure 1.

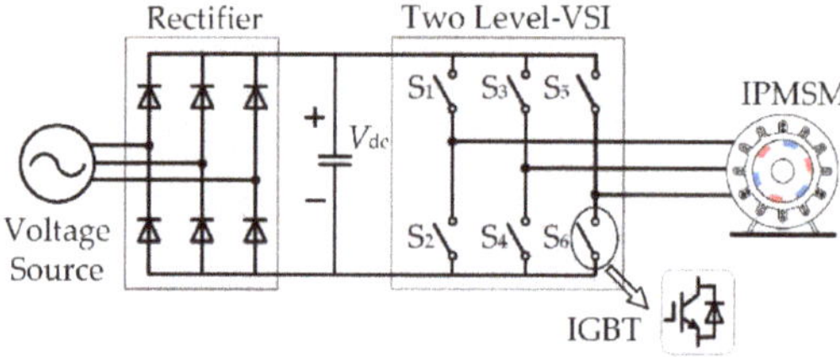

Figure 1. The topology of a two-level voltage source inverter (VSI) fed interior permanent magnet synchronous motor (IPMSM).

There are $2^3 = 8$ basic voltage vectors in the space diagram, corresponding to eight switching states, including six active basic vectors and two zero basic vectors. The control performance will be affected if one single vector is adopted in each control period. Consequently, if the control period is equally divided into three intervals, two adjacent active vectors and one zero vector are applied in each time interval, and more virtual vectors will be synthesized, as shown in Figure 2. For example, V_{100} can be synthesized by V_1 and V_0, which are applied in 1/3 control period and 2/3 control period, respectively, and V_{112} can be synthesized by V_1 and V_2, which are applied in a 2/3 control period and 1/3 control period, respectively. If more virtual vectors are applied, the control performance will be improved.

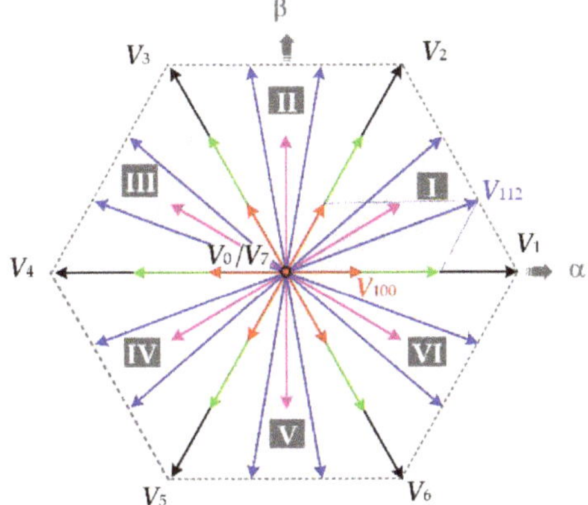

Figure 2. Diagram of a voltage vector synthesizer.

3. Model Predictive Torque Control for IPMSM

3.1. IPMSM Model

The stator voltage equation of the IPMSM in d–q axis rotation coordinate system based on the rotor magnetic field orientation can be expressed as Equation (1).

$$\begin{cases} u_d = R_s i_d + L_d \frac{di_d}{dt} - w_e L_q i_q \\ u_q = R_s i_q + L_q \frac{di_q}{dt} + w_e(\psi_f + L_q i_q) \end{cases}, \tag{1}$$

where, u_d and u_q represent the d-axis and q-axis components of the stator voltage, respectively; i_d and i_q represent the d-axis and q-axis components of the stator current, respectively; L_d and L_q represent the d-axis and q-axis the stator inductance; R_s represents the stator resistance; ω_e is the electrical rotor speed; and ψ_f is the permanent magnet flux linkage.

The expressions of the electromagnetic torque (T_e) and d–q axis stator flux components are as follows:

$$T_e = \frac{3p}{2}[\psi_f i_q + (L_d - L_q)i_d i_q], \tag{2}$$

$$\begin{cases} \psi_d = L_d i_d + \psi_f \\ \psi_q = L_q i_q \end{cases}, \tag{3}$$

where, ψ_d and ψ_q represent the d-axis and q-axis components of the stator flux linkage, respectively, and p is the number of pole pairs. As $L_d \neq L_q$ in IPMSM, the electromagnetic torque can be divided into two components. One is the excitation torque, and the other is the reluctance torque.

For IPMSM, the traditional $i_d = 0$ control algorithm is not suitable, because the reluctance torque is not considered. The maximum torque per ampere (MTPA) control is usually adopted. The d-axis and q-axis current are distributed according to Equation (2), so as to realize the maximization of the electromagnetic torque. As shown in Figure 3, the blue solid lines represent the constant torque loci for different values of T_e as a function of the d-axis and q-axis current components. So, there must be a certain point in each constant torque locus corresponding to the minimum amplitude of the stator current. The red solid line represents the minimum points for different values of T_e, which is usually referred to as the MTPA trajectory.

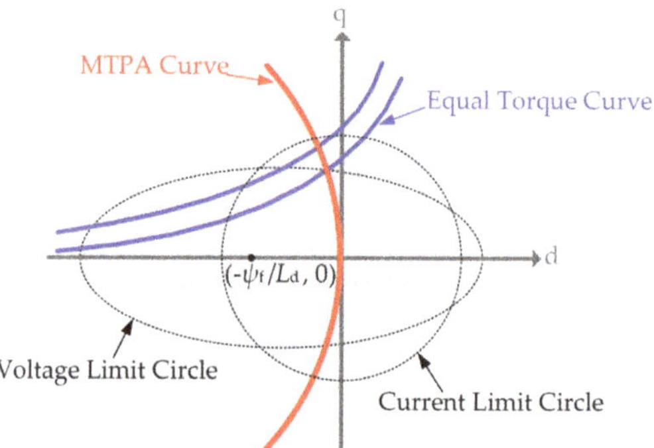

Figure 3. Trace of maximum torque per ampere (MTPA).

3.2. Model Predictive Torque Control

The feedback signals of the stator current, position, and speed of the rotor can be obtained by the current sensor and the encoder. Then, combined with the mathematical model of the motor, the impact of each vector on the stator current can be predicted. The relationship is as follows:

$$\begin{cases} i_{dn}(k+1) = i_d(k) + \frac{1}{L_d}[w_e(k)L_q i_q(k) - R_s i_d(k) + V_{dn}]T_s \\ i_{qn}(k+1) = i_q(k) + \frac{1}{L_q}[-w_e(k)L_d i_d(k) - R_s i_q(k) - w_e(k)\psi_f + V_{qn}]T_s \end{cases} \tag{4}$$

where, $n = 1, 2, 3 \ldots$ is the number of voltage vectors and its related variables; T_s is the control period; and $x(k)$ and $x(k + 1)$ are the values of variable x at the beginning of the kth and $(k + 1)$th control period, respectively. The electromagnetic torque ($T_{en}(k + 1)$) and the stator flux ($\psi_{sn}(k + 1)$) generated by the voltage vector (V_n) can be obtained by Equations (2) to (4).

The reference value and predictive value of the torque and flux are substituted in Equation (5). The optimal vector can be determined according to the value of the cost function.

$$\begin{cases} g(n) = \left| T_e^* - T_{en}(k+1) \right| + Q\left| \psi_s^* - \left| \psi_{sn}(k+1) \right| \right| \\ V_{opt} = V_{argming(n)} \end{cases} \tag{5}$$

where, V_{opt} is the optimal voltage vector, and Q is the weighted factor, which is the absolute value of the ratio of the rated torque to the rated flux, as shown in Equation (6).

$$Q = \left| \frac{T_N}{\psi_N} \right|. \tag{6}$$

However, the weighted factor (Q) obtained from Equation (6) cannot be used directly. It needs to be adjusted according to the operation condition. The optimal vector obtained in the kth control period can only be applied in the $(k + 1)$th control period. Therefore, the cost function can be further improved, as follows:

$$\begin{cases} g(n) = \left| T_e^* - T_{en}(k+2) \right| + Q\left| \psi_s^* - \left| \psi_{sn}(k+2) \right| \right| \\ V_{opt} = V_{argming(n)} \end{cases} \tag{7}$$

The predictive values of the torque and flux in the $(k + 2)$th control period in Equation (7) are obtained based on the predictive value of the $(k + 1)$th period, which could be obtained by Equation (4) and the output voltage vector of the kth period. According to the above principle, the MPTC block diagram is shown in Figure 4.

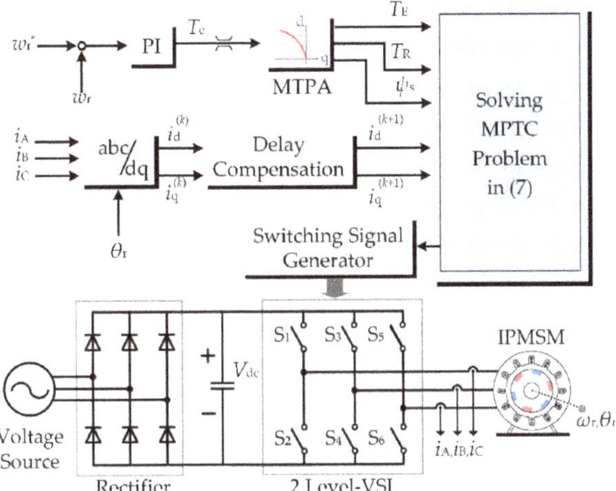

Figure 4. MPTC diagram for IPMSM drives.

3.3. Model Predictive Flux Control

The weighted factor will be eliminated in the model predictive flux control (MPFC) [17] if the control object is changed from the torque to the flux. The algorithm is simplified because only the prediction of the stator flux is needed. The implementation of MPFC is as follows.

The reference value (T_E^*), which is the output of the PI controller in the outer loop, is used as the input of the MTPA algorithm. The outputs of the MTPA algorithm are the d-axis and q-axis components of the reference flux of ψ_d^* and ψ_q^*, respectively. The d-axis and q-axis flux components, $\psi_{dn}(k+2)$ and $\psi_{qn}(k+2)$, respectively, with respect to the different voltage vectors, are deduced by substituting all of the voltage vectors in the finite control set into Equations (3) and (4), and then the cost function can be established as follows:

$$\begin{cases} g(\text{n}) = \left| \psi_s^* - \psi_{sn}(k+2) \right| = \left| \psi_d^* - \psi_{dn}(k+2) \right| + \left| \psi_q^* - \psi_{qn}(k+2) \right| \\ V_{\text{opt}} = V_{\text{argmin } g(n)} \end{cases} \tag{8}$$

The control objective of the MPFC is the stator flux, and the weighted factor is eliminated. Although the flux ripple is reduced, the torque ripple is increased. So, the cost function needs to be redesigned in order to improve the torque control performance.

4. Improved Model Predictive Torque Control

The torque control performances of both MPTC and MPFC are limited, because the cost function contains a flux component. However, the current will be unstable if the flux component is abandoned. So, a modified predictive control algorithm is presented in order to improve the torque control performance.

4.1. Improved Cost Function

For IPMSM, the torque can be regarded as the sum of the excitation torque and reluctance torque, and then Equation (2) can be rewritten as follows:

$$T_e = T_E + T_R, \tag{9}$$

where, $T_E = 3p\psi_f i_q/2$ is the excitation torque generated by the permanent magnet, which is proportional to the q-axis current; $T_R = 3p(L_d\text{-}L_q)i_d i_q/2$ is the reluctance torque generated by the magnetic reluctance, and is proportional to the product of d-axis and q-axis currents.

The d–q axis currents i_d and i_q can be restricted by controlling T_E and T_R, respectively. Accordingly, the modified cost function can be constructed as follows:

$$\begin{cases} g(n) = \left|T_E^* - T_{En}(k+2)\right| + \left|T_R^* - T_{Rn}(k+2)\right| \\ V_{opt} = V_{\mathrm{argmin}g(n)} \end{cases}. \tag{10}$$

It can be seen from Equation (10) that the units of both the excitation torque and reluctance torque are the same, so the weighted factor is unnecessary. Moreover, the torque control performance is directly influenced by the cost function, and the torque control performance can be improved.

For example, a motor with L_d = 0.200 mH, L_q = 0.555 mH, and ψ_f = 0.07574 Wb is analyzed. The data of T_E and T_R are shown in Figure 5. The simulation results of the stator current under a 10% rated load condition are shown in Figure 6.

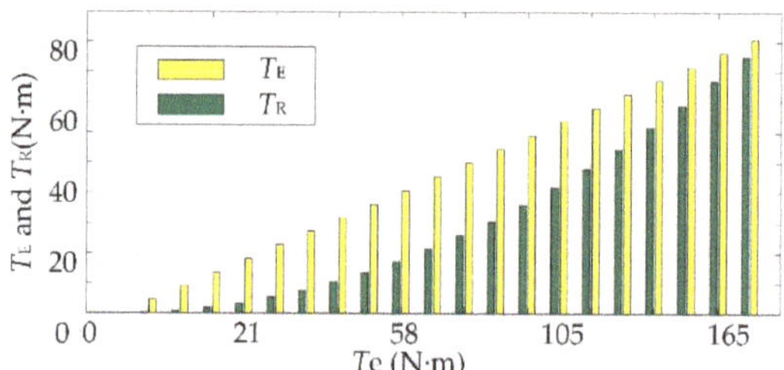

Figure 5. Torque characteristic of IPMSM.

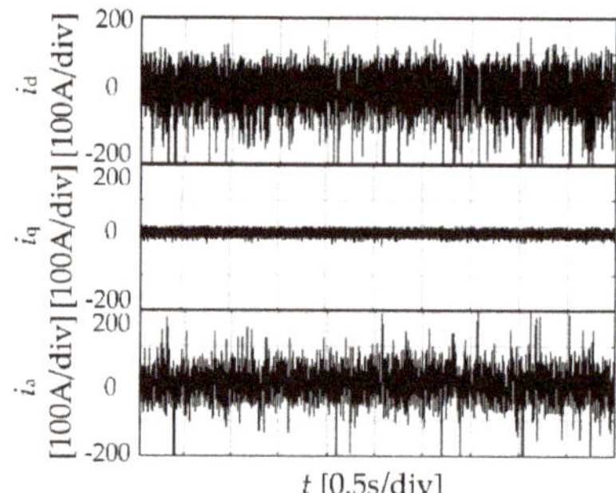

Figure 6. Current waveform under light load condition.

It can be concluded from Figures 3, 5 and 6 that the MTPA curve approximately coincides with the reference line of $i_d = 0$ when the motor is operated in conditions of no load and a light load. The torque is mainly composed of the excitation torque, which results in the fluctuation of the d-axis current. So, the cost function needs to be reconstructed in order to improve the torque performance in conditions of no load and a light load. The new cost function is as follows:

$$\begin{cases} g(n) = \begin{cases} \left|\psi_d^* - \psi_{dn}(k+2)\right| + \left|\psi_q^* - \psi_{qn}(k+2)\right| & \left|T_e^*\right| < T_X \\ \left|T_E^* - T_{En}(k+2)\right| + \left|T_R^* - T_{Rn}(k+2)\right| & \left|T_e^*\right| > T_X \end{cases} \\ V_{opt} = V_{\mathrm{argmin}(n)} \end{cases}, \tag{11}$$

where, T_X is the threshold value of the torque for switching the two cost functions in Equation (11), which is related to the motor's parameters.

The cost function will be switched frequently when T_X is close to the reference torque (T_e^*), so a hysteresis comparator with a reasonable width is adopted for eliminating unnecessary switches.

The proposed method combines the advantages of the traditional MPTC and MPFC. The weighted factor is eliminated, and the torque control performance is improved under a heavy load condition.

4.2. Finite Control Set

The number of virtual voltage vectors is increased to 40 by using the discrete voltage vector synthesis method, as shown in Figure 2. However, with the increase in virtual voltage vectors, the optimization process is more complicated. So, the three virtual voltage vectors nearest to the reference vector are preselected as the new finite control set. Then, the voltage vector that minimizes the value of the cost function can be obtained, and the switching signals can be generated by the PWM modulator, according to the selected voltage vector [18]. For example, when the optimal voltage vector is V_{120}, the switching signals are as shown in Figure 7.

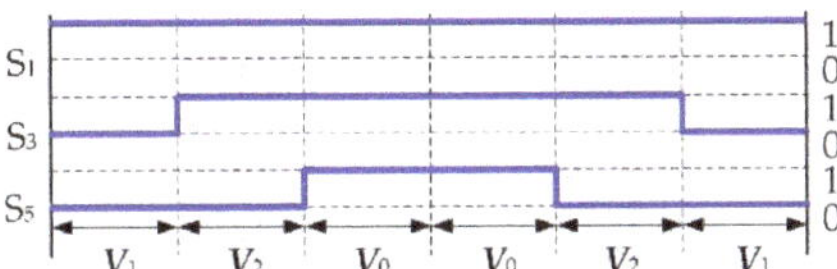

Figure 7. Switching state of S1, S3, and S5.

Compared with the traditional control algorithm, the calculation process is simplified and the performance of the system is improved. The control diagram of the proposed algorithm is illustrated in Figure 8.

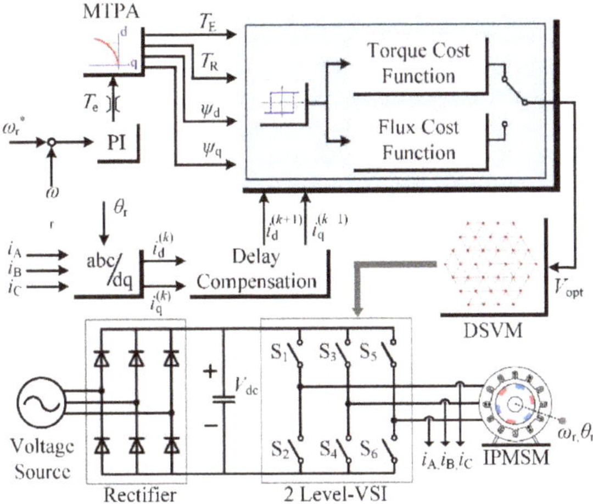

Figure 8. Control diagram of improved MPTC for the IPMSM drives.

It is worth mentioning that the mathematical model and parameters of the motor system are often nonlinear, time-varying, and strongly coupled. The basis of the rigorous optimization method is the parameter precision of the object model. Therefore, the rigorous optimization method often leads to a decrease in the control performance, or even results in control failure.

The model predictive torque control proposed in this paper belongs to the category of model predictive control. Model predictive control generally includes the following three parts: predictive model, rolling optimization, and feedback correction. The predictive model indicates the relationship between the inputs and outputs of the control system. Then, the cost function could be established so as to evaluate the impact of each possible control behavior on a certain control performance in the current control period. The control-behavior minimizing the cost function will be applied to the system in the next control period. Although the control performance may be affected by the accuracy of the predictive model, the rolling optimization could restrain the errors caused by the model mismatch, time-varying, and disturbance, and guarantee a good performance of the control system.

5. Simulation and Experimental Results

5.1. Simulation Analysis

The traditional and the proposed methods are simulated by MATLAB/Simulink, respectively. The sampling frequency of the single vector algorithm is 20 kHz, and the sampling frequency of the discrete voltage vector algorithm is 10 kHz. T_X is 40 N·m. The parameters of the motor are shown in Table 1.

Table 1. Parameters of an interior permanent magnet synchronous motor (IPMSM).

Parameter		Value
Rated voltage (U_{dc})	V	320
Number of pole-pairs (p)	-	4
Stator resistance (R_s)	Ω	0.0114
d-axis inductance (L_d)	mH	0.200
q-axis inductance (L_q)	mH	0.555
Permanent magnet flux linkage (ψ_f)	Wb	0.07574
Rated speed (n_N)	r/min	3000
Rated torque (T_N)	N·m	64
Maximum torque (T_{max})	N·m	180

The standard deviation is adopted for evaluating the control performance of different algorithms, which is defined as follows [15]:

$$
\begin{cases}
\sigma_x = \sqrt{\frac{1}{n-1} \sum_{i=1}^{n} \left(x(i) - \bar{x} \right)^2} \\
\bar{x} = \frac{1}{n} \sum_{i=1}^{n} x(i)
\end{cases}
\tag{12}
$$

The rotator speed (n_r), electromagnetic torque (T_e), stator flux amplitude ($|\psi_s|$), and stator current (i_a) of the traditional and the proposed algorithms are shown in Figure 9. The motor is accelerated from a static state to the rated speed. Figure 9a shows the results of the algorithm of the improved cost function of Equation (11) with a single vector modulation. Figure 9b shows the results of the algorithm of MPFC with discrete voltage vector modulation. Figure 9c shows the results of the proposed algorithm.

From Figure 9a, the speed of the motor is accelerated smoothly from 0 rpm to 3000 rpm (rated speed). Then, the torque reference is stepped up from 0 to 64 N·m (rated load), and the motor reaches the steady state rapidly. This indicates that the system has the ability of anti-disturbance. As can be seen from Figure 9b,c, with the adoption of discrete voltage vector modulation, the stability of the system remains. Moreover, the torque and flux ripples of both the proposed algorithm and MPFC are restrained.

The performances of the proposed algorithm and MPFC under a rated load/different speed conditions, both with discrete voltage vector modulation, are shown in Figures 10 and 11. For these two methods, the standard deviations of the torque are 2.54 and 2.03 N·m at a 10% rated speed, calculated according to Equation (12). While, the standard deviations of the torque are 2.56 and 2.31 N·m at the rated speed, respectively. Thus, the torque ripple of the proposed algorithm is lower than the MPFC under a rated load condition.

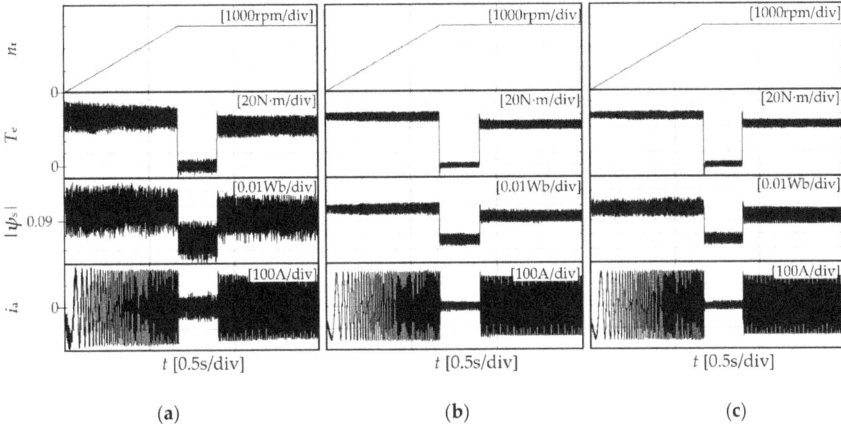

Figure 9. Simulation results: (**a**) improved method with a single vector; (**b**) MPFC with a discrete voltage vector; (**c**) improved method with a discrete voltage vector.

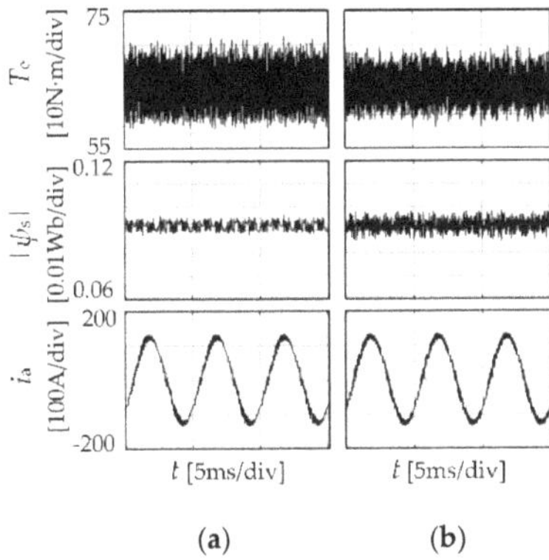

Figure 10. Simulation waveforms under the condition of 10% of the rated speed and rated load: (**a**) MPFC with discrete voltage vector; (**b**) improved method with a discrete voltage vector.

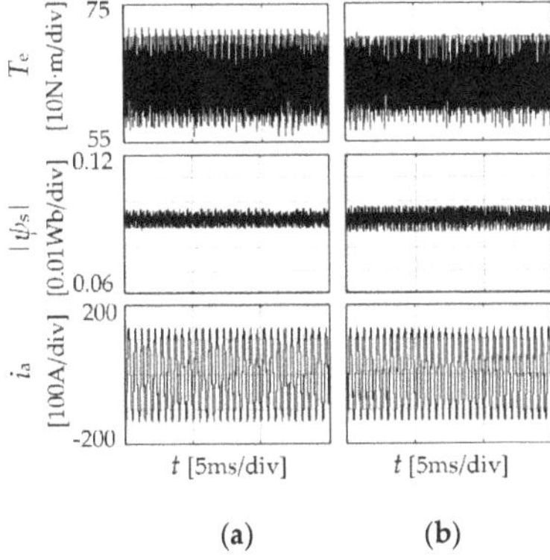

Figure 11. Simulation waveforms under the condition of rated speed and rated load: (**a**) MPFC with discrete voltage vector; (**b**) improved method with a discrete voltage vector.

5.2. Experimental Results

The proposed algorithm is implemented and evaluated on a 20-kW IPMSM driven by a two-level inverter, and the digital control unit is based on a Texas Instruments (Dallas, TX, USA) TMS320F28335 digital signal processor (company, city, country), both illustrated in Figure 12. The parameters of the motor are consistent with the simulation.

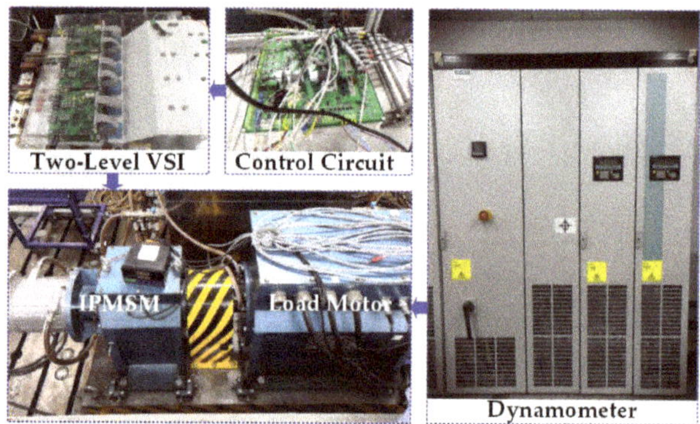

Figure 12. Experimental platform of IPMSM fed by a two-level VSI.

The speed of the motor is accelerated from 0 to 3000 rpm (rated speed) under a no-load condition. The torque and stator flux waveforms of the MPFC and proposed algorithm, both with a discrete voltage vector, are shown in Figure 13.

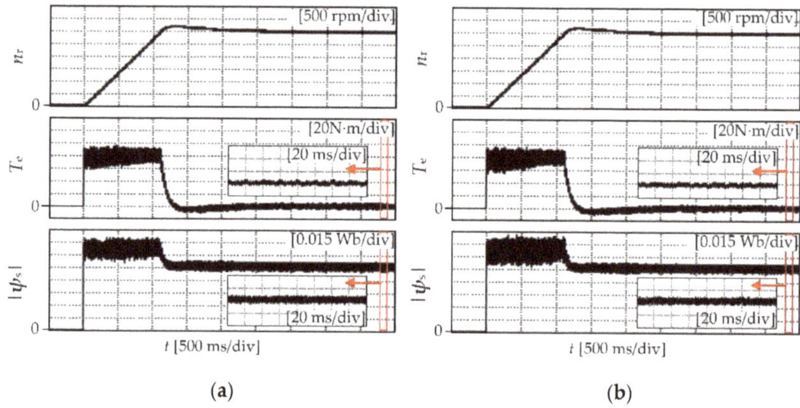

Figure 13. Experimental waveforms when the speed is accelerated from 0 to 3000 rpm: (**a**) MPFC with a discrete voltage vector; (**b**) improved method with a discrete voltage vector.

As can be seen from Figure 13, for both of the two algorithms, the speed of the motor is accelerated smoothly from 0 to 3000 rpm (rated speed). The torque and flux ripples under a no-load condition are almost the same.

Then, the torque reference is stepped from 0 to 64 N·m (rated load). The torque and stator flux waveforms of the MPFC and proposed algorithm, both with a discrete voltage vector, are shown in Figure 14.

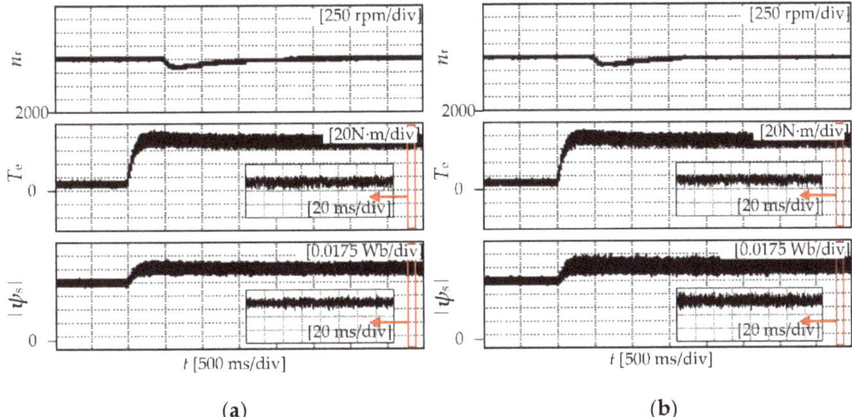

Figure 14. Experimental waveforms when the load is stepped up from 0 to 64 N·m: (**a**) MPFC with a discrete voltage vector; (**b**) improved method with a discrete voltage vector.

As can be seen, the motor reaches the steady state rapidly. It indicates that the system has the ability of anti-disturbance. Moreover, the torque ripple of the proposed algorithm is lower than that of MPFC.

The performance of the proposed method in the whole speed range is also verified. The speed reference is set to 10% of the rated speed, and the torque and stator flux waveforms at a steady state in conditions of no-load and a rated load are shown in Figures 15 and 16, respectively.

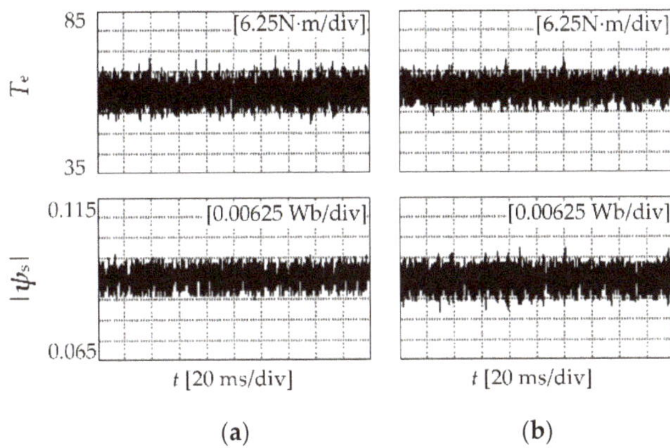

Figure 15. Experimental waveforms under the condition of a 10% rated speed/rated load: (**a**) MPFC with a discrete voltage vector; (**b**) proposed algorithm with a discrete voltage vector.

Under a low speed/rated load condition, the torque ripple of the proposed algorithm is lower than MPFC. While, under a low speed/no-load condition, the torque rippled of the proposed algorithm is the same as that of MPFC. The standard deviations of the above experimental results are calculated according to Equation (12), and shown in Table 2. As can be seen, the torque ripple of the proposed algorithm is lower than MPFC under heavy load conditions.

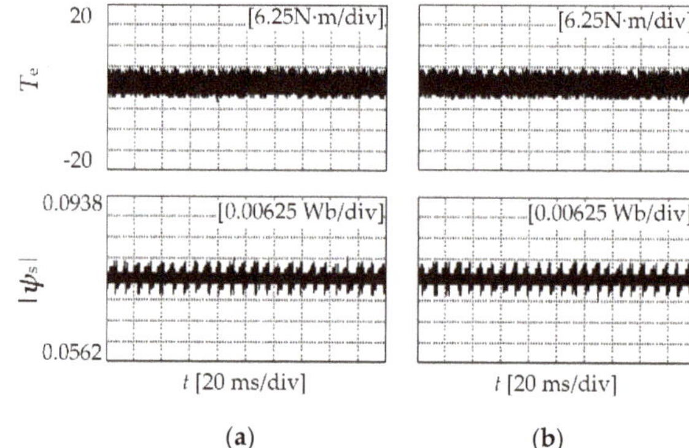

Figure 16. Experimental waveforms under the condition of a 10% rated speed/light load: (**a**) MPFC with a discrete voltage vector; (**b**) proposed algorithm with a discrete voltage vector.

Table 2. Comparison of standard deviation (STDEV) between two methods. MPFC—model predictive flux control.

Speed	STDEV	Load	MPFC	Improved Method
10% Rated speed	σ_T (Nm)	Light	1.508	1.502
		Rated	3.358	3.016
	σ_ψ (Wb)	Light	0.00148	0.00154
		Rated	0.00218	0.00284
Rated speed	σ_T (Nm)	Light	1.604	1.582
		Rated	3.624	3.094
	σ_ψ (Wb)	Light	0.00172	0.00178
		Rated	0.00267	0.00353

6. Conclusion

An improved MPTC algorithm is presented for IPMSM. The cost function is redesigned according to the load conditions. MPFC is adopted under light load conditions, so that there is no weighted factor in the cost function. While, under heavy load conditions, the weighted factor is eliminated by converting the torque and flux components to excitation torque and reluctance torque components, respectively, and the torque ripple is reduced. The simulation and experimental results verify the effectiveness of the proposed method.

Author Contributions: Conceptualization, X.G., X.L., and G.Z.; methodology, G.Z. and C.C.; software, Z.W.; validation, G.Z. and X.L.; formal analysis, G.Z. and C.C.; writing (original draft preparation), C.C.; writing (review and editing), X.G. and Z.W., and; funding acquisition, G.Z. and X.L.

Funding: This research was funded by the Youth Fund Projects of the National Natural Science Foundation of China, grant numbers 51807140 and 51807141.

Conflicts of Interest: The authors declare no conflict of interest.

References

1. Yuan, T.; Wang, D. Performance Improvement for PMSM DTC System through Composite Active Vectors Modulation. *Electronics* **2018**, *10*, 263. [CrossRef]
2. Kamel, T.; Abdelkader, D.; Said, B.; Padmanaban, S.; Iqbal, A. Extended Kalman Filter Based Sliding Mode Control of Parallel-Connected Two Five-Phase PMSM Drive System. *Electronics* **2018**, *2*, 14. [CrossRef]
3. Wang, M.; Hsieh, M.; Lin, H. Operational Improvement of Interior Permanent Magnet Synchronous Motor Using Fuzzy Field-Weakening Control. *Electronics* **2018**, *12*, 452. [CrossRef]

4. Bolognani, S.; Peretti, L.; Zigliotto, M. Online MTPA Control Strategy for DTC Synchronous-Reluctance-Motor Drives. *IEEE Trans. Power Electron.* **2011**, *26*, 20–28. [CrossRef]

5. Du, M.; Tian, Y.; Wang, W.; Ouyang, Z.; Wei, K. A Novel Finite-Control-Set Model Predictive Directive Torque Control Strategy of Permanent Magnet Synchronous Motor with Extended Output. *Electronics* **2019**, *4*, 388. [CrossRef]

6. Kirankumar, B.; Reddy, Y.V.S.; Vijayakumar, M. Multilevel inverter with space vector modulation: Intelligence direct torque control of induction motor. *IET Power Electron.* **2017**, *10*, 1129–1137. [CrossRef]

7. Pratibha, N.; Srinivas, S.; Ittamveettil, H. Five-level torque controller-based DTC method for a cascaded three-level inverter fed induction motor drive. *IET Power Electron.* **2017**, *10*, 1223–1230. [CrossRef]

8. Zhu, Z.Q.; Ren, Y.; Liu, J. Improved torque regulator to reduce steady-state error of torque response for direct torque control of permanent magnet synchronous machine drives. *IET Electr. Power Appl.* **2014**, *8*, 108–116. [CrossRef]

9. Ren, Y.; Zhu, Z.Q.; Liu, J. Direct Torque Control of Permanent-Magnet Synchronous Machine Drives With a Simple Duty Ratio Regulator. *IEEE Trans. Ind. Electron.* **2014**, *61*, 5249–5258. [CrossRef]

10. Zhang, Y.; Zhu, J. A Novel Duty Cycle Control Strategy to Reduce Both Torque and Flux Ripples for DTC of Permanent Magnet Synchronous Motor Drives with Switching Frequency Reduction. *IEEE Trans. Power Electron.* **2011**, *26*, 3055–3067. [CrossRef]

11. Masoud, A.; Farasat, M.; Jafarishiadeh, S. Model predictive current control of surface-mounted permanent magnet synchronous motor with low torque and current ripple. *IET Power Electron.* **2017**, *10*, 1120–1128. [CrossRef]

12. Zhang, Y.; Wei, X. Torque ripple RMS minimization in model predictive torque control of PMSM drives. In Proceedings of the 2013 International Conference on Electrical Machines and Systems (ICEMS), Busan, Korea, 26–29 October 2013; pp. 2183–2188. [CrossRef]

13. Cortes, P.; Rodriguez, J.; Vargas, R.; Ammann, U. Cost Function-Based Predictive Control for Power Converters. In Proceedings of the IECON 2006—32nd Annual Conference on IEEE Industrial Electronics, Paris, France, 6–10 November 2006; pp. 2268–2273. [CrossRef]

14. Rojas, C.A.; Rodriguez, J.; Villarroel, F.; Espinoza, J.R.; Silva, C.A.; Trincado, M. Predictive Torque and Flux Control Without Weighting Factors. *IEEE Trans. Ind. Electron.* **2013**, *60*, 681–690. [CrossRef]

15. Muddineni, V.P.; Bonala, A.K.; Sandepudi, S.R. Enhanced weighting factors selection for predictive torque control of induction motor drive based on VIKOR method. *IET Electr. Power Appl.* **2016**, *10*, 877–888. [CrossRef]

16. Zhang, Y.; Yang, H.; Xia, B. Model-Predictive Control of Induction Motor Drives: Torque Control Versus Flux Control. *IEEE Trans. Ind. Appl.* **2016**, *52*, 4050–4060. [CrossRef]

17. Zhang, Z.; Wei, C.; Qiao, W.; Qu, L. Adaptive Saturation Controller-Based Direct Torque Control for Permanent-Magnet Synchronous Machines. *IEEE Trans. Power Electron.* **2016**, *31*, 7112–7122. [CrossRef]

18. Xia, C.; Wang, S.; Gu, X.; Yan, Y.; Shi, T. Direct Torque Control for VSI-PMSM Using Vector Evaluation Factors Table. *IEEE Trans. Ind. Electron.* **2016**, *63*, 4571–4583. [CrossRef]

 *electronics*

Article

Feedforward Interpolation Error Compensation Method for Field Weakening Operation Region of PMSM Drive

Young-Bae Ji and Jung-Hyo Lee *

Department of Electrical Engineering, Kunsan National University, Gunsan, Jeollabuk-do 54150, Korea; dudqo848@naver.com
* Correspondence: jhlee82@kunsan.ac.kr; Tel.: +82-63-469-4707

Received: 9 August 2019; Accepted: 17 September 2019; Published: 18 September 2019

Abstract: This study proposes a field weakening control method with interpolation error compensation of the look-up table based permanent-magnet synchronous machine (PMSM) method. The look-up table (LUT) based control method has robust control characteristics compared to other control methods that use linear controllers for current reference generation. However, it is impossible to store all current references under all circumstances for torque commands. General LUT based control methods use two input parameters. In order to mitigate the effect of discretely stored data, two-dimensional interpolation is used to linearly interpolate values between discontinuous data. However, because the current trajectories of PMSMs are generally ellipsoidal, an error occurs between the linearly interpolated and controllable current references. This study proposes a method to compensate for this interpolation error using a feedforward controller for rapid compensation. The improvement using the proposed method is verified by experiment and simulation.

Keywords: look-up table; interpolation error; PMSM drive

---

1. Introduction

Recently, permanent-magnet synchronous motor (PMSM) drive systems have been widely adopted in vehicles to improve efficiency and convenience. Generally, for these PMSM drives, current command controls based on two-dimensional look-up tables (2D-LUT) are widely used [1–6]. Among these control methods, the flux-torque 2D-LUT based control method is the most generally used due to the reflection of DC-link voltage variation. This method is shown in Figure 1 [1].

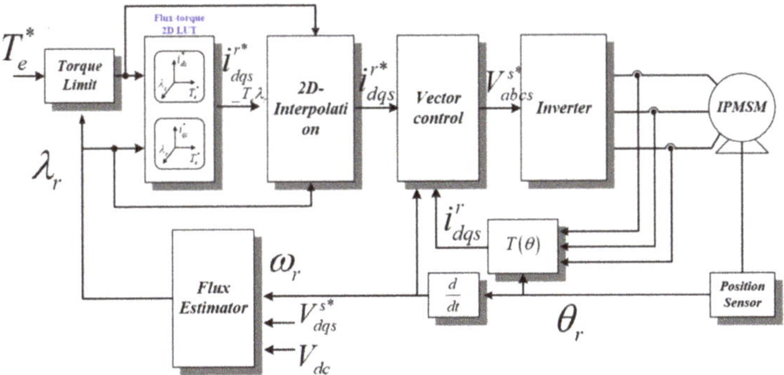

Figure 1. Flux-torque look-up table based permanent-magnet synchronous machine (PMSM) control method [1].

In this controller, the current data in the 2D-LUT is stored according to the respective memory addresses corresponding to flux-torque values. Because the memory address is a discontinuous value, only current reference data for a specific flux and torque can be stored in the memory. Therefore, it is impossible to store the appropriate current data for the entire driving region. To solve this problem, two-dimensional interpolation has been used to properly interpolate input data that have not been previously stored in the memory [7–9]. Two-dimensional interpolation produces the linear outputs of two continuous input parameters that are not defined at the specific torque and flux points of the look-up table address from the discreet look-up table data. For instance, when the input parameters are inserted into a look-up table, related outputs are generated from memory. These outputs are then linearly interpolated using first-order Newton's interpolation to calculate the approximate output for the input parameters.

Two-dimensional interpolation in PMSM control is only effective if the amount of stored data is sufficient to ignore the interpolation error. However, if the data is insufficient, the interpolation error cannot be ignored because all PMSM current trajectories are not linear, but ellipsoidal. Moreover, in automotive applications, in order to fulfill the international standard ISO26262, which aims to ensure driver safety, PMSM drives should have many fault diagnostic features and AUTOSAR software, which requires much memory to operate.

To solve this problem, only a few studies have investigated the problem of optimal memory use. Lenke et al. [10] used curve fitting to reduce the amount of memory used. However, they did not suggest the solution for the error between the interpolated curve fitting output and optimal output. A constant torque control method for PMSMs using any table was demonstrated in [11]. However, this method requires exact PMSM parameters. Moreover, if the target PMSM does not have sufficient saliency, its effectiveness is limited. Because current trajectories for PMSMs are ellipsoidal, the interpolation used in this application requires a second-order interpolation method, such as a second-order Lagrange or Spline interpolation with parameter modification for each operating condition [12]. However, these are seldom used in motor control because of heavy calculation burdens to digital signal processors (DSPs).

This study proposes a novel control method to reduce this interpolation error. First, we will analyze the cause of the interpolation error and define the problem characteristics. Next, we will illustrate the compensation method for the interpolation error using DC-link voltage feedforward.

2. Interpolation Error of the 2D Interpolation Technique for PMSM Drives

2.1. Two-Dimensional Interpolation Technique

Two-dimensional interpolation is a linear interpolation method for two input parameters. As mentioned before, 2D-LUT data on specific fluxes and torques have four different values: Outputs on the ceiling and floor of the flux input (λ_{max}, λ_{min}), and outputs on the ceiling and floor of torque input (T_{max}, T_{min}), as shown in Figure 2. To generate the interpolated output, the following linear interpolation formula is applied three times to the input parameters as shown in Figure 2b.

Research manuscripts reporting large datasets that are deposited in a publicly available database should specify where the data have been deposited and provide the relevant accession numbers. If the accession numbers have not yet been obtained at the time of submission, authors should state that they will be provided during review as they must be provided prior to publication.

$$f(x) = f(x_1) + \left(\frac{f(x_2) - f(x_1)}{x_2 - x_1} \right) \cdot (x - x_1) \tag{1}$$

For instance, in the linear interpolation of speed on T_{min}, x_2, and x_1 are the ceiling and floor of the speed input (λ_{max}, λ_{min}), respectively; and $f(x_2)$ and $f(x_1)$ are the current reference data on T_{min} corresponding to λ_{max} and λ_{min} (the notations in Figure 2b are $i^{r*}_{dqs\_T_{min}\lambda_{max}}$, $i^{r*}_{dqs\_T_{min}\lambda_{min}}$), respectively. x is the input of current speed (λ_r), and $f(x)$ is the interpolated output on T_{min} ($i^{r*}_{dqs\_T_{min}}$).

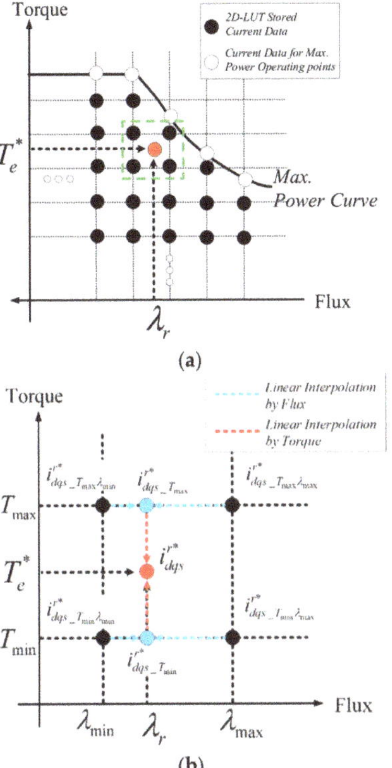

Figure 2. Two-dimensional interpolation procedure for current references outputs from two-dimensional look-up tables (2D-LUT). (**a**) Stored data in the 2D-LUT according to each speed and torque value. (**b**) The 2D interpolation procedure for specific speed and torque reference inputs.

To reduce the processing burden on microprocessors, the division of variables must be avoided as much as possible. This study uses the same fixed unit as that used in [9]. In this study, the current references from look-up tables were also established according to this unit value. Substituting the variable with the fixed unit value, Equation (1) can be changed as shown below.

$$f(x) = f(x_1) + \left(\frac{f(x_2) - f(x_1)}{x_{unit}} \right) \cdot (x - x_1) \tag{2}$$

where x_{unit} is the fixed unit value of the input parameter's deviation.

Note that the flux is an inverse value of the speed. If the speed is increasing, then the flux is decreasing, as shown in Figure 2a. Therefore, the calculated flux should be properly limited to avoid divergence.

2.2. Two-Dimensional Interpolation Error In Look-Up Table Based PMSM Control

Figure 3 shows the interpolated current trajectories using only six torque inputs. As shown in the figure, using 2D interpolation, two current reference points stored in the memory are linearly connected, whereas the operating condition that stored the data is absent. Although using 2D interpolation can reduce the error better than using only look-up table outputs, the generated power is lower than the capable maximum power. In addition, the effect of the interpolation error hardly exists in the constant torque operating region because current data are sufficient for the torque input. In contrast, the effect of the interpolation error is enhanced in the field weakening operating region because the current data are insufficient in high-speed operation.

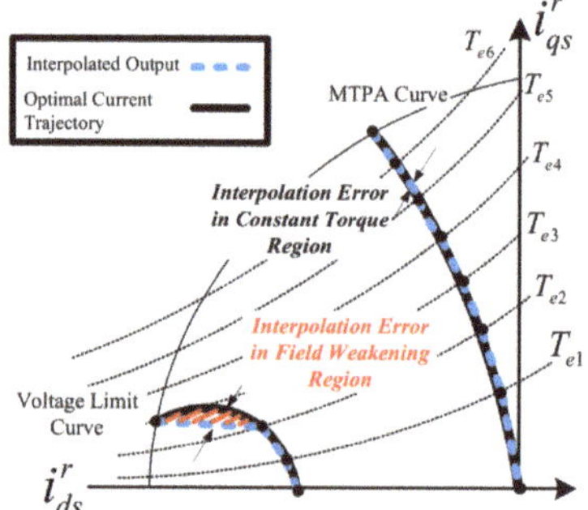

Figure 3. Interpolated current trajectories of PMSMs with a current table having the data of six torque inputs.

Figure 4 shows the magnified interpolation error described in Figure 3. The fundamental PMSM model equations to calculate this 2D interpolation error are shown below.

$$v_{ds}^r = R_s i_{ds}^r + L_d p i_{ds}^r - \omega_r L_q i_{qs}^r$$
$$v_{qs}^r = R_s i_{qs}^r + L_q p i_{qs}^r + \omega_r L_d i_{ds}^r + \omega_r \phi_f \tag{3}$$

$$T_e = 3P/4\left\{\phi_f i_{qs}^r - (L_q - L_d) i_{ds}^r i_{qs}^r\right\} \tag{4}$$

$$V_{\max} = \omega_r \sqrt{(L_d i_{ds}^r + \phi_f)^2 + (L_q i_{qs}^r)^2}$$
$$V_{\max} = \frac{V_{dc}}{\sqrt{3}} \tag{5}$$

where v_{ds}^r, v_{qs}^r are d- and q-axis voltages of the PMSM, respectively; T_e is torque, $V_{\max}$ is the voltage restriction of the PMSM; ω_r is electric angular speed; R_s is phase resistance; L_d, L_q are d- and q-axis inductances, respectively; ϕ_f is the permanent magnet's flux; P is the number of poles; and V_{dc} is DC-link voltage.

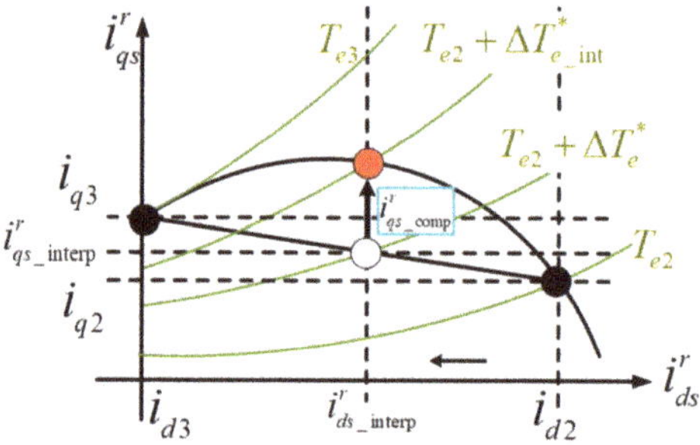

Figure 4. Magnified 2D interpolation error in field weakening region of Figure 3.

As shown in Figure 4, the interpolation error does not affect the d-axis current component ($i_{ds\_interp}^r$). However, it affects the q-axis current component ($i_{qs\_interp}^r$). This interpolated q-axis current reference can be expressed from Equation (1) as Equation (6) below.

$$i^r_{qs\_interp}(T_{e2} + \Delta T^*_e) = i_{q3} + \frac{i_{q3} - i_{q2}}{T_{e3} - T_{e2}}\Delta T^*_e$$
$$(\Delta T^*_e = T^*_e - T_{e2})$$

(6)

where T_{e2} and T_{e3} are floor and ceil of torque references in Figure 4, respectively.

Using Equation (6), the magnitude of the back-EMF (electric motive force) using the interpolated current reference can be expressed as Equation (7):

$$V_{mag} = \omega_r \sqrt{(L_d i^r_{ds\_interp}(T_{e2} + \Delta T^*_e) + \phi_f)^2 + (L_q i^r_{qs\_interp})^2}$$
$$V_{mag} < V_{max}$$

(7)

where V_{mag} is the back EMF magnitude for the variation of torque reference.

If the interpolated d-axis current is according to the varied torque in Equation (7), then the magnitude of the maximum torque generated by this d-axis current can be obtained through the intersection of the voltage restriction curve and the torque curve. The equation is expressed in quadratic form as shown in Equation (8), which has four roots (two imaginary, two real). Of the two real roots, the effective d-axis current is the magnitude with the minimum value.

$$
\begin{aligned}
&-\left(L_d - L_q\right)^2 L_q{}^2 i^r_{ds\_interp}{}^4 \\
&+\left\{-2L_d{}^2\left(L_d - L_q\right) - 2\phi_f L_d\left(L_d - L_q\right)^2\right\} i^r_{ds\_interp}{}^3 \\
&+\left\{-2L_d{}^2\phi_f{}^2 - 4\phi_f L_d\left(L_d - L_q\right) - \left(L_d - L_q\right)^2\phi_f{}^2 + \left(L_d - L_q\right)^2\left(\tfrac{V_{max}}{\omega_r}\right)^2\right\} i^r_{ds\_interp}{}^2 \\
&+\left\{-2\phi_f{}^3 L_d - 2\left(L_d - L_q\right)\phi_f{}^2 + 2\left(L_d - L_q\right)^2\left(\tfrac{V_{max}}{\omega_r}\right)^2\right\} i^r_{ds\_interp} \\
&+\left\{\phi_f{}^2\left(\tfrac{V_{max}}{\omega_r}\right)^2 - \left(\tfrac{L_q(T_{e2} + \Delta T_{e\_int})}{P_n}\right)^2 - \phi_f{}^4\right\} = 0
\end{aligned}
$$

(8)

where P_n is pole-pair and $\Delta T^*_{e\_int}$ is increased torque output from compensated current.

The compensation value of the q-axis current can be obtained by substituting the d-axis current value from Equation (8) into Equation (4).

$$i^r_{qs\_comp} = \frac{4}{3P}\frac{T_{e2} + \Delta T^*_{e\_int}}{\left(\phi_f + (L_d - L_q)i^r_{ds\_interp}\right)} - i^r_{qs\_interp}$$

(9)

From Equation (9), the increased torque output using the compensated q-axis current ($\Delta T^*_{e\_int}$) is always higher than the increased torque output using the interpolated q-axis current (ΔT^*_e), as shown in Figure 4.

This proves that a torque error exists between the interpolated output and optimal output. In addition, the torque generated from the interpolated output is always lower than that generated from the optimal output.

3. An Interpolation Error Compensation Method with DC-Link Voltage Feedforward Controller

Generally, because the magnitude of the current in the field weakening region is not large, the magnitude of the back EMF has a major influence on the components in the permanent magnet's flux. Therefore, the variation in the torque in this region is largely affected by the variation in the q-axis current and the voltage restriction is greatly affected by the speed and the d-axis current. Among these, the speed is not a direct control target of the electric motor, but a restrictive condition that is influenced by the machine-driving environment. Therefore, to compensate for the reduced output caused by the interpolation, the q-axis current must be compensated to raise the torque.

This torque error occurs because the magnitude of the back EMF of the interpolated q-axis current is smaller than the voltage limit, which is determined by DC-link voltage. Therefore, the compensated q-axis current can be obtained from the voltage error between the back EMF and the voltage limit calculated from the DC-link voltage. From this error, the proportional-integral (PI) controller in this study deducts the compensated q-axis current [13,14].

$$i^r_{qs\_comp} = \frac{sK_p + K_i}{s}(V_{\max} - V_{mag})$$

(10)

where K_p and K_i are P and I gains of PI controller, respectively.

However, as all current trajectories for PMSM operation are ellipsoidal, the PI controller is insufficient to calculate the compensation current. A PI controller is widely known to be effective with fixed references. However, if the references are ellipsoidal, the final output value always has an error from delayed response.

The easiest way to respond suitably to ellipsoidal references is to use a proportional-integral-differential (PID) controller to increase the poles of the controller's characteristic equation or to adapt numerous higher gains of PI controller to reduce this final output error. However, the D controller requires suitable filters to avoid divergent outputs and a higher gain PI controller can be very unstable operation for disturbance.

In this work, a feedforward controller was added to improve the dynamics of the PI controller. To obtain the feedforward compensation current, the voltage error between the back EMF and the voltage limit can be obtained as Equation (11) from Equation (7).

$$\begin{aligned}
V_{s_err} &= V_{\max} - V_{mag} \\
&= \frac{V_{dc}}{\sqrt{3}} - \sqrt{(\omega_r L_q i^r_{qs_linear})^2 + (\omega_r L_d i^r_{ds_linear} + \phi_f)^2}
\end{aligned}$$

(11)

where V_{s_err} is voltage error from the interpolation.

As shown in Figure 4, because the interpolated d-axis current reference generates a proper current for the voltage limit, Equation (11) can be changed to Equation (12).

$$\begin{aligned}
V_{s_err} &= V_{\max} - V_{mag} \\
&= \frac{V_{dc}}{\sqrt{3}} - \omega_r L_q \left| i^r_{qs_linear} \right|
\end{aligned}$$

(12)

To reduce this voltage error to zero, the compensation feedforward q-axis current can be established using Equation (13).

$$i^r_{qs\_ff} = \frac{V_{s\_err}}{\omega_r L_q}$$

(13)

where, $i^r_{qs\_ff}$ is compensation feedforward q-axis current.

Using Equation (13), the proposed overall control block for q-axis compensation current is established in Figure 5.

As mentioned, an interpolation error hardly exists in the constant torque operating region. Moreover, the back EMF in this region has its own value at each operating point. Compensating for interpolation error in this region requires other back EMF look-up table data for each flux and torque values. Therefore, in this work, the compensation block was deactivated while the PMSM operated in the constant torque region.

To reduce the amount of stored memory for voltage magnitude, the start points for the field weakening operation of each speed were obtained experimentally instead of from voltage magnitudes all over the operating region. Figure 6 shows the stored data of field weakening start points. As can be seen from the figure, as speed increases, the generated torque reduces, generating a constant power output. In addition, because an error hardly occurs between the practical torque-speed curve and the linearly interpolated curve, the field weakening start torque, which is generated from flux-torque LUT of Figure 5, can determine whether to enable or disable the compensation block. Instead of using speed inputs, estimated flux is used to reflect input voltage variation. Conversion from speed to flux can simply be obtained using Equation (14).

$$\lambda_r = V_{\max}/\omega_r \tag{14}$$

The activation condition of the q-axis current compensation can be determined by the following equation.

$$if(T_e^* - T_{e\_FW} > 0) : \text{sgn}(T_r^* - T_{e\_FW}) = 1$$
$$if(T_e^* - T_{e\_FW} \leq 0) : \text{sgn}(T_r^* - T_{e\_FW}) = 0 \tag{15}$$

With this enable-disable control block, the proposed compensation block is only activated when the effect of compensation is maximized.

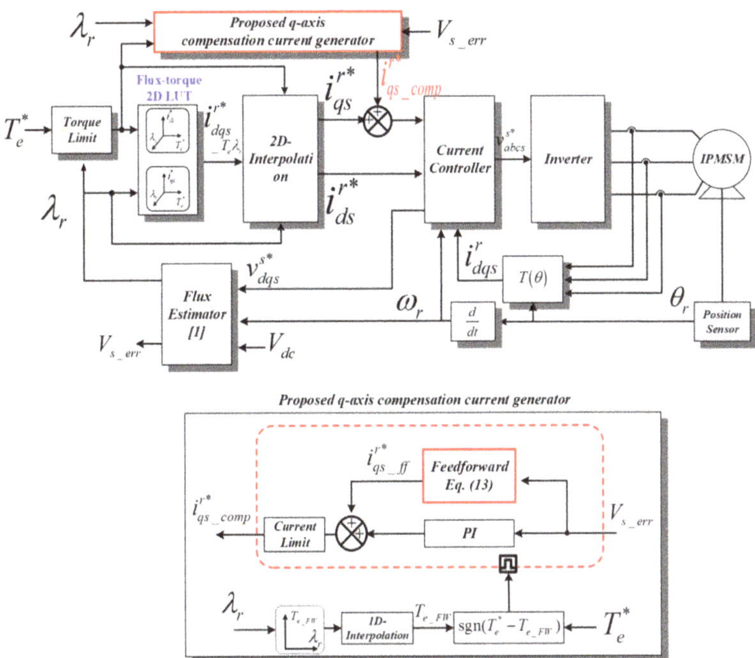

Figure 5. Overall proposed control block for q-axis current compensation.

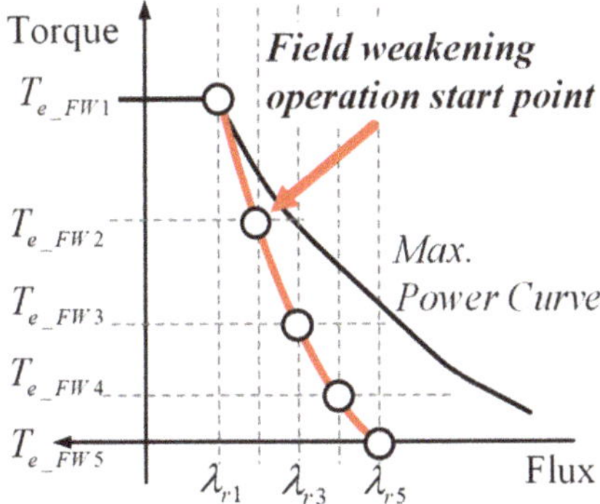

Figure 6. Field weakening operation start points for each speed.

4. Experiment

To verify the proposed control algorithm, the experiment was set up as shown in Figure 7. The test motor parameters are described in Table 1. The controller of the inverter used the DSP TMS320F28335 from Texas Instruments Corporation.

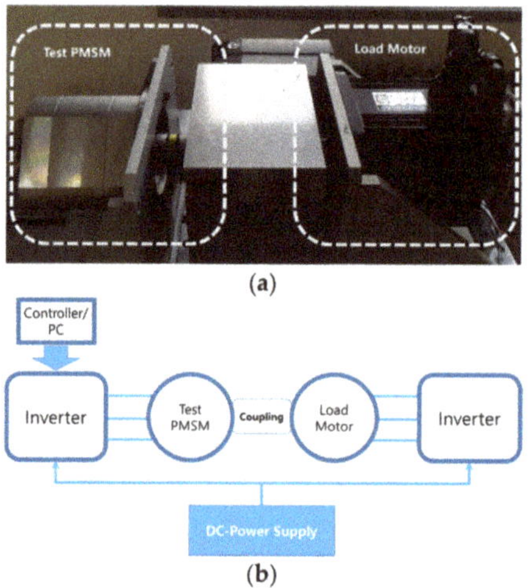

Figure 7. Experimental setup: (**a**) Picture of the setup; (**b**) composition of the setup.

Table 1. Test motor parameters.

Pole	8
Phase Resistance [mΩ]	20
d-axis Inductance [mH]	2.03
q-axis Inductance [mH]	2.13
Permanent Magnet Flux [Wb]	0.1439
Rated DC-Link Voltage [V]	48
Rated Torque [Nm]	15
Rated Speed [rpm]	200
Max. Speed [rpm]	1500
Max. Phase Current [A]	30

Although this DSP has sufficient memory to store current 2D-LUT data, for the worse environment configuration, the current map is constructed using very little memory compared to the conventional method.

The d-q-axis current map for each speed-torque was measured through experiments [6]. The results are shown in Figure 8. In this study, the experimental current map was stored at one-third of the rated torque and at one-quarter of the maximum driving speed. The resulting speed-torque current map was transformed into a flux-torque map using Equation (14). As a result, it can be seen that only one torque data appears in the d-q-axis current value for the torque command change at the maximum speed, except for the d-q-axis current value during the zero-torque control. This is because the inductance expressed by the nominal value in the actual motor and the permanent magnet flux decreased due to the influence of saturation in the actual experiment.

Figure 9 shows the comparison of experimental results with or without the proposed compensation method. Without the compensation method, the d-q-axis currents are linearly controlled due to the 2D interpolation according to the increase of torque references. Despite applying 2D interpolation, a considerable error exists between the maximum controllable output and linearly controlled output because the stored current references are insufficient in the high-speed field weakening operating region. However, with the proposed compensation method, the d-q-axis current trajectory follows the voltage limit ellipse, which means that the target PMSM generates the maximum controllable output at current motor speed.

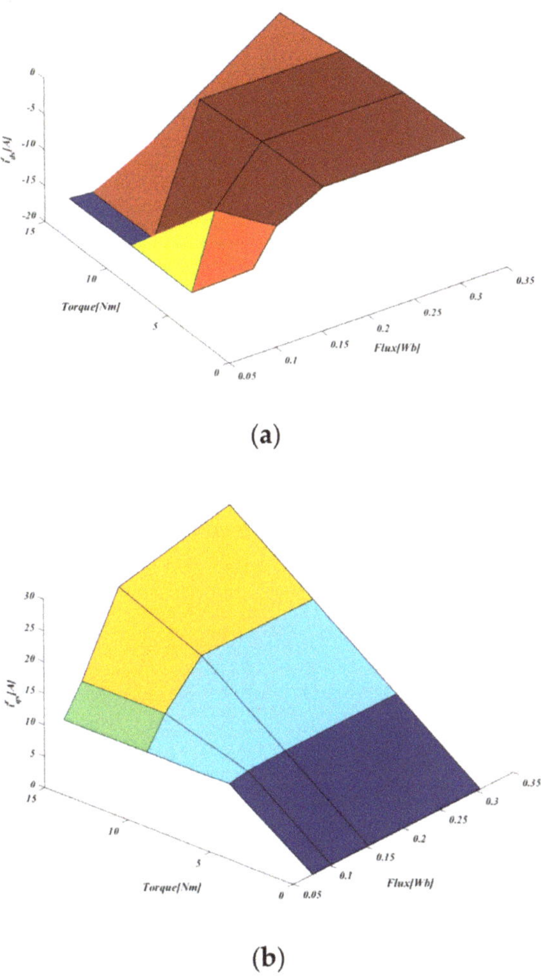

(a)

(b)

Figure 8. *Cont.*

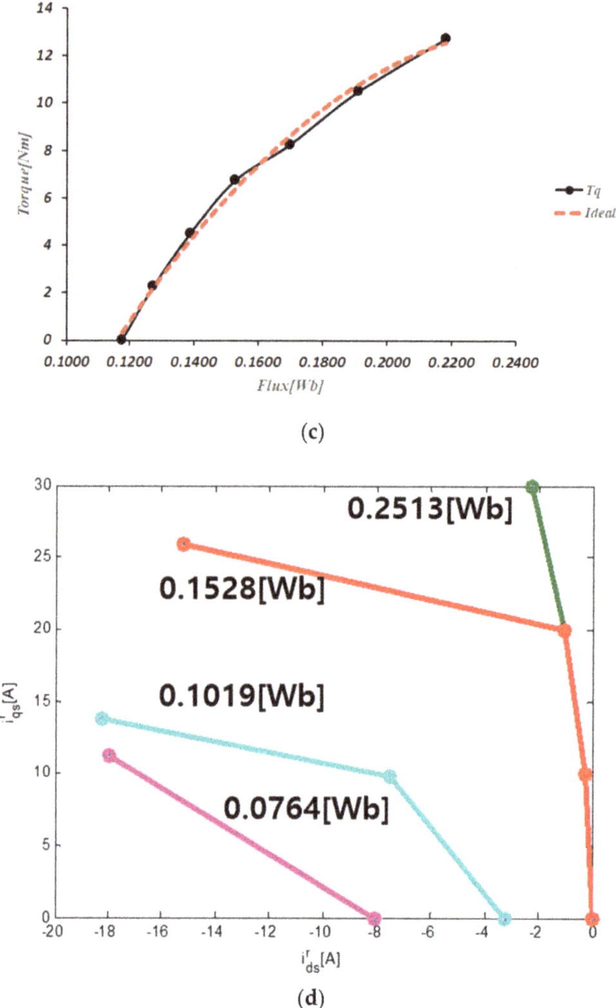

(c)

(d)

Figure 8. Current maps in the experiment: (**a**) D-axis current map, (**b**) q-axis current map, (**c**) look-up table for the field weakening operation start point, (**d**) d-q-axis current maps according to variable flux.

Figure 10 shows the compared experimental results for operation at 1500 rpm. The applied torque reference is from 0 to 15 Nm. The shape of the torque reference is a ramp with a slope of 0.75 Nm/ms. As shown in Figure 10a, without any compensation, the d-q-axis current references are linearly straight lines because the stored currents data have only two points at this speed. In Figure 10b, for the compensation algorithm using only a PI controller [14], the compensated amount is small and delayed because the voltage error is not fixed, but is varied according to the torque reference. The compensator requires time to generate suitable compensation currents. If the torque reference is changed rapidly, the response time of the PI controller must be considered because excessive q-axis current occurs and saturates the back EMF. In Figure 10c, with the proposed feedforward compensator, the response time of the PI controller can be improved. Therefore, a suitable compensation current can be obtained, even if a rapid varied torque reference is applied.

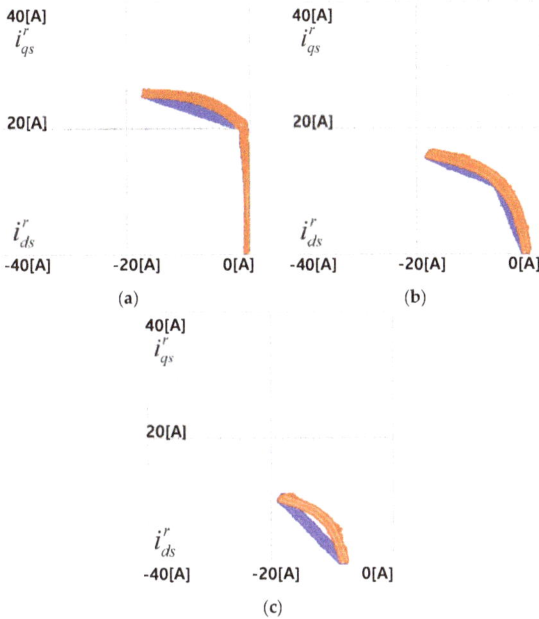

Figure 9. Comparison of experimental results of the conventional and the proposed method with and without 2D interpolation compensation. (**a**) Operation at 750 rpm, (**b**) operation at 1150 rpm, (**c**) operation at 1500 rpm.

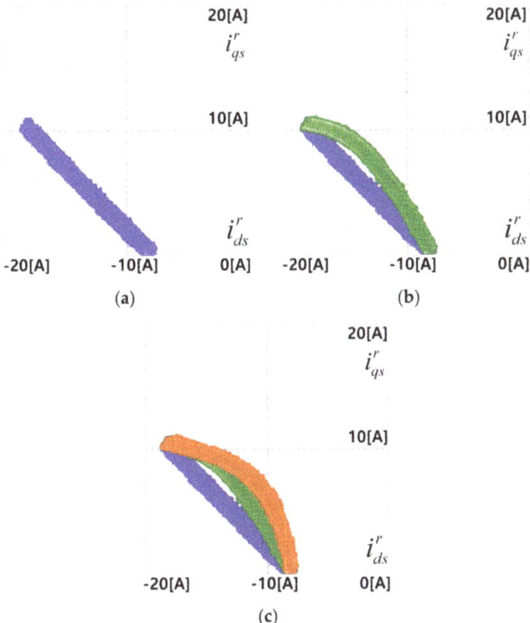

Figure 10. Magnified comparison of the experimental results at 1500 rpm using (**a**) only 2D interpolation, (**b**) compensator with a PI controller [14], and (**c**) the proposed compensation method. (Slope of the torque reference: 0.75 Nm/ms).

5. Conclusions

This study proposed a compensation method for the interpolation error in the general flux-torque 2D-LUT based PMSM control method. To achieve this, we first calculated the error between the back EMF and the voltage limit in the field weakening operating region. We then obtained the q-axis compensation current from PMSM voltage equations. To improve the compensation control dynamic for ellipsoidal references in the feedback controller, a feedforward controller based on DC-link voltage was attached to the general PI controller.

With the proposed compensation blocks, the generated power was enhanced beyond the power obtained using only 2D interpolation, although very restricted numbers of memory are allowed for flux-torque 2D-LUT. The proposed control method was verified by the experiment with a test setup.

Author Contributions: J.-H.L. designed and performed the experiment, analyzed the theory, and wrote the manuscript. J.-H.L. and Y.-B.J. participated in research plan development and revised the manuscript. All authors have contributed to the manuscript.

Funding: This research was supported by Basic Science Research Program through the National Research Foundation of Korea (NRF) funded by the Ministry of Education (NRF-2018R1C1B6008895). This work is supported the Human Resources Development Program (Grant No. 20174010201350) by the Korea Institute of Energy Technology Evaluation and Planning (KETEP) grants.

Conflicts of Interest: The authors declare no conflict of interest.

References

1. Jung, S.-Y.; Hong, J.; Nam, K. Current minimizing torque control of the IPMSM using Ferrari's method. *IEEE Trans. Power Electron.* **2013**, *28*, 5603–5617. [CrossRef]
2. Lee, J.H.; Lee, J.H.; Park, J.H.; Won, C.Y. Field-weakening strategy in condition of DC-link voltage variation using on electric vehicle of IPMSM. In Proceedings of the 2011 International Conference on Electrical Machines and Systems, Beijing, China, 20–23 August 2011; pp. 1–6.
3. Yang, N.; Luo, G.; Liu, W.; Wang, K. Interior permanent magnet synchronous motor control for electric vehicle using look-up table. In Proceedings of the 7th International Power Electronics and Motion Control Conference, Harbin, China, 2–5 June 2012; pp. 1015–1019.
4. Kwon, T.-S.; Choi, G.-Y.; Kwak, M.-S.; Sul, S.-K. Novel Flux-Weakening Control of an IPMSM for Quasi-Six-Step Operation. *IEEE Trans. Ind. Appl.* **2008**, *44*, 1722–1731. [CrossRef]
5. Cheng, B.; Tesch, T.R. Torque Feedforward Control Technique for Permanent-Magnet Synchronous Motors. *IEEE Trans. Ind. Electron.* **2010**, *57*, 969–974. [CrossRef]
6. Lee, J.-H.; Won, C.-Y.; Lee, B.-K.; Kim, H.-B.; Baek, J.-H.; Han, K.-B.; Chung, U.-I. IPMSM Torque Control Method Considering DC-link Voltage Variation and Friction Torque for EV/HEV applications. In Proceedings of the 2012 IEEE Vehicle Power and Propulsion Conference, Seoul, Korea, 9–12 October 2012; pp. 1063–1069.
7. Tursini, M.; Chiricozzi, E.; Petrella, R. Feedforward Flux-Weakening Control of Surface-Mounted Permanent-Magnet Synchronous Motors Accounting for Resistive Voltage Drop. *IEEE Trans. Ind. Electron.* **2010**, *57*, 440–448. [CrossRef]
8. Park, J.-H.; Lee, J.-H.; Lee, J.-H.; Won, C.-Y. Current Control Method of IPMSM in Constant Power Region for HEV. In Proceedings of the 2011 International Conference on Electrical Machines and Systems, Beijing, China, 20–23 August 2011; pp. 1015–1019.
9. Lenke, R.U.; de Doncker, R.W.; Kwak, M.-S.; Kwon, T.-S.; Sul, S.-K. Field Weakening Control of Interior Permanent Magnet Machine using Improved Current Interpolation Technique. In Proceedings of the 2006 37th IEEE Power Electronics Specialists Conference, Jeju, Korea, 18–22 June 2006; pp. 1–5.
10. Cintron-Rivera, J.G.; Foster, S.N.; Nino-Baron, C.A.; Strangas, E.G. High performance controllers for Interior Permanent Magnet Synchronous Machines using look-up tables and curve-fitting methods. In Proceedings of the 2013 International Electric Machines & Drives Conference, Chicago, IL, USA, 12–15 May 2013; pp. 268–275.
11. Yoon, Y.-D.; Lee, W.-J.; Sul, S.-K. New flux weakening control for high saliency interior permanent magnet synchronous machine without any tables. In Proceedings of the 2017 European Conference on Power Electronics and Applications (ECPEA), Aalborg, Denmark, 2–5 September 2007; pp. 1–7.

12. Huang, S.; Chen, Z.; Huang, K.; Gao, J. Maximum Torque Per Ampere and Flux-weakening Control for PMSM Based on Curve Fitting. In Proceedings of the 2010 IEEE Vehicle Power and Propulsion Conference (VPPC), Lille, France, 1–3 September 2010; pp. 1–5.
13. Lee, J.-H. DC-link voltage feedforwarded interpolation error compensation method for field weakening operation region of look-up table based PMSM drive. In Proceedings of the IOP Conference Series: Materials Science and Engineering, Tokyo, Japan, 22–25 May 2019; Volume 600.
14. Lee, J.-H. Interpolation Error Compensation Method for PMSM Torque Control. *Trans. Korean Inst. Electron. Eng.* **2018**, *67*, 391–397. [CrossRef]

 *electronics* 

Article

Novel Efficacious Utilization of Fuzzy-Logic Controller-Based Two-Quadrant Operation of PMBLDC Motor Drive Systems for Multipass Hot-Steel Rolling Processes

Mohanraj Nandakumar [1,*], Sankaran Ramalingam [1], Subashini Nallusamy [1] and Shriram Srinivasarangan Rangarajan [1,2,*]

[1] Department of EEE, SASTRA Deemed to be University, Thanjavur, Tamil Nadu 613401, India; rs@eee.sastra.ac.in (S.R.); nmnsi@eee.sastra.ac.in (S.N.)

[2] Department of Electrical & Computer Engineering, Clemson University, Clemson, SC 29634, USA

* Correspondence: snehammohan@eee.sastra.ac.in (M.N.); shriras@g.clemson.edu (S.S.R.)

Received: 8 May 2020; Accepted: 13 June 2020; Published: 16 June 2020

Abstract: This study investigates the rough steel-rolling process, which requires repeated and rapid bidirectional hot-rolling operations and proposes a fuzzy-logic-controller-based brushless electric DC (BLDC) motor drive system for the same. We present a modeling of the hot-steel rough-rolling process using a set of metallurgical parameters and mechanical equations based on their operating conditions, specific features and characteristics, all obtained from actual data. The above equations and related parameters were modeled in MATLAB/Simulink schematic under variations in temperature and slab thickness corresponding using three different hot-rolled (HR) steel specimens. This led to the creation of a pair of speed and torque- profiles with alternate polarities for successive passes covering the entire rolling process for each steel specimen. A fuzzy logic controller utilized the above profiles on the motor shaft by incorporating speed and current feedback loops to attain reference speed and calculation of instantaneous stator currents of the BLDC motor with respective phase sequences, so as to satisfy the torque-profile. Simulation results showing the detailed performance of the drive system are presented. Further, experimental work on a BLD-motor-drive system is presented, along with loading arrangements and an arm controller embedded with control algorithm for the multi-loop feedback system used for the closed loop speed control. The efficacy of the new applications proposed in this study for the first time can be seen from the validation of the results from the BLDC motor with its fuzzy-based controller in terms of simulation and hardware, thereby serving to be an attractive alternative to conventional induction motor drive systems for steel rolling.

Keywords: PMBLDC motor; power electronics; fuzzy-logic controller design

---

1. Introduction

The process of steel making is highly power-intensive. About eight percent of total energy consumption is required in hot-steel rolling. Steel products in the form of steel plates, welded pipes and wired products are utilized by various industries, such as automobiles, ship building and constructional areas. The first stage in hot rolling is the rough-rolling mill where steel ingots are rolled to produce products like slabs [1]. Stages use these intermediate specimens for production of plates or bars or rods and similar products. Prior to the above two phases of rolling, the incoming hot steel slab specimen must be surface-cleansed by scrubbing to remove layers of oxides.

The rolling process consists of a set of work and backup rolls driven by two identical high-power electric motors through a set of gear train arrangements. Each backup roll is driven by an individual drive system, such that the two work rolls rotate in opposite directions for forward/reverse feeding of the work piece at hand. The mechanical calculations [2] of rolling force and torque at the rollers for multipass operation involve taking into account various metallurgical properties and physical parameters of roll material, work temperature and roller dimensions. The rough-rolling process is aimed at successive reduction of thickness over multiple passes of rolling. Calculations over repeated forward-reverse rolling passes leads to the creation of profiles that define a series of operating points of the drive motors, in terms of shaft speed and developed torque.

Different from conventional DC electric motors, AC synchronous motors and induction motors, a new entrant is the permanent magnet brushless DC (PMBLDC) motor [3–5], which possess certain advantages like a high torque-to-current ratio, a high power-to-weight ratio as well as a quick response to load fluctuations and speed reversal. PMBLDC motors can operate over a wide range of speeds efficiently, while delivering nearly-full-load-rated torque and are amendable to precise control using feedback signals.

A significant contribution has been made [6] regarding the quick achievement of downloading an algorithm onto the controller board being used, but the output waveforms quick settlement in steady state has not achieved. One can read the study [7] concerned with proportional integral derivative (PID) self-tuning and model reference adaptive control, but it is not focused on experimental implementation. The main focus of this study [8] is incorporation of fuzzy PID for a closed-loop speed control, but it was utilized only for speed control-not on inner-current-control required for torque control. This study is focused on achieving both better torque and speed control in steady state, as well as several power electronic devices involved in hardware implementation.

In this study, the multipass rolling process of hot rough-steel-rolling milling was mathematically modeled starting with basic set of equations, characterized by draft, roller contact length, temperature dependent strength coefficient and strain-hardening exponent. The above model was integrated with a motor drive system through reduction gears and alternate control algorithms using speed and current feedback. An intelligent computing technique called fuzzy logic control was implemented for controlling the BLDC motor drive system [9–13]. Several controllers like conventional PI whose parameter values were selected by trial and error approach—and a tuned PI controller whose parameter values were selected by using Ziegler-Nichols approach—were also tested. Later, a comparison of all three controllers was done to ensure the quickness in reaching the steady state among the controllers. The overall drive system was simulated in the MATLAB/Simulink platform for performance evaluation. This work was carried out for multipass, bidirectional rolling of 3 different HR steel materials. It was found that the rolling operation closely follows the respective work profile. The performance of the drive system covering variables like motor shaft speed, electromagnetic torque, back-EMF and stator currents were evaluated for ten passes of hot-steel rolling over a duration of ten seconds, and the simulation results are presented. An experimental setup for speed control of the entire drive system was developed and the control algorithm for implementing the same was embedded as hex file on the arm microcontroller. The arm microcontroller calculated the speed error based on reference input profile values and actual output speed values sensed from the BLDC motor and provided necessary control action based on the developed control algorithm. Thereby, it regulated the speed of the BLDC motor.

This study is organized as follows. Section 2 describes the model of a hot-roughing steel-rolling mill. In Section 3 the profiles created for the rolling operation are discussed. The Section 4 deals with the mathematical model developed for the BLDC motor. In Section 5 the design of a fuzzy-logic controller employed for BLDC motor is introduced. In Section 6 the controller configuration with feedback design is incorporated. In Section 7 simulation schematic is dealt with. Section 8 describes experimental set up for the simulations created in simulation schematic. Finally, in Section 9 the results are discussed in the conclusion.

2. Model of A Hot-Roughing Steel-Rolling Mill

A hot-roughing steel-rolling mill is formed by integrating several mechanical components to produce the desired forces and forward movement of roll material for thickness reduction of incoming steel slabs. A typical hot-strip mill stand employs two identical BLDC motor drive systems coupled to a hot-roughing steel-rolling mill through reduction gears on either side (Figure 1). The work described in this study assumes multipass rolling, in which the hot material specimen is compressed between two work rolls, and each alternate pass is for a duration of one second—either in forward and reverse direction—over a total run of ten passes to obtain the desired reduction in thickness from an initial thickness (t_o) to final thickness (t_f) by the lamination force of the two parallel rolls.

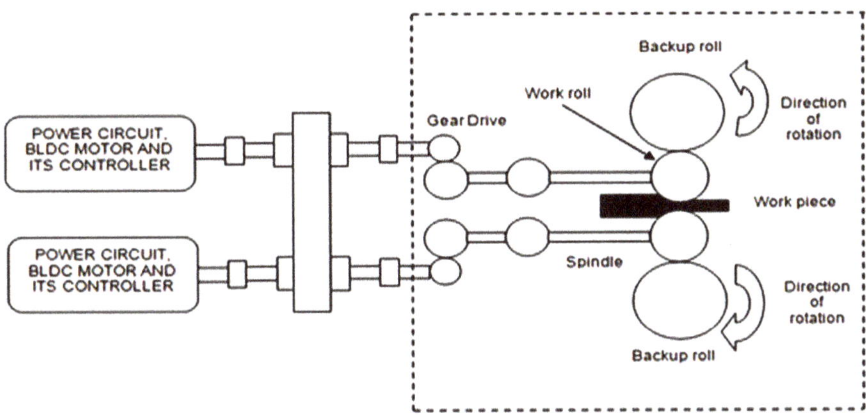

Figure 1. Schematic of two identical brushless electric DC (BLDC) motor drive systems coupled to steel rolling mill through reduction gears.

A simplified schematic portraying the modeling of BLDC motor drive system incorporating speed and current feedback loops using fuzzy controller is shown in Figure 2. The essential components taking part in the rolling operation are a pair of work rolls, backup rolls and a gear train operated by a pair of BLDC motors. For simplicity, the power transmission from one BLDC motor alone is shown in Figure 2. In actual practice, another identical motor drive system is available to operate the work roller and back up roller on the opposite side of the work piece.

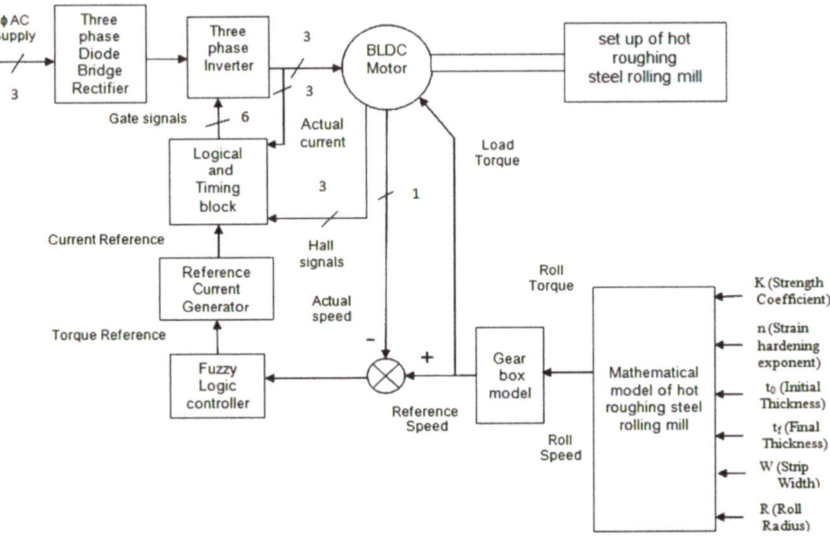

Figure 2. Schematic of steel rolling mill model and BLDC motor drive system.

The electrical components of the system consist of three phased AC sources, three phase-uncontrolled bridge rectifiers, a 6-pulse inverter, a 4-pole BLDC motor and a speed error loop, along with reference torque and reference current generation modules, which are implemented by using fuzzy logic-based control algorithm [14–24].

2.1. Mathematical Modeling of Steel Rolling Process

Draft is a term specified for thickness reduction in steel-material rolling under specified operating conditions from hot slab to steel plates. This phenomenon—otherwise known as deformation—is governed by the following Equation (1):

$$d = t_o - t_f = 2R \left(1 - \cos \alpha\right) \tag{1}$$

where d = draft, mm, t_o = initial thickness, mm and t_f = final thickness, mm, R = roll radius, mm and α = bite angle in degrees.

The true strain experienced by the work piece while rolling is based on the thickness of the work material before and after rolling and is given by the Equation (2):

$$\varepsilon = \ln\left(\frac{t_o}{t_f}\right) \tag{2}$$

During flat rolling, average flow stress $\overline{Y_f}$ on the work material can be determined by the following Equation (3):

$$\overline{Y_f} = \frac{K\varepsilon^n}{1+n} \tag{3}$$

where K = strength coefficient, MPa and n = strain-hardening exponent.

The contact length L in mm is related to the rolling operation is defined as in Equation (4):

$$L = \left[R \left(t_o - t_f\right)\right]^{0.5} \tag{4}$$

Roll force F required for flat rolling is estimated by the following relation (5):

$$F = \overline{Y_f}w \times L \tag{5}$$

where w = width of the steel slab, mm

In hot rolling, load torque required on each work roll is given by the Equation (6):

$$T = 0.5FL \text{ N-m} \tag{6}$$

The PMBLDC motor is coupled to the steel-rolling system through an arrangement of reduction gears. The material properties and operational parameters of the rolling process are discussed in Section 2.2.

2.2. Rolling Process Parameters

Considering the low-carbon specimens of 1040 HR steel, 1080 HR steel and 12L14 HR steel, the metallurgical data and parameters were obtained from standard ASM Metals Handbook and shown in Table 1.

Figures 3–5 show the variation of the strength coefficient K and the strain-hardening coefficient n with respect to temperature in °C.

Table 1. Properties of three different hot-rolled (HR) steel materials.

Properties	1040 HR Steel	1080 HR Steel	12L14 HR Steel
Tensile strength (MPa)	620	772	540
Yield strength (MPa)	415	425	415
Elastic modulus (GPa)	190–210	205	190–210
Poisson ratio	0.27–0.30	0.29	0.27–0.30
Vickers hardness	211	241	170
Density (gr/cm^3)	7.845	7.7–8.03	7.87

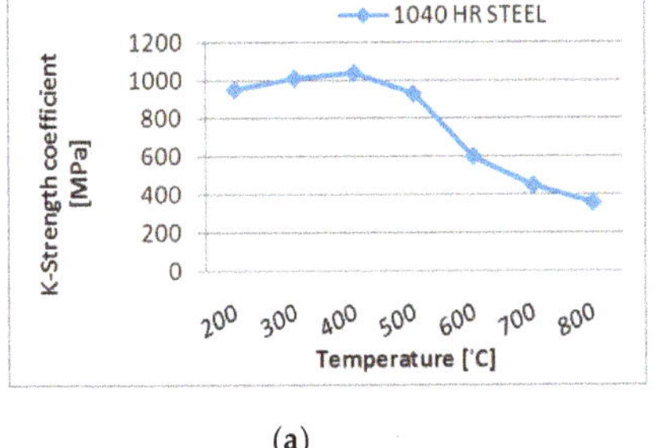

(**a**)

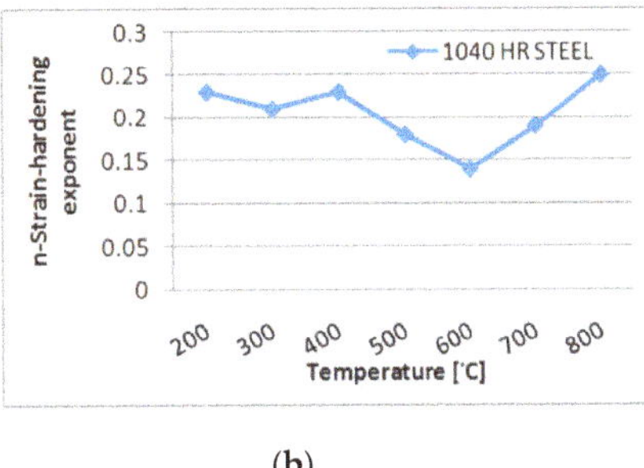

(**b**)

Figure 3. 1040 HR steel characteristics (**a**) K at different temperatures; (**b**) n at different temperatures.

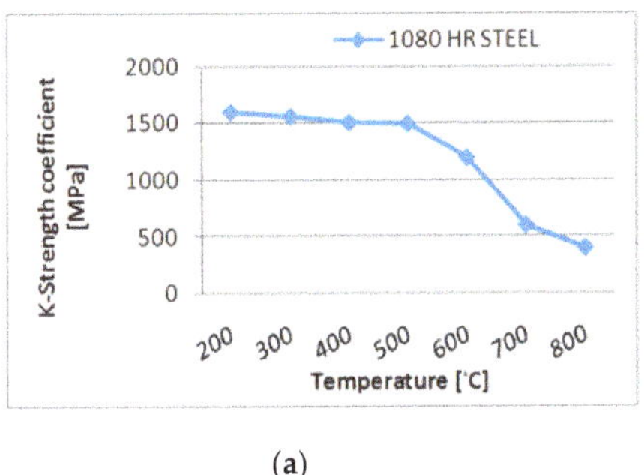

(**a**)

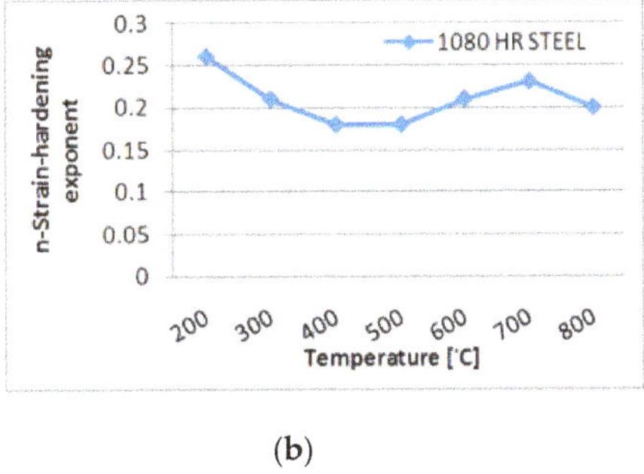

(**b**)

Figure 4. 1080 HR steel characteristics (**a**) K at different temperatures; (**b**) n at different temperatures.

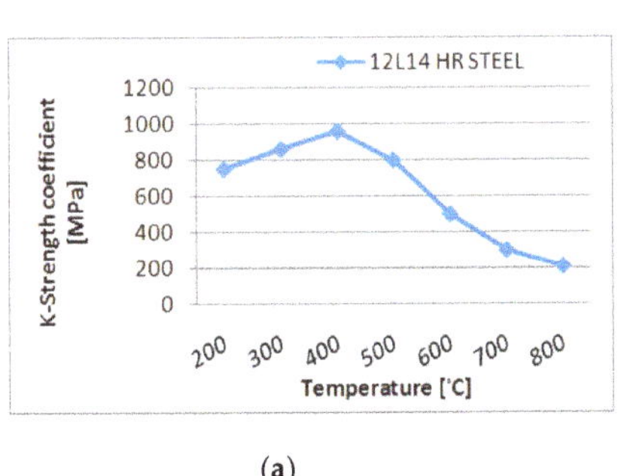

(**a**)

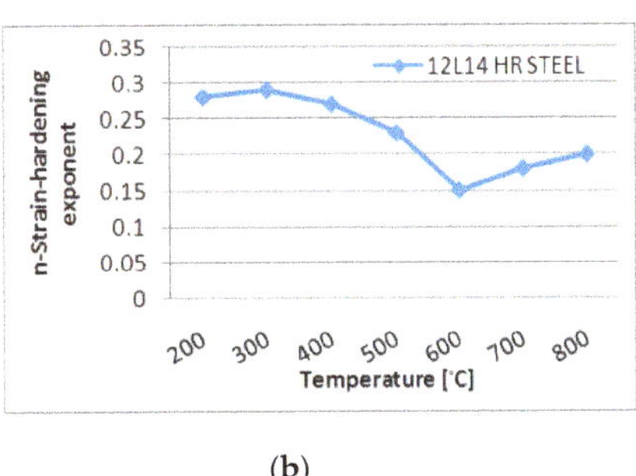

(**b**)

Figure 5. 12L14 HR steel characteristics (**a**) K at different temperatures; (**b**) n at different temperatures.

The ten rolling passes carried out to reduce the overall thickness from 100 mm to 70 mm are characterized by the range of parameters in Table 2.

Table 2. Rolling parameters at 10 passes for 3 different HR steel materials.

Pass	Temperature °C	1040 HR		1080 HR		12L14 HR	
		K, MPa	n	K, MPa	n	K, MPa	n
1	815	345	0.25	360	0.21	230	0.18
2	810	346	0.258	364	0.2102	231	0.179
3	805	348	0.256	369	0.2104	232	0.1792
4	800	349	0.255	372	0.2106	233	0.1788
5	795	350	0.253	376	0.2108	234	0.1784
6	790	351	0.252	380	0.2110	235	0.1780
7	785	352	0.250	384	0.2112	236	0.1776
8	780	354	0.248	388	0.2114	237	0.1772
9	775	355	0.247	392	0.2116	238	0.1768
10	770	358	0.246	396	0.2118	239	0.1764

3. Profile of Rolling Operation

For the rapid and efficient rolling of the hot material, it is advisable to employ multiple bidirectional rolling. Here, the rolling if the work slab was carried out in the opposite direction by reversing the rotation of the work rolls. This action was repeated several times to realize successive reduction of slab thickness. Accordingly, both the drive speed and torque became alternatively positive and negative, which corresponded to two-quadrant operation of the BLDC motor.

Calculation of successive rolling force and roll torque over ten passes yielded a torque-profile covering total time duration of one second. Considering a roll radius 250 mm and width 250 mm—and assuming a typical work roller speed range of 20 RPM to 30 RPM and gear reduction ratio of 40:1—the BLDC motor shaft speed varied from 800 RPM to 1200 RPM. The complete profile showing the motor speed and torque covering the 10 passes of rolling for three different HR steel materials is given in Table 3.

Table 3. Speed and torque-profile values for 10 passes of 3 different HR steel materials.

Time Duration (s)	Reduced Thickness (cm)	Roller Speed (RPM)	Motor Speed (RPM)	Load Torque for 3 Different HR Steel Materials		
				1040 HR (N-m)	1080 HR (N-m)	12l14 HR (N-m)
1	9.7	20	800	258.89	334.97	243.68
1	9.4	−20	−800	−263.98	−340.60	−246.53
1	9.1	22.5	900	269.97	346.37	249.45
1	8.8	−22.5	−900	−274.19	−352.28	−252.43
1	8.5	25	1000	279.69	358.33	255.50
1	8.2	−25	−1000	−284.18	−364.55	−258.64
1	7.9	27.5	1100	289.99	370.93	261.87
1	7.6	−27.5	−1100	−296.81	−377.51	−265.20
1	7.3	30	1200	301.79	384.29	268.62
1	7.0	−30	−1200	−309.16	−391.29	−272.16

4. Mathematical Model of BLDC Drive System

The ideal operation of the BLDC motor is characterized by trapezoidal waveforms of back-EMF and 120° conduction of stator-phase currents that have rectangular waveforms. Assuming balanced stator windings and constant self-inductances and mutual inductances, the equations of phase voltages take the matrix form as shown below in Equation (7):

$$\begin{bmatrix} V_{an} \\ V_{bn} \\ V_{cn} \end{bmatrix} = \begin{bmatrix} R & 0 & 0 \\ 0 & R & 0 \\ 0 & 0 & R \end{bmatrix} \begin{bmatrix} i_a \\ i_b \\ i_c \end{bmatrix} + \begin{bmatrix} L-M & 0 & 0 \\ 0 & L-M & 0 \\ 0 & 0 & L-M \end{bmatrix} \frac{d}{dt} \begin{bmatrix} i_a \\ i_b \\ i_c \end{bmatrix} + \begin{bmatrix} e_a \\ e_b \\ e_c \end{bmatrix} \tag{7}$$

where L is self induced EMF, M is mutually-induced EMF, e_a, e_b, e_c are back emf corresponding to phase a, b, c and i_a, i_b, i_c are current in phase a, b, c, respectively.

The interaction between the stator currents and the magnetic fields of the permanent magnets in the rotor gives rise to a unidirectional electromagnetic torque T_e in N-m is given by the following Equation (8):

$$T_e = \frac{e_a i_a + e_b i_b + e_c i_c}{\omega} \tag{8}$$

where ω corresponds to angular speed in rad/s.

The motor was fed from a 6-pulse voltage source inverter, with the gate control of the IGBT's satisfying 120 degree conduction mode. The required stator frequency was slaved to the rotor speed by sensing the rotor position using Hall sensors, which implement the logic, sequence and timing of the gate-triggering scheme.

5. Design of Fuzzy-Logic Controller

The fuzzy-logic controller is a rule-based controller which consists of input, processing and output stages. The input—or fuzzification stage—maps sensor or other inputs to the appropriate membership functions and logic levels. The processing stage invokes appropriate if–then rules for linking the fuzzified variables. Finally, the output—or defuzzification stage—converts the processed fuzzy variables into specific, crisp variables. A triangular membership function was chosen for implementing the "max–min" inference rules. The simulation diagram of fuzzy-logic controller is shown in Figure 6, where the inputs are the speed error (e) and its derivative (Δe), while the output generated is the reference torque.

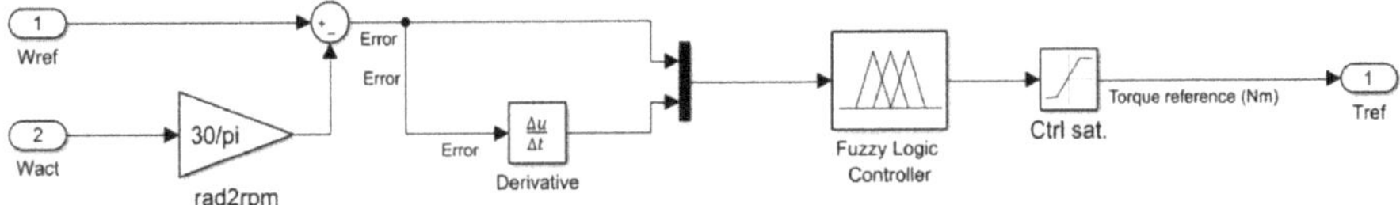

Figure 6. Fuzzy logic controller in MATLAB/Simulink.

Fuzzy rule has a 7 × 7 decision table with two input variables viz, error (e) and rate of change of error (Δe) and one output variable. The look-up Table 4 presents the input and output rules defined for seven fuzzified variables (NB, NM, NS, Z, PS, PM, PB) that stand for negative big, negative medium, negative small, zero, positive small, positive medium and positive big, respectively.

Table 4. Rule Table for fuzzy logic controller.

e \ Δe	NB	NM	NS	Z	PS	PM	PB
NB	NB	NB	NB	NM	NS	NS	Z
NM	NB	NM	NM	NM	NS	Z	PS
NS	NB	NM	NS	NS	Z	PS	PM
Z	NB	NM	NS	Z	PS	PM	PB
PS	NM	NS	Z	PS	PS	PM	PB
PM	NS	Z	PS	PM	PM	PM	PB
PB	Z	PS	PS	PM	PM	PB	PB

6. Controller Configuration with Feedback Design

Since a steel-rolling mill operation requires particular values of speed and torque over successive passes as defined in the respective profiles, the controller must ensure the operation of the BLDC drive system at the required frequency and specific stator current. This required a feedback control configuration consisting of an outer speed-control loop and an inner current-control loop. The outer loop sensed the speed error for processing through a fuzzy controller, so as to produce a reference torque. A set of three reference-phase-current waveforms were generated by making use of the reference torque. A current-controller block used these reference currents in comparison with the actual motor-phase-current waveforms to implement an error based hysteresis control algorithm in each phase through the gate triggering of the 6-pulse IGBT inverter-feeding the BLDC motor, so that the actual phase currents follow the reference values closely. Further, the alternate polarity of speed and torque values in the respective profiles enabled the 2-quadrant operation of the BLDC motor in successive passes of the rolling operation. The controller was chosen so as to meet the above requirements of generating alternate phase sequences at the output of the inverter along with desired current values.

7. Simulation Schematic

The MATLAB/Simulink software was used for modeling the overall proposed scheme for the operation of a hot-roughing steel-rolling mill integrated with BLDC motor along with gear box arrangement (Figure 7). Further, in Section 2.1 the hot-roughing steel-rolling process as modeled by Equations (1)–(6) was implemented using Simulink functional blocks and is shown in Figure 8. This schematic required a fairly large set of primary data like work roll radius, width, initial and final thickness of specimen and properties of 3 types of HR steel material and calculated the roll force and roll torque. Simulations run over 10 successive time slots of 1 s each generated the roll torque and speed profiles covering a total duration of 10 s. This profile represents the operating conditions of the multi-cyclic, forward-reverse rolling process. Figure 7 shows the overall schematic of the BLDC motor based drive system incorporating power circuit and control modules, so as to implement speed and current feedback algorithm [25–29]. The mechanical-load demand on the motor shaft in terms of shaft speed and torque were deduced from the profile of the rolling process by considering parameters of the gear train arrangement. This arrangement consisted of gear box modeled with a gear reduction ratio of 40:1 with gear box efficiency of 0.98. Figure 9 presents the MATLAB/Simulink model in closed-loop operation of BLDC motor.

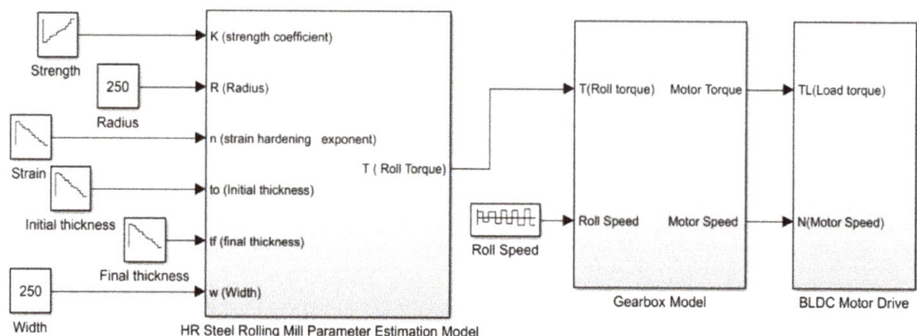

Figure 7. MATLAB/Simulink model of BLDC motor-drive system coupled with rolling process.

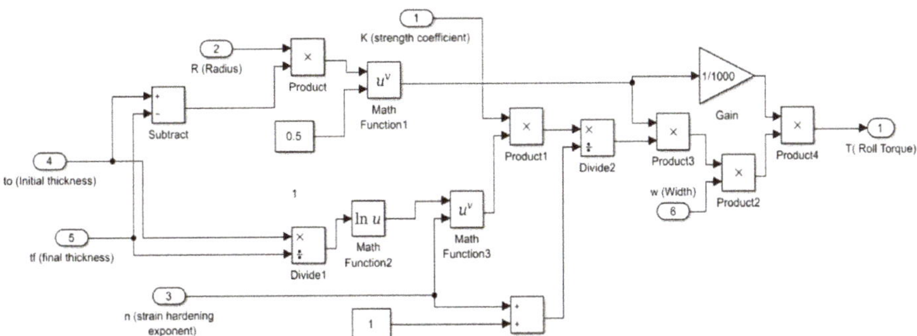

Figure 8. MATLAB/Simulink model of steel rolling process.

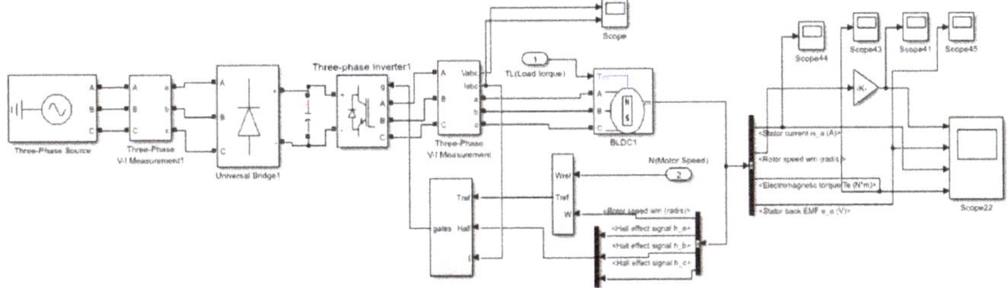

Figure 9. MATLAB/Simulink model in closed-loop operation of BLDC motor.

Simulation Results

Figure 10 shows the simulation results of the profiles corresponding to the speed and shaft torque values calculated for 3 different HR steel specimens viz. 1040 HR, 1080 HR, 12L14 HR. The profile values indicate the geared-down values of roll torque and the corresponding speed of the BLDC motors.

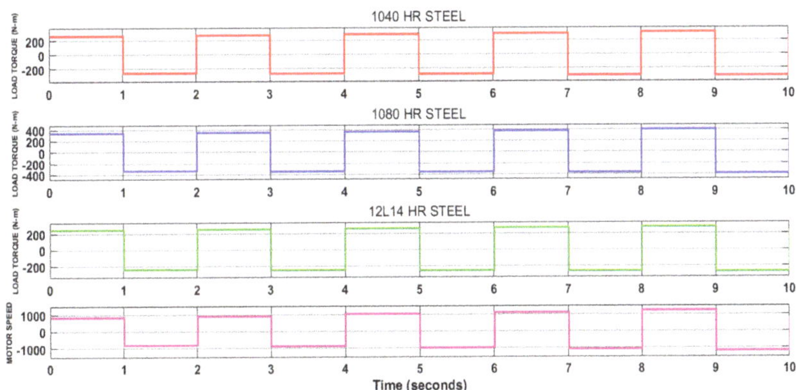

Figure 10. Speed and load torque demand at the work roll for 3 different HR steel.

The operation of the entire drive system to meet the speed and torque-profiles was simulated covering a total simulation history of ten seconds. The forward-reverse operation indicated in the profiles requires running of the motor in Quadrant I and Quadrant III alternatively in successive passes. The results show the variation of back-EMF, rotor speed, stator current and electromagnetic torque.

The expanded view of the simulation results of two passes for forward-reverse rolling is shown in Figure 11 depicting 2-quadrant operation of the BLDC drive system. It was seen that the rapid reversal of speed and torque as required in the rolling operation was achieved by the action of the fuzzy logic-based controller in changing the phase sequence of the inverter output. An expanded view of the transition from clockwise to the counter-clockwise rotation of the motor shaft along with the corresponding back-EMF transition is shown in Figure 11. As can be seen in the graph, the response of the drive system for bidirectional steel-rolling was fast. Figure 12 depicts the overall performance of the drive system covering five forward-reverse cycles totaling ten passed of rolling over the duration of 10 s. As can be seen in the graph, the response of the drive system for bidirectional steel rolling was fast.

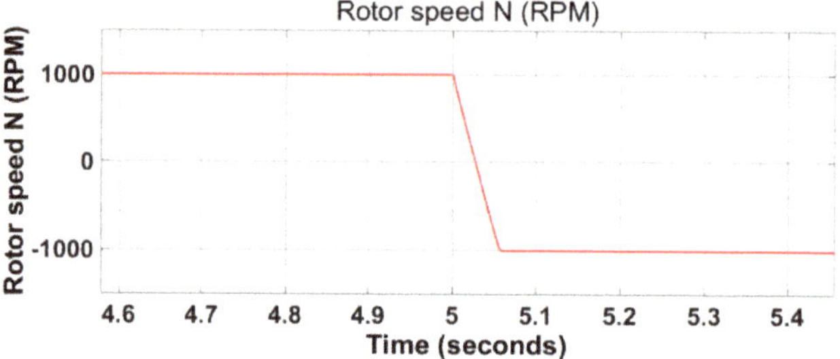

Figure 11. Expanded view of first two passes for forward-reverse rolling.

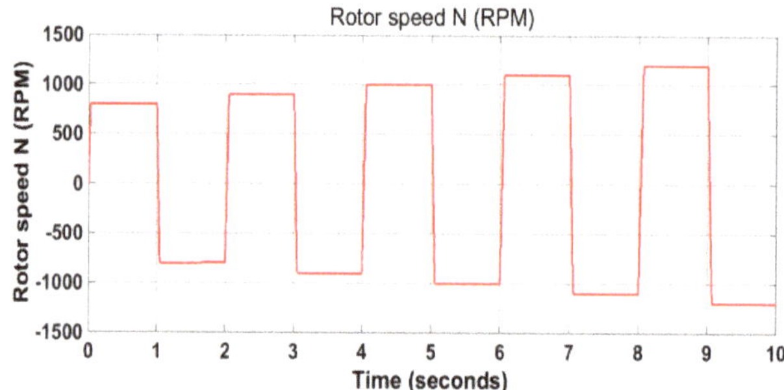

Figure 12. Actual speed of the BLDC motor for ten seconds.

The stator back-EMF waveform spanning ten seconds is shown in Figure 13; for each motor speed, the corresponding change in the back-EMF waveform can also be observed. The expanded view of back-EMF is shown in Figure 14 to indicate that the back-EMF of the BLDC motor was trapezoidal.

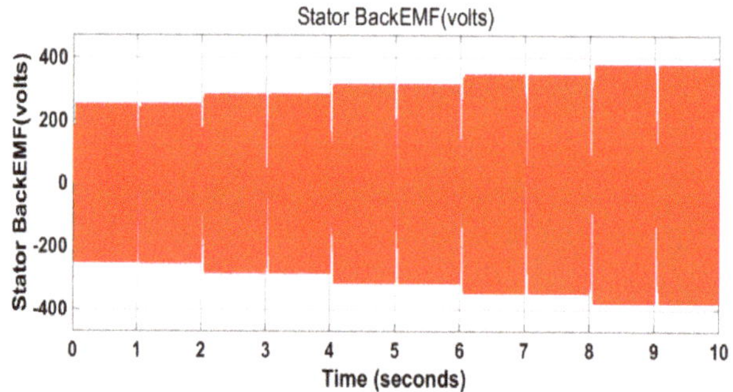

Figure 13. Stator back-EMF of BLDC motor for ten seconds.

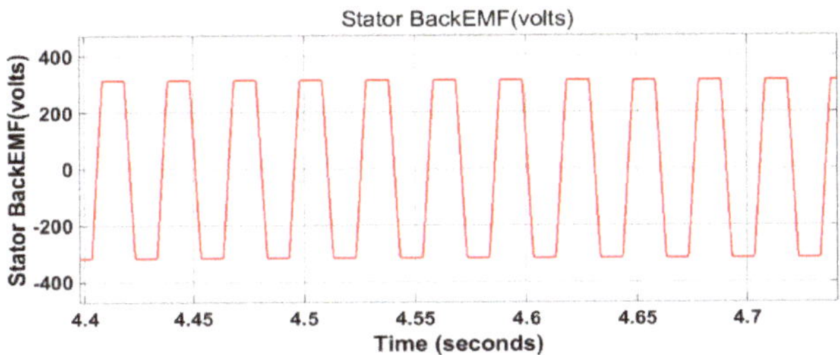

Figure 14. Expanded view of the stator back-EMF of the BLDC motor.

Figures 15–17 presents the overall performance of the drive system spanning 5 forward-reverse cycles totaling 10 passes of rolling over a duration of 10 s for 3 different HR steels viz. 1040 HR, 1080 HR and 12L14 HR steels. This clearly indicates a series of stepping up of all variables over successive cycles, as the rolling process progresses indicating the 2-quadrant operation of the BLDC drive system. It is seen that the rapid reversal of speed and torque as required in the rolling operation was achieved by the action of the fuzzy logic-based controller in changing the phase sequence of the inverter output. The applied-load torque alternated every second and was bidirectional in shape, which was termed as the reference torque. The developed electromagnetic torque followed the reference torque continuously. Figures 15–17 displays the variation and expanded views of transition of shaft torque and stator current in Phase A. Table 5 shows the ratings and parameters of the BLDC motor used in this study.

Table 5. BLDC motor parameters.

Motor Mating	**62 HP**
Voltage	500 V
Rated speed	1500 RPM
Phase resistance	0.2 Ω
Phase inductance	8.5 mH
Number of pole pairs	2
Back EMF	Trapezoidal

Several controllers were employed for checking the performance of the developed system. Its ability to track the reference speed could be observed from the expanded view (Figure 18a,b). The uniqueness of the fuzzy-logic controller performance was superior in terms of steady-state tracking, lesser overshoot and undershoots rather than the other two controllers employed. The overall system provided better performance in achieving the profile values rather than motor drive system using neural network compensation.

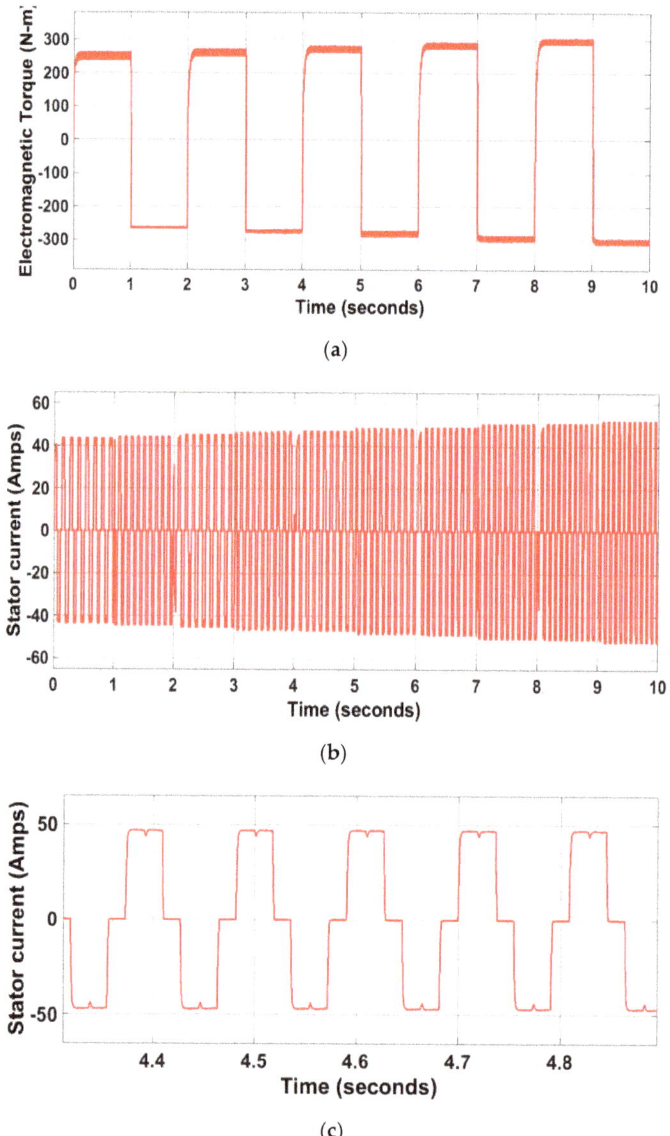

Figure 15. (**a**) Electromagnetic torque developed by a motor using the parameters of 1040 HR steel; (**b**) stator current in Phase A of a motor using the parameters of 1040 HR steel; (**c**) expanded stator current waveform in Phase A of motor corresponding to 1040 HR steel.

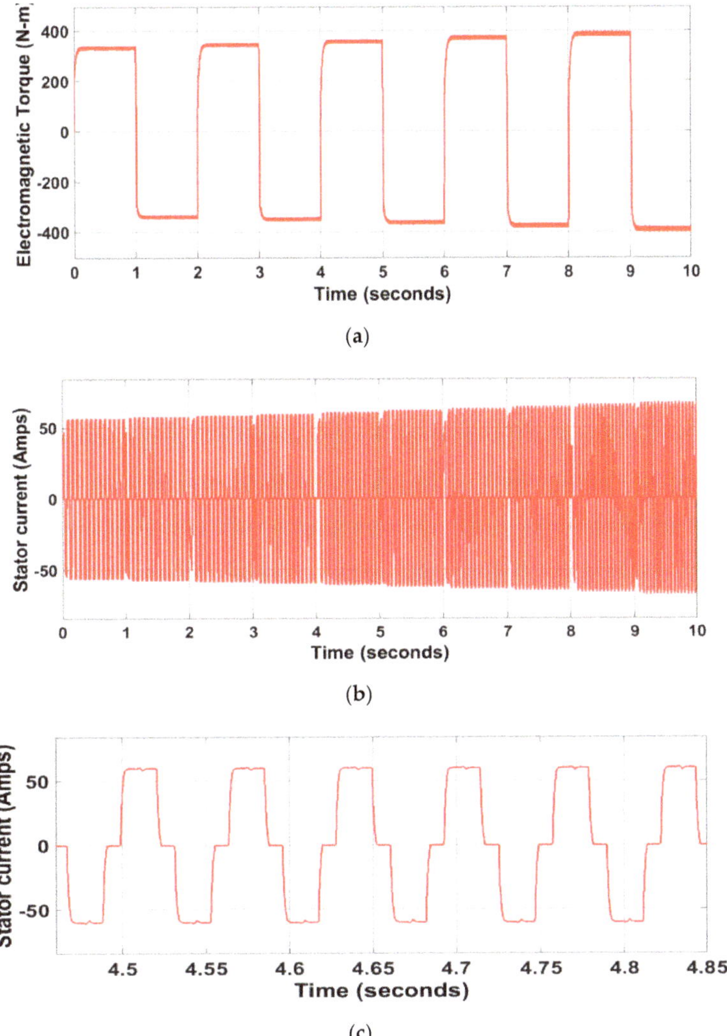

(a)

(b)

(c)

Figure 16. (**a**) Electromagnetic torque developed by a motor using the parameters of 1040 HR; (**b**) stator current in Phase A of a motor using the parameters of 1080 HR steel; (**c**) expanded stator current waveform in Phase A of motor corresponding to 1080 HR steel.

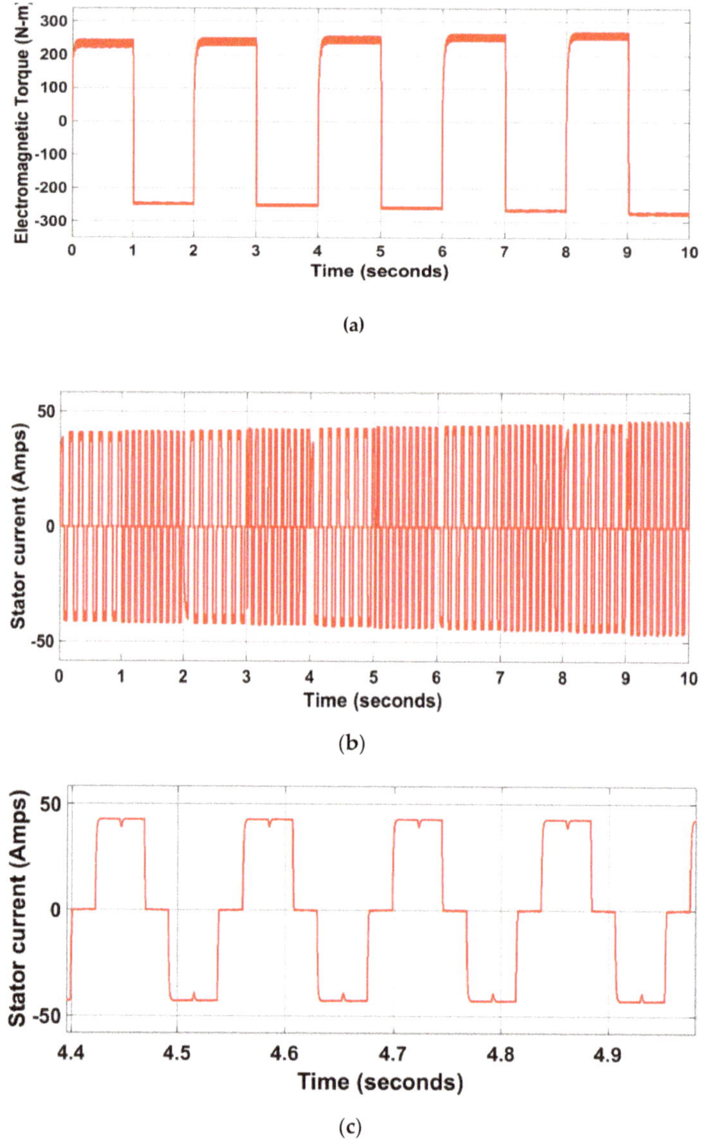

(a)

(b)

(c)

Figure 17. (**a**) Electromagnetic torque developed by a motor using the parameters of 12L14 HR steel; (**b**) stator current in Phase A of a motor using the parameters of 12L14 HR steel; (**c**) expanded waveform of stator current in Phase A of motor corresponding to 12L14 HR steel.

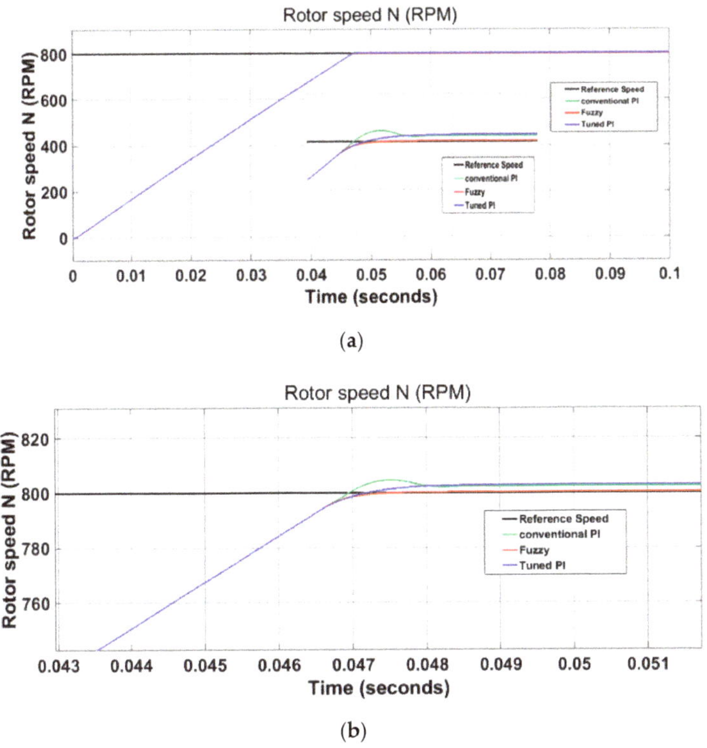

Figure 18. (**a**) Comparison of speeds for various controllers; (**b**) expanded view for various controllers.

8. Experimental Setup

The block diagram consisted of a BLDC motor with loading arrangements, an IGBT-based intelligent power module (IPM) and arm processor development board—along with speed sensors and interfaces required for the closed-loop operation of the drive system (Figure 19). Here, three-phase power from the AC mains was supplied to the diode bridge rectifier, which was converted into DC supply and fed to the 3-phase inverter supplying power to the BLDC motor. The actual speed of the motor was sensed and compared with the set or reference speed and accordingly; a control algorithm was employed for the generation of reference torque. The reference torque responsible for the production of three-phase reference current was compared with actual motor current and hysteresis control algorithm was employed for the calculation of gate signals to trigger the controlled switches of the three-phase inverter at various timing instants. The computation performed from sensing to the gate signal production at various instants, the control algorithm was developed and embedded into the arm microcontroller. Based on this action, the arm microcontroller provided the triggering signals at different instants which were interfaced through gate drivers to the controlled devices essential for running the BLDC motor. The BLDC motor drive set up was interfaced with CPU for monitoring the actual and reference speed and displayed the speed waveforms through a graphical user interface (GUI). This facility allowed monitoring of the actual speed of the BLDC motor and reference speed in the Simulink scope window. The experimental set up of the entire drive system is shown in Figure 20.

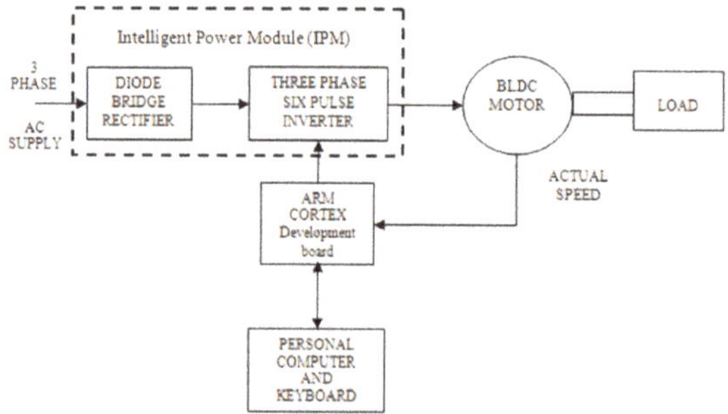

Figure 19. Block diagram of real-time BLDC motor drive system.

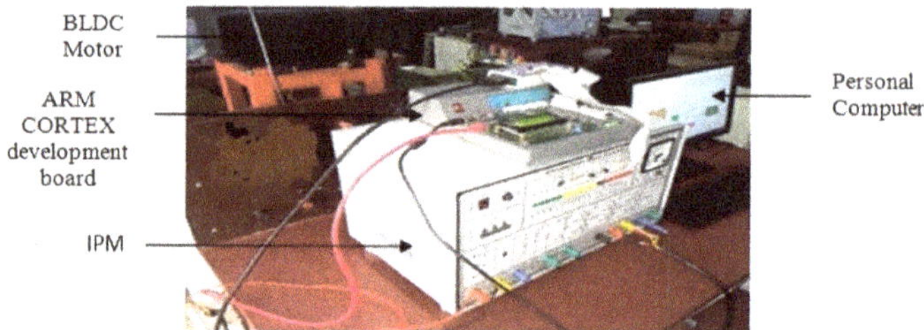

Figure 20. Experimental set up of BLDC motor drive system.

The online monitoring of the actual motor speed based on the set speed adjustment displayed in GUI was captured and shown in Figure 21. The set or reference speed indicated in red color spans over a range of 800 RPM to 1300 RPM in steps of 100 RPM. The actual motor speed indicated in blue color follows the set speed, even though we changed the speed value at various instants of time. The control algorithm embedded in the arm microcontroller was much efficient to adopt changes in following the set speed.

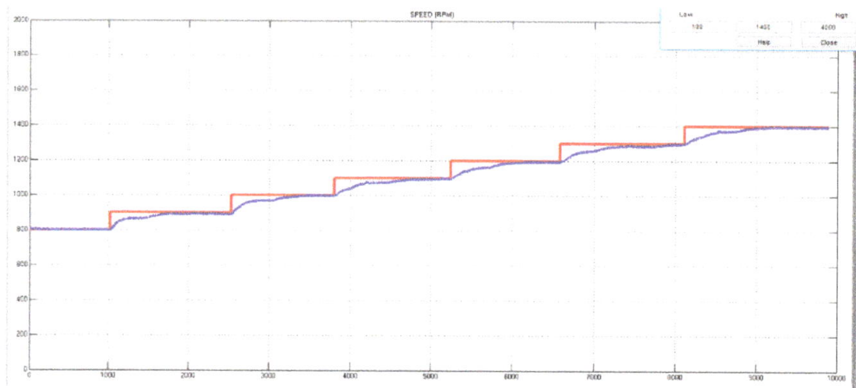

Figure 21. Online monitoring of set speed and the actual speed of motor through a graphical user interface (GUI).

9. Conclusions

The work described in this study commences with the mathematical modeling of hot rough-steel-rolling mill taking into account the metallurgical and mechanical aspects of the process for three different specimens of steel. A pair of profiles for the variation of roll speed and roll torque covering multipass forward–reverse rolling was generated in MATLAB/Simulink using the above model. The overall Simulink schematic also included the BLDC motor model and fuzzy logic-based controller model. The profile calculations obtained from steel rolling process were utilized as command inputs by the controller, which was configured with an outer speed and an inner current feedback loop. Performance of the whole system was evaluated by simulating the overall schematic and the resulted covering a set of electrical and mechanical variables are presented. The results confirm that the range of speed and torque as required over successive forward–reverse passes in the rolling operation were successfully met by the drive system. The results indicate satisfying rapid reversal of rolling speed and the corresponding roll torque over successive passes suggested an energy-efficient solution. It could be well witnessed that the fuzzy-based controller for BLDC motor could definitely play a vital game changing role as an alternate solution to induction motor drive for steel rolling.

It could be clearly concluded from the validation of hardware and simulation results that the novel application of a BLDC motor with a fuzzy logic-based controller was a highly effective approach. The efficacy of novel contribution proposed in this study for the first time could be witnessed from the validation of the resulted pertaining to BLDC motor with its fuzzy-based controller thereby serving to be an attractive alternative to conventional induction motor-drive systems for steel rolling applications. It could be well witnessed that the proposed fuzzy-based controller responds rapidly within a time period of 0.0001 s, compared to conventional PI and tuned PI that takes more than 0.003 s. The superiority in the performance in terms of faster response gives it a major edge as a part of the proposed application.

In contrast to conventional induction motor drive, the most quantifiable finding of this study is that the novel approach of employing fuzzy-based BLDC motor was far superior in terms of performance and a highly energy-efficient approach. This study will serve as a reference for all researchers and engineers working in this domain to effectively employ this methodology in steel rolling mill applications.

Author Contributions: M.N. performed the work under the guidance and direction of S.R., S.N. and S.S.R. All authors analyzed the data and contributed towards the paper. M.N. wrote the paper and executed the experiments under the guidance, supervision and suggestions from S.R., S.N. and S.S.R. All authors have read and agreed to the published version of the manuscript.

Funding: This research received no external funding.

Acknowledgments: Authors would like to thank the management of SASTRA Deemed University and the R& D lab at Power electronics and drives laboratory of SEEE at SASTRA University for providing necessary facilities to carry out the research.

References

1. Lee, D.; Moon, C.; Moon, S.C.; Park, H. Development of healing control technology for reducing breakout in thin slab casters. *Control Eng. Pract.* **2009**, *17*, 3–13. [CrossRef]
2. Yamada, F.; Sekiguchi, K.; Tsugeno, M.; Anbe, Y.; Andoh, Y.; Forse, C.; Guernier, M.; Coleman, T. Hot Strip Mill Mathematical models and Set Up Calculation. *Ind. Appl.* **1991**, *27*, 131–139. [CrossRef]
3. Bose, B. Power Electronics and Motor Drives Recent Progress and Perspective. *IEEE Trans. Ind. Electron.* **2008**, *56*, 581–588. [CrossRef]
4. Xu, F.P.; Li, T.C.; Tang, P.H. A low cost drive strategy for BLDC motor with low torque ripples. In Proceedings of the IEEE Conference on Industrial Electronics and Applications, Singapore, 3–5 June 2008.
5. Lianbing, L.; Hui, J.; LiQiang, Z.; Hexu, S. Study on Torque Ripple Attenuation for BLDCM Based on Vector Control Method. In Proceedings of the 2009 Second International Conference on Intelligent Networks and Intelligent Systems, Tianjin, China, 1–3 November 2009.
6. Matlab, Version 7.1. 2005. Available online: http://www.mathworks.com (accessed on 1 February 2012).

7. Potnuru, D.; Ch, S. Design and implementation methodology for rapid control prototyping of closed loop speed control for BLDC motor. *J. Electr. Syst. Inf. Technol.* **2018**, *5*, 99–111. [CrossRef]
8. El-Samahy, A.A.; Shamseldin, M.A. Brushless DC motor tracking control using self-tuning fuzzy PID control and model reference adaptive control. *Ain Shams Eng. J.* **2018**, *9*, 341–352. [CrossRef]
9. Ravell, D.A.M.; Maia, M.M.; Diez, F.J. Modeling and control of unmanned aerial/underwater vehicles using hybrid control. *Control Eng. Pract.* **2018**, *76*, 112–122. [CrossRef]
10. Pillay, P.; Krishnan, R. Modeling, simulation, and analysis of permanent- magnet motor drives, part II: The brushless dc motor drive. *IEEE Trans. Ind. Appl* **1989**, *IA-25*, 274–279. [CrossRef]
11. Kim, I.; Nakazawa, N.; Kim, S.; Park, C.; Yu, C. Compensation of torque ripple in high performance BLDC motor drives. *Control Eng. Pract.* **2010**, *18*, 1166–1172. [CrossRef]
12. Mahfouf, M.; Yang, Y.Y.; Gama, M.A.; Linkens, D.A. Roll Speed and Roll Gap Control with Neural Network Compensation. *ISIJ Int.* **2005**, *45*, 841–850. [CrossRef]
13. Niapour, S.M.; Tabarraie, M.; Feyzi, M. A new robust speed-sensorless control strategy for high-performance brushless DC motor drives with reduced torque ripple. *Control Eng. Pract.* **2014**, *24*, 42–54. [CrossRef]
14. Mazaheri, A.; Radan, A. Performance evaluation of nonlinear Kalman filtering techniques in low speed brushless DC motors driven sensor-less positioning systems. *Control Eng. Pract.* **2017**, *60*, 148–156. [CrossRef]
15. Lu, H.; Zhang, L.; Qu, W. A New Torque Control Method for Torque Ripple Minimization of BLDC Motors With Un-Ideal Back EMF. *IEEE Trans. Power Electron.* **2008**, *23*, 950–958. [CrossRef]
16. Shi, T.; Guo, Y.; Song, P.; Xia, C. A New Approach of Minimizing Commutation Torque Ripple for Brushless DC Motor Based on DC–DC Converter. *IEEE Trans. Ind. Electron.* **2009**, *57*, 3483–3490. [CrossRef]
17. Salah, W.A.; Ishak, D.; Hammadi, K. PWM Switching Strategy for Torque Ripple Minimization in BLDC Motor. *J. Electr. Eng.* **2011**, *62*, 141–146. [CrossRef]
18. Sathyan, A.; Milivojevic, N.; Lee, Y.J.; Krishnamurthy, M.; Emadi, A. An FPGA-Based Novel Digital PWM Control Scheme for BLDC Motor Drives. *IEEE Trans. Ind. Electron.* **2009**, *56*, 3040–3049. [CrossRef]
19. Zhao, J.; Wang, W.; Liu, Q.; Wang, Z.; Shi, P. A two-stage scheduling method for hot rolling and its application. *Control Eng. Pract.* **2009**, *17*, 629–641. [CrossRef]
20. Ma, L.; Dong, J.; Peng, K.; Zhang, K. A novel data-based quality-related fault diagnosis scheme for fault detection and root cause diagnosis with application to hot strip mill process. *Control Eng. Pract.* **2017**, *67*, 43–51. [CrossRef]
21. Steinboeck, A.; Wild, D.; Kugi, A. Nonlinear model predictive control of a continuous slab reheating furnace. *Control Eng. Pract.* **2013**, *21*, 495–508. [CrossRef]
22. Lotter, U.; Schmitz, H.P.; Zhang, L. Application of the Metallurgically Oriented Simulation System "TKS-StripCam" to Predict the Properties of Hot Strip Steels from the Rolling Conditions. *Adv. Eng. Mater.* **2002**, *4*, 207–213. [CrossRef]
23. Ko, J.S. Robust position control of BLDC motors using Integral-Proportional-Plus Fuzzy logic controller. *IEEE Trans. IE* **1994**, *41*, 308–315.
24. Li, W. Design of a hybrid fuzzy logic proportional plus conventional integral-derivative controller. *IEEE Trans. Fuzzy Syst.* **1998**, *6*, 449–463. [CrossRef]
25. ASM Metals Handbook, Forming and Forging. *ASM Int.* **1996**, *14*, 351–355.
26. Mikell, P. Groover Fundamentals of Modern Manufacturing. In *Materials, Processes and Systems*, 4th ed.; John Wiley & Sons: Hoboken, NJ, USA, 2010; pp. 395–403.
27. AISI/SAE Standard (American Iron and Steel Institute/Society of Automotive Engineers). *SAE* **2014**, *J403*, 1010.
28. AISI/SAE Standard (American Iron and Steel Institute/Society of Automotive Engineers). *SAE* **2014**, *J403*.
29. Varshney, A.; Gupta, D.; Dwivedi, B. Speed response of brushless DC motor using fuzzy PID controller under varying load condition. *J. Electr. Syst. Inf. Technol.* **2017**, *4*, 310–321. [CrossRef]

Article

A Wide-Frequency Constant-Amplitude Transmitting Circuit for Frequency Domain Electromagnetic Detection Transmitter

Xiujuan Wang [1,*]**, Zhihong Fu** [1,*]**, Yao Wang** [2]**, Wendong Wang** [1]**, Wei Liu** [1] **and Junli Zhao** [1]

[1]   School of Electrical Engineering, Chongqing University, State Key Laboratory of Power Transmission Equipment and System Security, Chongqing 400044, China; cquwwd@cqu.edu.cn (W.W.); 20181101027@cqu.edu.cn (W.L.); lijunli_65@163.com (J.Z.)

[2]   Yellow River Survey Planning Design and Research Institute Co. LTD., Zhengzhou 450003, China; 20161101003@cqu.edu.cn

*   Correspondence: shanshui510@163.com (X.W.); fuzhihong@cqu.edu.cn (Z.F.); Tel.: +86-13062352738 (Z.F.)

Received: 7 May 2019; Accepted: 5 June 2019; Published: 6 June 2019

Abstract: In this paper, a novel AC magnetic transmitter current source circuit is proposed for application of frequency domain electromagnetic method (FEM) prospecting. The proposed current source circuit is capable of generating high frequency and high constant amplitude currents, which are key technical problems for FEM. It is suitable for very wide frequencies. The main circuit of the proposed current source consists of a rising-edge enhancing unit, a constant current control unit, and a high voltage clamping unit. Large constant clamping voltage is applied during the rising edge and the falling edge of the alternating square current to obtain a high frequency and high linearity current source. On the current flat stage, the constant current unit provides the energy to the load to ensure the constant amplitude of the output current. Detailed operations of the proposed magnetic current source are given. Simulation and experimental results demonstrate that the proposed circuit achieves short reversal time, the linearity of the rising/falling edge, constant amplitude and low power loss. These are the desired characteristics of the ac square current source probing transmitter for the magnetic FEM applications.

Keywords: current source circuit; voltage clamping; constant current control; frequency domain electromagnetic

---

1. Introduction

The electromagnetic detection method has the advantages of deep exploration depth, high horizontal resolution, strong penetration capability and anti-interference, and has been widely used in geophysical exploration such as mineral exploration, water resources exploration and engineering geological survey [1–4]. The research of the excitation source is an important topic in geoscientific instrument research. However, in most power supply situations, there are large inductive loads, small loop resistance, low power supply voltage, etc. Therefore, it is very difficult to obtain ideal high-frequency AC square wave current source.

The linearity of the commutation edge, short commutation time, high stability are important characteristics of the pulse current source, and also the core problems for researchers. When the turn-off time is constant, increasing the voltage pulse amplitude can improve the linearity of the current falling edge and improve the waveform quality. Based on the idea of high voltage clamping, [5,6] proposed a method to control the turn-off time and improve the slope of the front and rear edges of the current by adjusting the voltage of the clamping voltage source. In [7], the solid-state DC/DC boost front-end and h-bridge output are used to achieve the purpose of rapid current rise. In order to realize the output current steep pulse, literatures [8] and [9] respectively put forward the method of using high-voltage dc power supply and increasing boost charging circuit to charge capacitor during pulse current commutation. In order to improve the quality stability of transmitting current waveform, a method for suppressing the ripple current of the flat-top section of the emission current by introducing a passive bypass circuit in [10], and reference [11] proposed a high precision mixed-signal (analog and digital) plan for output current stabilization. A model predictive current control scheme with phase-shifted pulse-width modulation is introduced in [12], which has the advantages of small tracking error and high steady-state performance. What's more, reference [13] proposed a high-precision pulse current source control system can get good dynamic response by improving switching characteristics, and the flat top current can be regulated with precisely defined precision. A converter based on switched-capacitor units with high voltage gain and high efficiency was proposed in [14], the proposed topology can generate repetitive high-voltage pulses. The literature [13] proposed a pulse current source control system with high-precision and fast dynamic response through the improvement of control strategy and algorithm.

Recently, due to the need for engineering geophysical prospecting, detection of geological information in the shallow layer and the superficial layer of the ground has received a great deal of attention [15,16], this requires the transmitter to have a higher frequency of current. The frequency domain electromagnetic transmitter needs to solve the problems of fast commutation of current, high linearity of commutation, wide frequency transmission (500 Hz ~ 100 KHz) and constant current amplitude. For the key technical problems of frequency domain electromagnetic transmitter, this paper proposes bipolar broadband communication pulse current source circuit, the highest transmission frequency can reach 100 KHz, the maximum emission current is 15 A, the current rises rapidly to the set value with the slope of 10^7 A/s, the current is quickly turned off from the positive polarity amplitude to the negative polarity amplitude, and there is no dead-time in the middle, it can maintain the current value at 15 A in the flat stage, which achieves wide-band constant amplitude and the current amplitude is greatly improved at high frequencies. The proposed approach provides a solution to the high frequency and high current emission problems encountered in frequency domain transmission systems.

2. Magnetic Source Frequency Domain Electromagnetic Transmitter

This section describes the schematic, operation principle, and realization details of the proposed magnetic source FEM probing circuit.

2.1. Operation Principle

The traditional full bridge circuit model is shown in Figure 1a, and the schematic of the proposed current source circuit is depicted in Figure 1b. The circuit consists of three major units according to its functionalities. They are rising-edge enhancing unit, constant current control unit, and high voltage clamping unit. In order to achieve linearity of the load current commutation edge, it is necessary to ensure that the load voltage is constant during current commutation. Using the existing high-speed turn-off technique, the linearity of the entire commutation edge can be achieved by clamping the load during load current rise. In this paper, in order to avoid the influence of supply voltage and output frequency on the current amplitude, the constant current source is used as the load power supply. By clamping the load voltage during the rising edge, leading edge linearity is achieved. To simplify the circuit, leading-edge boost circuit and fast turn off circuit share the same clamp circuit and clamp voltage source to achieve the consistency of rising and falling edges of the square wave current. The main circuit is shown in Figure 1b.

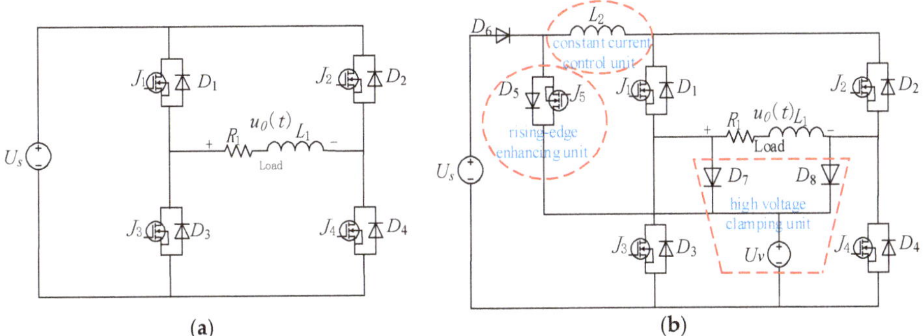

Figure 1. Circuit model: (**a**) Traditional full bridge circuit model. (**b**) Constant current and voltage clamping control circuit.

The energy supplement circuit is made up of diode D_5, J_5 and clamping voltage source U_v; the AC square wave current generator is a full bridge transmitter circuit; clamp circuit is composed of diodes D_7, D_8 and clamping voltage source U_v, Constant current inductor L_2 is used as a constant current source. The waveforms of the load current and voltage are shown in Figure 2.

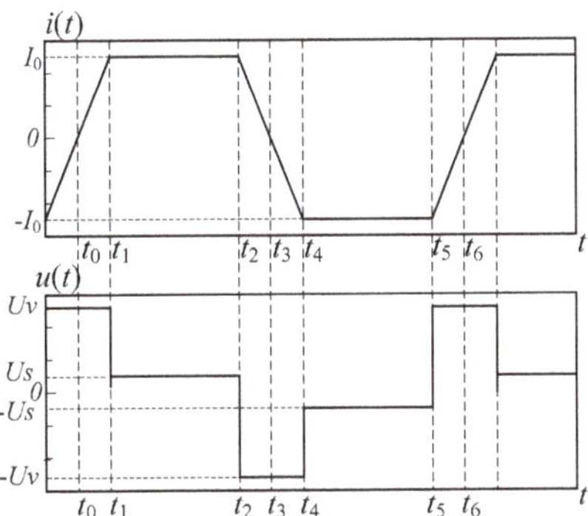

Figure 2. Expected waveforms of load current and voltage.

2.2. Process of the Constant Current and Voltage Clamping Control Circuit

One charge cycle of the load current is divided into two equal time periods, that is, positive power supply time and reverse power supply time. The basic idea is that during the current commutation period, the load voltage is always been clamped to the clamp voltage source to achieve commutation direction edge linear. While during the stable positive and reverse power supply, the load inductor is powered by the constant current source in order to make sure the output current amplitude is constant. The implementation process is as follows:

1. In $t_0 \sim t_1$ period: switches J_1 and J_4 are turned on, J_2 and J_3 are turned off. The constant current inductor L_2 and the load inductance L_1 are connected in series, which have unequal initial currents. High voltage is produced at the load inductance. D_7 is turned on, the current is less than the set value I_0 during commutation. J_5 is turned on, so the current of L_2 flows through J_1, D_7, J_5 to keep the circuit unblocked, at the same time the current of load inductor flows past R_1, L_1, J_4, U_v, D_7, the load voltage is clamped to the clamp voltage U_v, where $u(t) = U_v$, $L_1 di(t)/dt = U_v$, and the load current increases linearly. U_v is much larger than the DC supply voltage U_s and AC square wave current linearly rises from zero to the set amplitude of load current I_0.

2. In $t_1 \sim t_2$ period: switches J_1 and J_4 are turned on, J_2 and J_3 are turned off. The load inductance L_1 and constant current inductance L_2 are connected in series. When the load inductance L_1 current rises to be equaled to the constant current inductance of L_2, $u(t) = U_s$, the output current is the square wave, $i(t) = I_0$. During this period, when the load current is lower than the expected set minimum value I_{min}, J_5 is turned on, the clamped voltage source replenishes the energy to the constant current inductance, U_v, J_5, L_2, J_1, R_1, L_1, J_4 form a loop, the current of constant inductor rises rapidly. When the load current exceeds the set maximum value I_{max}, J_5 is turned off and the circuit continues to be supplied by the power supply.

3. In $t_2 \sim t_3$ period: switches J_2 and J_3 are turned on, J_1 and J_4 are turned off, the load current suffers a sudden change, D_8 is turned on, the load current flows through R_1, L_1, D_8, U_v, J_3, $u(t)$ changes from positive to negative, $u(t) = -U_v$. The voltage of the load is clamped to the clamped voltage source, $L_1 di(t)/dt = -U_v$, the square wave current begins to drop linearly, the load current $i\,(t_3)$ decays to zero at the time of t_3, the constant current of inductor L_2 flows through J_2, D_8, U_v, U_s, D_6 to charge the clamped voltage source.

For the same reason, the reverse cycle of the load current is similar to that of the positive supply cycle, which is not described here.

Analysis of the circuit shows that, the rising edge waveform of the load current of the circuit designed in this paper is linear, and when $U_v \gg R_1 I$, the absolute values of the rising edge slope and the falling edge slope are equal, the waveform is symmetrical completely. The load current is constant maintained by the constant current inductance, and it has nothing to do with the supply voltage, the frequency and the load device. The main function of the diode D_6 is to block the path of U_v and U_s. Because of D_6, when U_v is greater than U_s, J_5 is turned on, the clamp voltage source U_v works. Therefore, D_6 is indispensable.

3. Results Main Technical Index Analysis and Parameter Selection

3.1. Main Technical Index Analysis

Suppose the voltage $u(t)$ is constant, denote as U_v, deduce the formulas of square wave's positive edge and negative edge in time domain.

$$i(t) = -\frac{U_v}{R_1} + \left[i(t_2) + \frac{U_v}{R_1} \right] e^{-\frac{R_1}{L_1}(t-t_2)} \tag{1}$$

$$i(t) = \frac{U_v}{R_1} - \left[i(t_5) + \frac{U_v}{R_1} \right] e^{-\frac{R_1}{L_1}(t-t_5)} \tag{2}$$

Suppose formula (1) and (2) equals zero, the falling edge and the negative edge can be expressed as follows.

$$t_{d1} = t_3 - t_2 = -\frac{L_1}{R_1} \ln\left(\frac{U_v}{U_v + IR_1} \right) \tag{3}$$

$$t_{d2} = t_6 - t_5 = -\frac{L_1}{R_1} \ln\left(\frac{U_v}{U_v - IR_1} \right) \tag{4}$$

The commutation time of square wave current is $t_{d1} + t_{d2}$ or

$$t_d = -\frac{L_1}{R_1} \left(\ln\left(\frac{U_v}{U_v + IR_1} \right) + \ln\left(\frac{U_v}{U_v - IR_1} \right) \right) \tag{5}$$

where I is the amplitude of load current, U_v is the voltage of the clamp voltage source, L_1 and R_1 are the load inductance and the equivalent resistance, respectively. The absolute value of negative edge and positive edge slopes of AC square wave current $i(t)$ are:

$$K_1 = \left|\frac{di(t)}{dt}\right| = \left|\frac{R_1 I + U_v}{L_1}\right| \tag{6}$$

$$K_2 = \left|\frac{di(t)}{dt}\right| = \left|\frac{R_1 I - U_v}{L_1}\right| \tag{7}$$

When $U_v \gg R_1 I$, the commutation time t_d and the edge slope K are:

$$t_d = t_{d1} + t_{d2} \approx \frac{2IL_1}{U_v} \tag{8}$$

$$K = K_1 = K_2 \approx \frac{U_v}{L_1} \tag{9}$$

Since $I = Kt$, deduce the following expressions:

$$I = K_2 \cdot t_{d2} = \frac{U_v}{R_1}\left(\ln\frac{U_v}{U_v - IR_1}\right) \tag{10}$$

According to the analysis above, the main specifications of the square wave current source are:

1. Shorten the commutation time, heighten the output frequency. From formula (8), the commutation time can be adjusted by changing U_v; By increasing U_v, the commutation time t_d will decrease. Decreasing the ratio of t_d over one cycle helps to increase the output frequency.
2. Increase current amplitude, output remains constant. Form formula (10), the square wave current can be adjusted by changing U_v; heighten U_v, the output current will increase. When $U_v \gg U_s$, the instantaneous value of load current will increase rapidly to I with a slope of K and stabilize in a short time. Since I has nothing to do with source voltage, output frequency and load inductor, the load current will remain constant.
3. The emission current waveform is symmetrical. During the load current commutation, load current is clamped at a constant high voltage, the commutation edge performs in linear. When $U_v \gg R_1 I$, the slope of positive edge equals the slope of negative edge, the commutation edge is a line with a slope K, the positive and negative edges remain conformity and share a clamping circuit, the positivity and negativity of load current also remain conformity, there is no current reverse overshoot, and the current in flat region is constant.
4. Circuit energy loss is low. When the load current decreases, the load energy will store in the clamp voltage source, which lowers the circuit energy loss.

3.2. Circuit Parameters Selection

Let the steady state load current be $I_0 = 15$ A and the clamp voltage be $U_v = 1000$ V. Optimize the load inductor, capacitor and constant current inductor L_2.

When $U_s = 12$ V, $I = 15$ A, $R_1 = 0.3\ \Omega$, $L_1 = 00\ \mu$H, $\Delta I = 0.5$ A, the current fluctuation range should consider the hysteresis output, which is no more than the switch diode's maximum frequency $f_{max} = 100$ kHz. The constant current inductance is:

$$L_{min} = \frac{R_1}{f_{max} \cdot \ln\frac{R_1 I - U_s}{R_1 I + R_1 \Delta I - U_s}} - L_1 \tag{11}$$

where ΔI is the amplitude fluctuation of the square wave current; f_{max} is the maximum frequency of switch diode J_5. Given that the lower frequency of J_5 is, the larger constant current inductance is required, L_1 should be no less than L_{min}.

4. Simulation of Circuit Model

The wide-band banner current source circuit proposed in this paper adds constant current control and high voltage clamping measures to the traditional full-bridge circuit model. It can be realized that the amplitude of the output current is constant during the stable positive/negative power supply, and the waveform of the output current is linear during the current commutation.

In order to prove that the method can achieve the expected results, the circuit model is simulated and verified by Power Simulation software. In order to compare and analyze, this paper simulates the output waveform of the proposed circuit model and the traditional full-bridge circuit model. Figures 3 and 4 show the schematic diagram of the traditional full-bridge circuit model and the proposed circuit model in the PSIM software simulation environment.

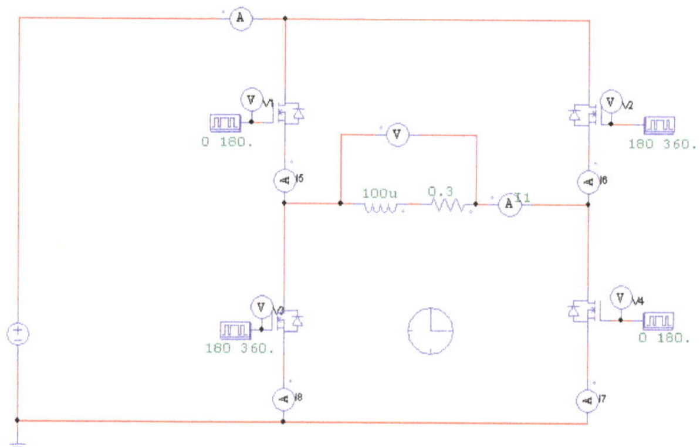

Figure 3. Simulation model of traditional full bridge circuit.

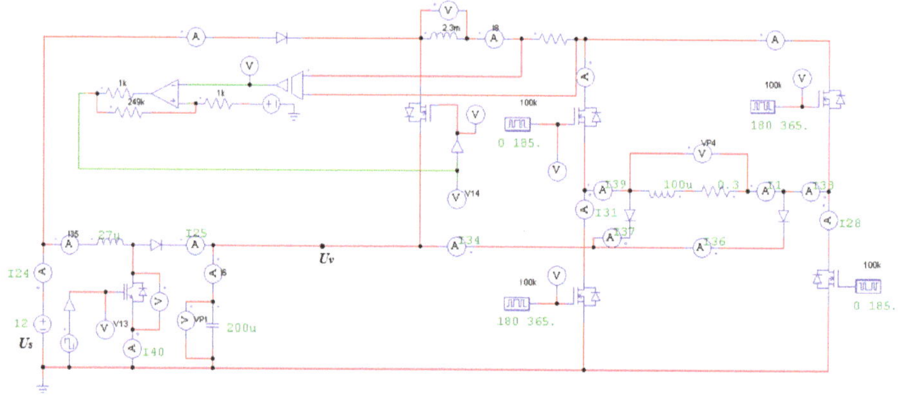

Figure 4. Simulation model of constant current and voltage clamping control circuit.

The setting parameters are as follows: supply voltage U_s is 12 V, clamp voltage U_v is 1000 V, as shown in Figure 5. load inductance L_1 is 100 μH, R_1 is 0.3 Ω. In the case of the same parameters, we compared the circuit proposed in this paper with the traditional full bridge circuit through simulation. L_2 is selected to tolerate a constant current below the maximum value of 15 A. At the same time, set the upper limit of current hysteresis band control as 15.5 A and the lower limit as 14.5 A. To prove that the proposed circuit is suitable for a very wide frequency band, five frequency points are selected between 500 Hz and 100 KHz for simulation, which are 500 Hz, 1 KHz, 10 KHz, 50 KHz, 100 KHz. The comparisons of the simulation load responses between the full-bridge circuit and the proposed circuit are shown in Figures 6–10.

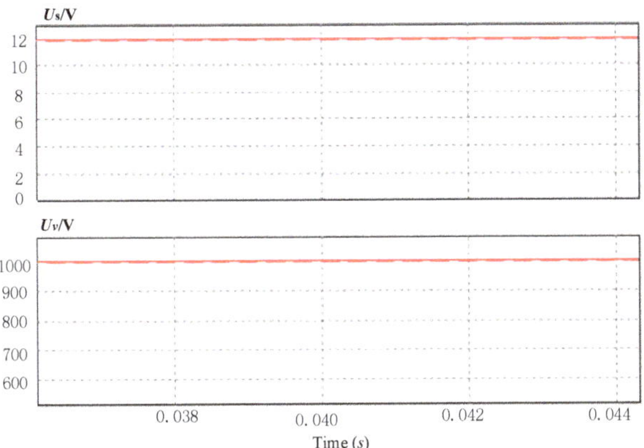

Figure 5. Simulation results of supply voltage U_s is 12 V, clamp voltage U_v is 1000 V.

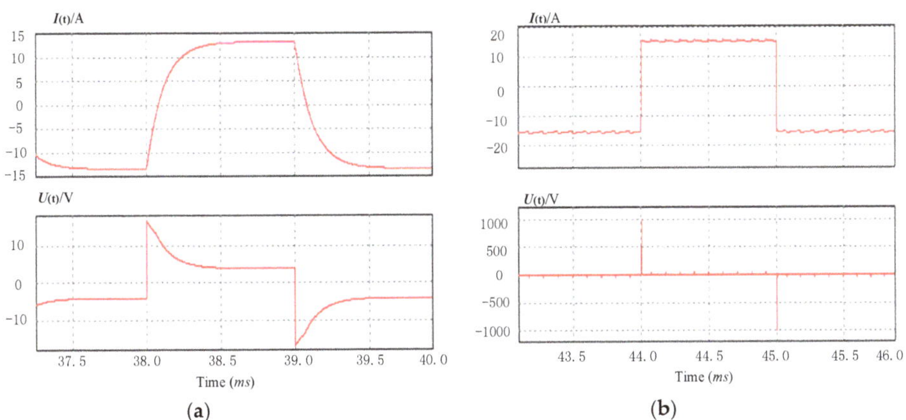

Figure 6. Waveforms of the output current and voltage, 500 Hz. (**a**) Traditional full-bridge circuit; (**b**) Constant current and voltage clamping control circuit.

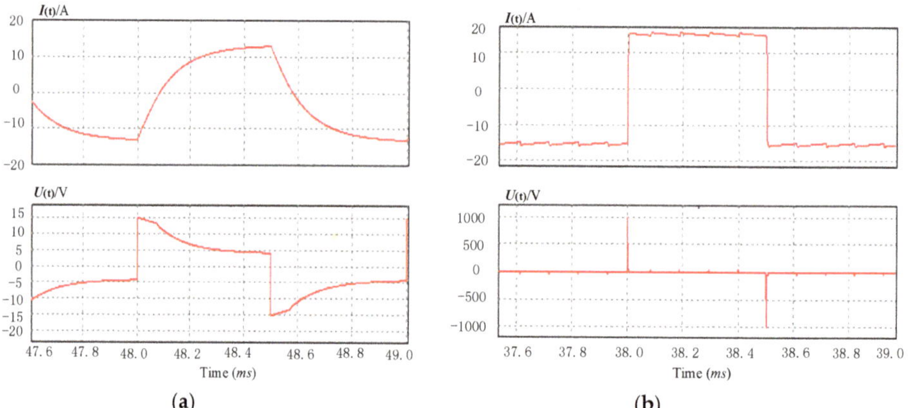

Figure 7. Waveforms of the output current and voltage, 1 KHz. (**a**) Traditional full-bridge circuit; (**b**) Constant current and voltage clamping control circuit.

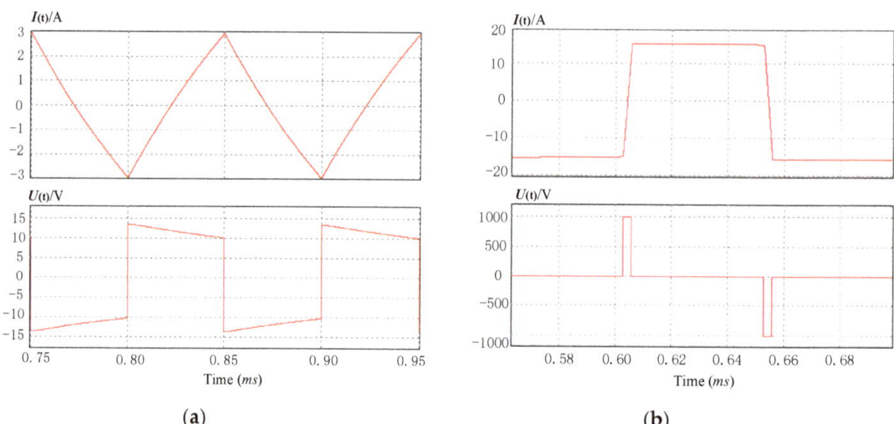

Figure 8. Waveforms of the output current and voltage, 10 KHz. (**a**) Traditional full-bridge circuit; (**b**) Constant current and voltage clamping control circuit.

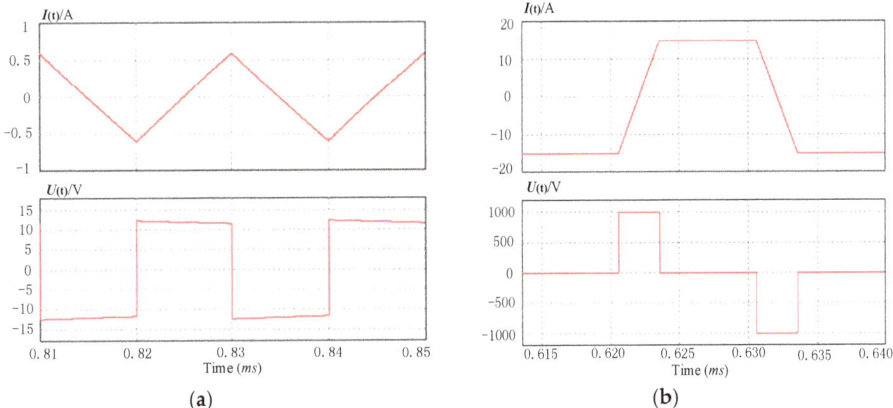

Figure 9. Waveforms of the output current and voltage, 50 KHz. (**a**) Traditional full-bridge circuit, (**b**) Constant current and voltage clamping control circuit.

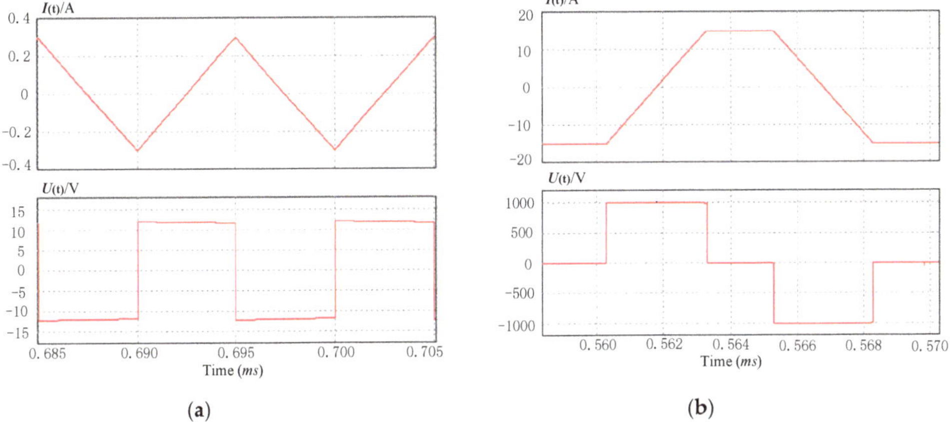

Figure 10. Waveforms of the output current and voltage, 100 KHz. (**a**) Traditional full-bridge circuit; (**b**) Constant current and voltage clamping control circuit.

It can be seen from Figures 6 and 7 that at the lower frequencies of 500 Hz and 1 KHz, the current waveform of the output of the traditional full bridge circuit is irregular, and the amplitudes are 15 A and 12 A. The output current waveform of the circuit presented in this paper is basically stable at 15 A in the flat region. Figure 8 shows that when the frequency is 10 KHz, the output waveform of the traditional full-bridge circuit is similar to that of the triangular wave, and the amplitude is 3 A. The circuit output current waveform proposed in this paper is linear during the rising/falling period, the flat region is stable, no fluctuation, and the amplitude is stable at 15 A. Figures 9 and 10 show that at the higher frequency of 50 KHz and 100 KHz, the current waveform of the output of the traditional full bridge circuit is triangular wave with amplitudes of 0.6 A and 0.3 A respectively. The current waveform of the circuit output proposed in this paper is stable in the flat region, and the amplitude is stable at 15 A. The linearity is high during the rising/falling period.

As shown in Figures 6–10, as the frequency increases, the output current amplitude of the traditional full-bridge circuit gradually decreases, and the waveform gradually changes from irregular to triangular wave; the output voltage amplitude is equal to the supply voltage Us of 12 V, and the waveform changes from irregular to square wave, but at the flat region always unstable.

In the case of the same circuit parameters, since the circuit proposed in this paper adds constant current control and voltage clamping measures, the output current waveform does not fluctuate with changes in frequency, source voltage and load. While the clamp voltage is 1000 V DC voltage, the current of the circuit I rises to the set value with fast rising edge at a rate of 10^7 A/s and maintains at 15 A when the current is in steady state. This means that this kind of circuit can well operate in a wide frequency range and simultaneously maintain constant output current, from 500 Hz to 100 kHz.

According to $di(t)/dt = -U_v/L_1$, in the case of the inductance, the circuit was defined and the rise steepness of the current was proportional to the value of the clamp voltage U_v. The higher the clamping voltage, the faster the rate of change of the rising edge of the current, the shorter the time it takes for the current to rise, and thus the cycle decreases and the operating frequency of the circuit increases. On the contrary, the lower the clamp voltage, the slower the rate of change in the rising edge of the current, the longer the current rise process, and the longer the cycle, the lower the operating frequency of the circuit. E.g., if the clamp voltage U_v is changed from 1000 V to 500 V, the slope of the current change will be half of the original, it is needed much more time to reach the specified current I_0, the cycle becomes larger, the operating frequency is reduced. In the case of equipment conditions permitting, the higher clamp voltage U_v, the better. Of course, in reality, the voltage cannot be increased indefinitely due to the limitation of the performance of the circuit device.

5. Experimental Results

In order to experimentally verify the constant current and voltage clamping control circuit proposed in this paper, the experimental circuit is established according to Figure 1b. The main circuit of the experiment is shown in Figure 11. The supply voltage U_s is 12 V, clamp voltage U_v is 1000 V, load inductance L_1 is 100 μH, constant current inductor L_2 is 2 mH. The maximum current in the circuit is about 15 A, and the voltage stress of the switching device is about 1000 V. Considering a certain voltage and current margin, Mosfet (C2M0045170D) and Dioxide (DSDI60-16A) are selected according to the datasheet. The main device parameters of the circuit are shown in Table 1.

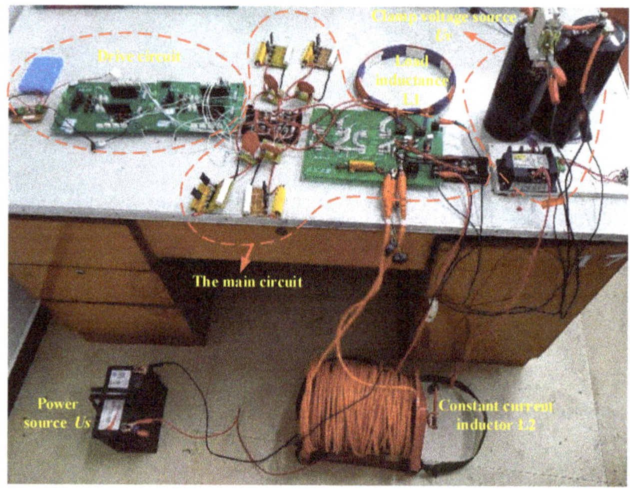

Figure 11. The photo of the main circuit in experiment.

Table 1. Main device parameters of the circuit.

DEVICE NAME	PARAMETER
MOSFET (1 ~ J5)	C2M0045170D (1700 V, 72 A)
DIOXIDE (D6 ~ D8)	DSDI60-16A (1600 V, 63 A)
INPUT VOLTAGE (U$_S$)	12 V
INDUCTANCE (L2)	2 mH
LOAD (L1)	100 µH

Limited by the performance of existing electronic devices, our experimental circuit can output stable waveforms of voltage and current with the maximum frequency of 50 KHz. The experimental results waveform of the voltage and current are shown in Figures 12 and 13 respectively. It is shown in Figure 12, during the experiments, that the load voltage is always been clamped between 998 V and 1002 V, and the waveform quality is very stable. Figure 13 shows that during the experiments the maximum output current of the circuit can reach up to 15 A, the current variation range in the flat phase is only 0.2 A. Figure 13 also shows that the output current waveform of the circuit maintains with high linearity during commutation, the commutation delay is less than 3 µs and the slope of the commutation edge is linearly stable. The experimental results shown in Figures 12 and 13 indicate that the constant current and voltage clamping control circuit proposed in this paper has reached the intended purpose of fast rising edge and amplitude constant.

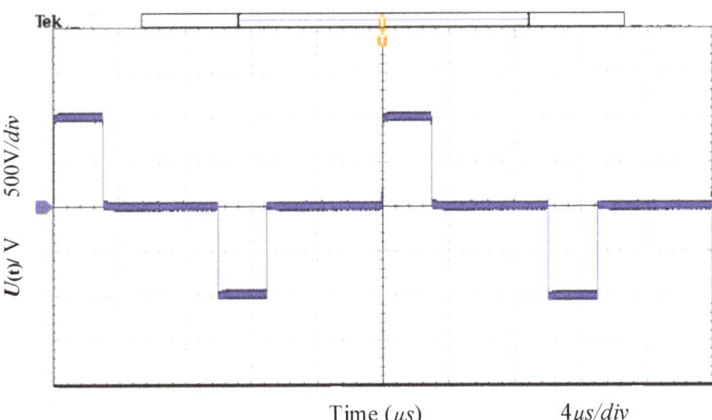

Figure 12. Experimental voltage waveform.

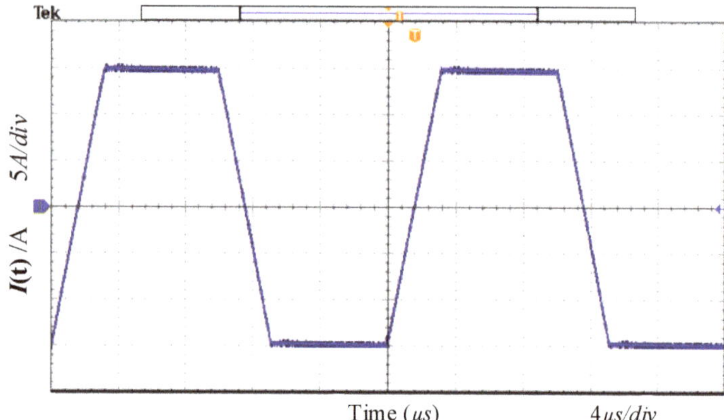

Figure 13. Experimental current waveform.

The experimental results show that the circuit proposed in this paper provides energy for the load by the constant current source. During the current commutation, the load voltage is clamped in a constant voltage, and the linearity of the commutation of the square wave current is realized. At the same time, in order to simplify the circuit, the leading-edge boost circuit and fast shutdown circuit share the same clamp circuit and clamp voltage source, while achieve the consistency of rising/falling edge of the square wave current. In the case of constant current control, the paper achieves that the output current does not change with the change of the power supply voltage, the frequency and the load device. The current rise (down) is fast, there is no dead zone in the middle section, the duty ratio is high, and the amplitude is constant in the flat phase. Controlling the clamped voltage source can control the steepness of the positive and negative edge, increase the clamp voltage, reduce commutation delay and improve the switching frequency.

6. Conclusions

In this paper, a bipolar broadband AC pulse current source emission circuit has been developed for the key technical problems such as rapid commutation of transmission current, constant linearity, wide frequency transmission and constant current amplitude in the existing frequency domain electromagnetic transmitter. The leading-edge linear technique and the constant-current control strategy of load current are proposed to realize wide-band constant-amplitude AC square-wave current emission of inductive load. By applying a constant high voltage clamp to the load during the current commutation to achieve the commutation edge linearity. Controlling the clamped voltage source can control the steepness of the positive and negative edge, increase the clamping voltage, reduce commutation delay and improve the switching frequency. In the case of constant current control, the paper achieves that the output current does not change with the changes of the power supply voltage, the frequency and the load device. The maximum transmission frequency is 100 KHz, the maximum emission current is 15 A; the current can rise quickly to the set value with very large slope of 10^7 A/s. The current can be quickly turned off from the amplitude of the positive polarity to the amplitude of the negative polarity. Without a dead zone in the middle section, the amplitude is constant in the flat stage; the amplitude of the current in the high frequency range is increased from 3 A to 15 A.

Author Contributions: Conceptualization, Z.F.; Data curation, W.L.; Formal analysis, X.W., W.W. and J.Z.; Investigation, Z.F.; Methodology, X.W.; Project administration, X.W., Z.F. and Y.W.; Software, X.W., W.W. and W.L.; Supervision, Z.F.; Writing—original draft, X.W.

Funding: This research was funded by National Key Research and Development program "Long distance linear engineering integrated intelligent detection technology and equipment research development", grant number

2018YFC0406904, National Natural Science Foundation of China; "Study on magnetic induced polarization effect and corrosion diagnosis of grounding grid", grant number 51777017, Chongqing Key Industry Generic Key Technology Innovation Special Project, grant number CSTC2017ZDCY-ZDYFX0045.

Acknowledgments: This work was supported in part by National Key Research and Development program, grant number 2018YFC0406904, National Natural Science Foundation of China, grant number 51777017, Chongqing Key Industry Generic Key Technology Innovation Special Project, grant number CSTC2017ZDCY-ZDYFX0045.

Conflicts of Interest: The authors declare no conflict of interest.

References

1. Sheard, S.N.; Ritchie, T.J.; Christopherson, K.R.; Brand, E. Mining, environmental, petroleum, and engineering indus-try applications of electromagnetic techniques in geophysics. *Surv. Geophys.* **2005**, *26*, 653–669. [CrossRef]

2. Xue, G.; Chen, W.; Yan, S. Research study on the short offset time-domain electromagnetic method for deep exploration. *J. Appl. Geophys.* **2018**, *155*, 31–137. [CrossRef]

3. Jang, H.; Kim, H.J.; Nam, M.J. In-Loop transient electromagnetic responses with induced polarization effects of deep-Sea hydrothermal deposits. *IEEE Trans. Geosci. Remote Sens.* **2016**, *54*, 7272–7278. [CrossRef]

4. Chen, W.Y.; Xue, G.Q.; Muhammad, Y.K.; Gelius, L.J.; Zhou, N.N.; Zhong, H.-S. Application of Short-Offset TEM (SOTEM) Technique in Mapping Water-Enriched Zones of Coal Stratum, an Example from East China. *Pure Appl. Geophys.* **2015**, *172*, 1643–1651. [CrossRef]

5. Wang, S.L.; Yin, C.C.; Lin, J.; Yang, Y.; Hu, X.Y. Bipolar square-wave current source for transient electromagnetic systems based on constant shutdown time. *Rev. Sci. Instrum.* **2016**, *87*, 034707. [CrossRef] [PubMed]

6. Fu, Z.; Zhou, L.; Tai, H.-M. Current Pulse Generation for Transient Electromagnetic Applications. *Electr. Power Compon. Syst.* **2007**, *35*, 1201–1218. [CrossRef]

7. Bae, S.; Kwasinski, A.; Flynn, M.M.; Hebner, R.E. High-Power Pulse Generator with Flexible Output Pattern. *IEEE Trans. Power Electron.* **2010**, *25*, 1675–1684. [CrossRef]

8. Mi, Y.; Yao, C.; Jiang, C.; Li, C.; Sun, C.; Tang, L.; Liu, H. A ns-μs duration, millitesla, exponential decay pulsed magnetic fields generator for tumor treatment. *IEEE Trans. Dielectr. Electr. Insul.* **2011**, *18*, 1111–1118. [CrossRef]

9. Zhu, X.; Su, X.; Tai, H.M.; Fu, Z.; Yu, C. Bipolar Steep Pulse Current Source for Highly Inductive Load. *IEEE Trans. Power Electron.* **2016**, *31*, 6169–6175. [CrossRef]

10. Xiao, H.X.; Ma, Y.; Lv, Y.; Ding, T.; Zhang, S.; Hu, F.; Li, L.; Pan, Y. Development of a high-stability flat-top pulsed magnetic field facility. *IEEE Trans. Power Electron.* **2014**, *29*, 4532–4537. [CrossRef]

11. Liu, L.; Liang, Z.; Liang, H.; Yin, H.; Liu, Y. A study on current stabilization technology of transmitter for controlled-source frequency electromagnetic sounding. In Proceedings of the IEEE 2011 11th International Conference on Electronic Measurement & Instruments, Chengdu, China, 16–19 August 2011.

12. Zhou, D.; Yang, S.; Tang, Y. Model Predictive Current Control of Modular Multilevel Converters with Phase-Shifted Pulse-Width Modulation. *IEEE Trans. Ind. Electron.* **2019**, *66*, 4368–4378. [CrossRef]

13. Penovi, E.; Maestri, S.; Retegui, R.G.; Wassinger, N.; Benedetti, M. Event-based control system suitable for high-precision pulsed current source applications with improved switching behavior. *IEEE Trans. Ind. Inform.* **2015**, *11*, 987–996. [CrossRef]

14. Khosravi, R.; Rezanejad, M. A New Pulse Generator with High Voltage Gain and Reduced Components. *IEEE Trans. Ind. Electron.* **2019**, *66*, 2795–2802. [CrossRef]

15. Christiansen, A.V.; Auken, E.; Sørensen, K. The transient electromagnetic method. In *Groundwater Geophysics: A Tool for Hydrogeology*; Springer: Berlin, Germany, 2009; pp. 179–225.

16. Dennis, Z.R.; Cull, J.P. Transient electromagnetic surveys for the measurement of near-surface electrical anisotropy. *J. Appl. Geophys.* **2012**, *76*, 64–73. [CrossRef]

Article

A Topology-Based Approach to Improve Vehicle-Level Electromagnetic Radiation

Cunxue Wu [1], Feng Gao [1,*], Hanzhe Dai [1] and Zilong Wang [2]

[1] School of Automotive Engineering, Chongqing University, Chongqing 400000, Chnia; wucx1@changan.com.cn (C.W.); 20161102021@cqu.edu.cn (H.D.)
[2] China Automotive Technology and Research Center, Tianjin 300300, China; wangzilong@catarc.ac.cn
* Correspondence: gaofeng1@cqu.edu.cn; Tel.: +86-189-9618-8196

Received: 27 January 2019; Accepted: 20 March 2019; Published: 25 March 2019

Abstract: The popularity of the electric vehicle (EV) brings us many challenges of electromagnetic compatibility (EMC). Automotive manufacturers are obliged to keep their products in compliance with EMC regulations. However, the EV is a complex system composed of various electromagnetic interferences (EMI), sensitive equipment and complicated coupling paths, which pose great challenges to the efficient troubleshooting of EMC problems. This paper presents an electromagnetic topology (EMT) based model and analysis method for vehicle-level EMI prediction, which decomposes an EV into multi-subsystems and transforms electromagnetic coupling paths into network parameters. This way, each part could be modelled separately with different technologies and vehicle-level EMI was able to be predicted by algebra calculations. The effectiveness of the proposed method was validated by comparing predicted vehicle-radiated emissions at low frequency with experimental results, and application to the troubleshooting of emission problems.

Keywords: electric vehicle; electromagnetic compatibility; electromagnetic topology; radiated emission

---

1. Introduction

With the rising demand of the electric vehicle (EV), the electromagnetic environment deteriorates rapidly because of the big voltage, great current and high frequency in a limited vehicle space [1,2]. The EV is a source of radiation, and where human beings travel as passengers or drivers, so human exposure conditions are always taken into account in the study of electromagnetic compatibility (EMC) conditions. Due to the body shielding effect [3], a certain amount of electromagnetic radiation is absorbed depending on several factors, such as the frequency and polarization of signal, and the features of environment [4]. To prevent electromagnetic interference (EMI) generated by onboard electronic/electrical components from breaking down the operation of other equipment components, EVs are obliged to satisfy special component and vehicle level EMC regulations, such as CISPR 12 and SAE J551-5 [5]. However, such tests cannot be conducted until the components are confirmed and the vehicle layout is established, which is at a very late stage in the vehicle development process, and many resources are required to find out and solve electromagnetic compatibility (EMC) problems. To improve the development efficiency, numerical simulation technologies have been gradually adopted for early stages by automotive manufacturers to predict and troubleshoot according to some improvement measures, such as change of vehicle layout, usage of special wires, shielding bodies, circuit filters, etc. [6,7]

Many efforts have been attempted and presented to improve EV EMC by controlling the interference sources [8–12]. For example, the authors of Reference [9] studied the internal noise transmission path of a high-power DC/DC converter based on which some interference suppression methods were suggested. A new active filter for the high frequency common mode (CM) currents of a motor driver was designed to reduce the radiated emission [10]. Nevertheless, vehicle-level EMC problems are associated with not only components but also factors such as body structure, vehicle layout, cable routing, etc. [2]. Moreover, even if all modules satisfy their EMC requirements, there may still exist vehicle-level EMC problems. It is necessary to model and simulate the vehicle-level EMC from early development stages.

There are primarily two ways to model and simulate EMC problems of such complex systems as EV. One is realized by transforming the electromagnetic coupling, parasitic effects, generation of interferences, etc., into equivalent circuits with lumped parameters [13–17]. Reference [13] extracted the conducted electromagnetic disturbance parameters of the wiring system using the partial element equivalent circuit method. Additionally, the connector was further considered [14]. Reference [15] presented a transmission line theory-based approach to predict the EMI induced on the communication network. Because of the difficulty in fitting with the equivalent circuits, especially when the considered frequency is very high, this method is limited to the low frequency range and used in conducted problems [16,17]. Moreover, extensive experience is required to determine the parts needed to be modeled exactly, especially when the analyzed system is very complicated. Reference [18] divided the vehicle-level EMC problem into electrical large and small parts, and then the multi-port network was adopted to predict the emission. This technique can be used in the high frequency range, but the whole system is required to be modeled to acquire the network parameter. Reference [19] studied the EMI generated by the power inverter system in EV based on series of a two-port network.

The other ignores the internal details and adopts transfer functions to describe both conducted and radiated process, which makes it easy to use in practice [20]. Among these types of methods, electromagnetic topology (EMT) is representative. It decomposes a large and complex system into multi-subsystems according to the electromagnetic shielding level [21–23], and can be successfully applied to evaluate the EMC performance of a Boeing 707 airplane [21]. The confidence level of prediction relies heavily on the accuracy of the transfer function, which varies with the impedance of connected ports. This implies that all actual components are required to be present and interconnected correctly to obtain accurate transfer functions and any change may result in a new model process of the whole system.

To overcome these problems, this paper proposes a topology-based method to predict vehicle-level EMI from 150 kHz to 30 MHz. This method adopts multi-port networks to describe the subsystem of EMT, which decouples the characteristic coupling among ports and transfer paths, and enables each part to be modeled separately with different technologies. The algebra equation of this topological model is derived to solve the radiated EMI analytically. Furthermore, the main interference source is found by the sensitivity analysis using this model. This paper is organized as follows: Section 2 introduces the methods and materials of modelling for the studied EMC problems of EV. Section 3 analyzes the predicted and diagnostic results and Section 4 concludes the paper.

2. Methods and Materials

The studied sedan was electric and obliged to satisfy the regulation SAE J551-5. The vehicle running state and test system layout were consistent with this regulation during the testing. The antenna was positioned 1 m above the ground and 3 m away from EV body. Figure 1 shows the layout of the loop antenna when measuring the radiated magnetic emission at the front side of the vehicle as an example.

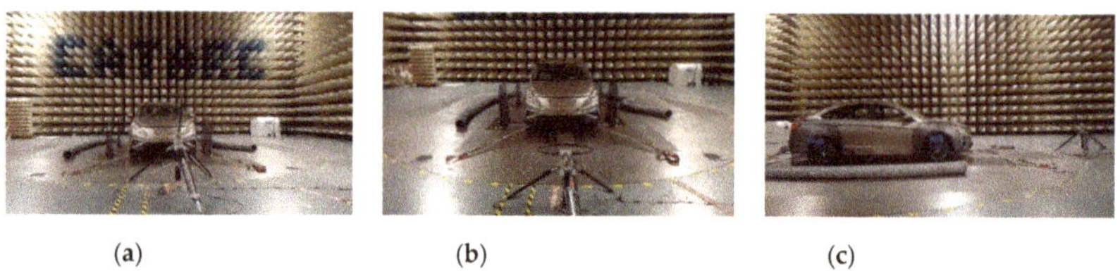

(a)   (b)   (c)

Figure 1. Layout of loop antenna at front side. (**a**) y polarization direction; (**b**) z polarization direction; (**c**) x polarization direction.

Its radiated electric field was below the regulation limit, while the magnetic emission exceeded as shown in Figure 2, which only shows some results. During testing, the radiated emission was measured at four directions around the vehicle, the electric field was measured by a rod antenna with vertical polarization, and x, y, z polarization directions for loop antenna when measuring the magnetic field (Figure 1). From the test results, the radiated magnetic field on both left and right side exceeded the limit by about 10 dB at the frequency around 18 MHz. Since the vehicle layout was delivered and fixed, it was almost impossible to change the component arrangement and electrical system, and one feasible way would have been to find out and improve the main interference source. One way to identify the source is by comparing the emissions under different combinational running states by switching on/off components. It is very resource-consuming and cannot provide an efficient improvement measure. In this paper, a numerical way is presented to identify the main interference source and reduce the radiated emission of EV.

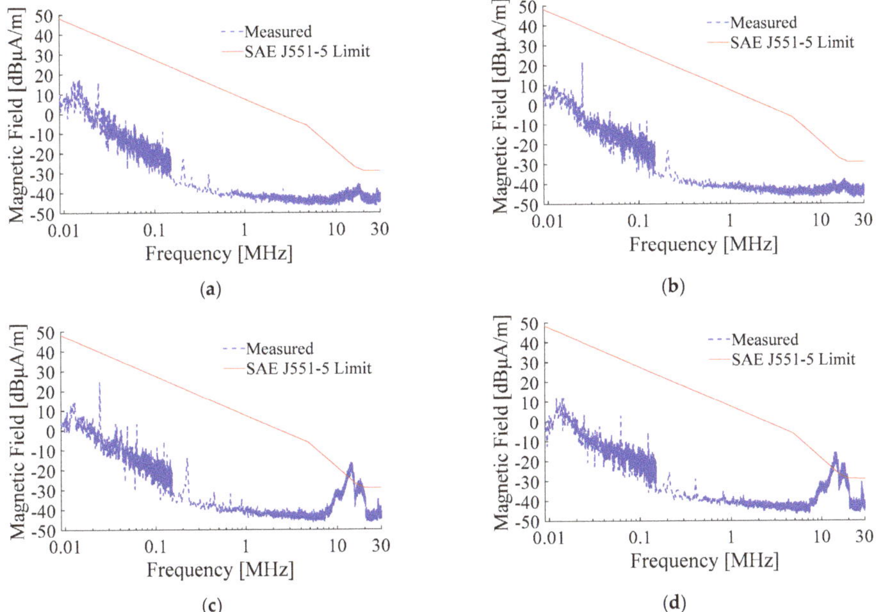

Figure 2. Radiated magnetic field. (**a**) x polarization direction at front side; (**b**) x polarization direction at rear side; (**c**) x polarization direction at left side; (**d**) x polarization direction at right side.

2.1. Modelling of Vehicle Level Radiation

The radiated emission at low frequency was mainly caused by the high-voltage system of the EV, and so first the high-voltage system was analyzed to find out what and how to model.

2.1.1. High-Voltage System Analysis

The high-voltage system of the EV consisted of DC/DC, on board charger (OBC), positive temperature coefficient heater (PTC), battery heater, compressor, traction system, power battery, cables and power distribution unit (PDU) as shown in Figure 3. Under the test condition, the components PTC, OBC and compressor are inactive. The radiated emission is mainly caused by DC/DC and traction system due to their internal power electronic devices with high frequency switching. The cables combined with the vehicle body act as the transmitting antenna. Since the positive and negative high-voltage cables were routed parallel and closely, the emission generated by the differential mode noise was ignorable and only CM was considered here.

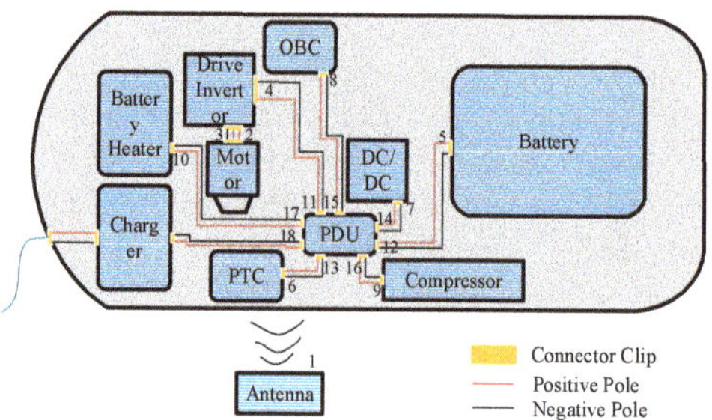

Figure 3. Schematic of high-voltage system layout.

Then the considered system was divided into two types of sub-volumes according to their shielding levels, depicted in Figure 4a [21]. The vehicle's metal body allowed electromagnetic waves to travel through apertures such as windows, door gaps, etc., so the internal space of EV was classified as an unscreened volume V1-1. Since the metal cases of the components including DC/DC, OBC, traction system, etc., were excellent shielding, they were defined as shielded volumes indexed from V2-1 to V2-10. Different from these components connected with only one pair of high-voltage cables, PDU (indexed by V1-2) was a conjunction of all high-voltage cables, in which the cables were connected with bus–bar resulting in conducted and near field couplings (see Figure 4b).

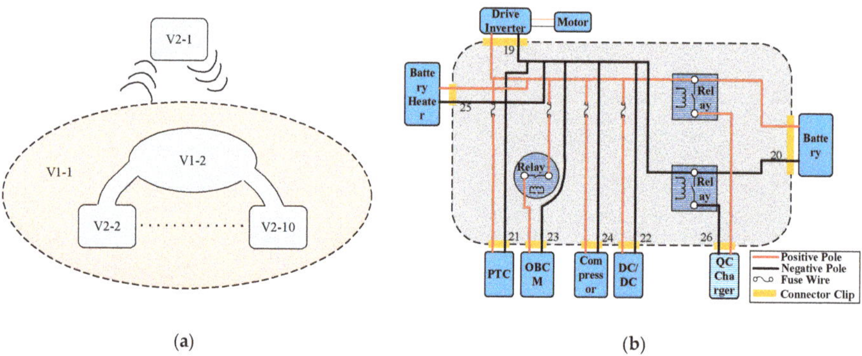

Figure 4. Topological model of high-voltage system. (**a**) Topological diagram; (**b**) internal schematic of power distribution unit (PDU).

The principle of the modelling process of a high-voltage system is shown in Figure 5. Good shielding isolated the radiated noise generated by inside sources and the vehicle-level emissions were mostly radiated by the connecting cables. Such shielded components were modeled by the Thevenin's equivalent circuits. On the contrary, the parts in an unscreened volume had unneglectable couplings with the outside. To model with less experience, a multi-port network was adopted to describe such complicated relations. Though multi-port networks mainly dealt with the linear coupling, from the above analysis it was realized that the emissions were mostly radiated by the cables and bodies, which were linear. The nonlinear parts were integrated in the shielded sub-volumes described by the equivalent circuits.

2.1.2. Modelling of Electromagnetic Radiation

Based on the analysis in Section 3.1, the EV was split into two multi-port networks as shown in Figure 6, where Network 1 was made up of the considered components including cables, measuring antennas, body, etc., i.e., V1-1 and V2-1~V2-10. Network 2 only consisted of the PDU, i.e., V1-2. The number of network ports was the same as that in Figures 3 and 4.

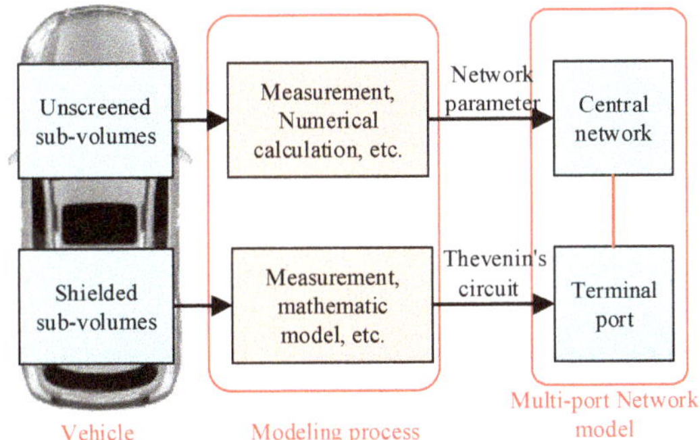

Figure 5. Conversion principle of electromagnetic path to equivalent circuit.

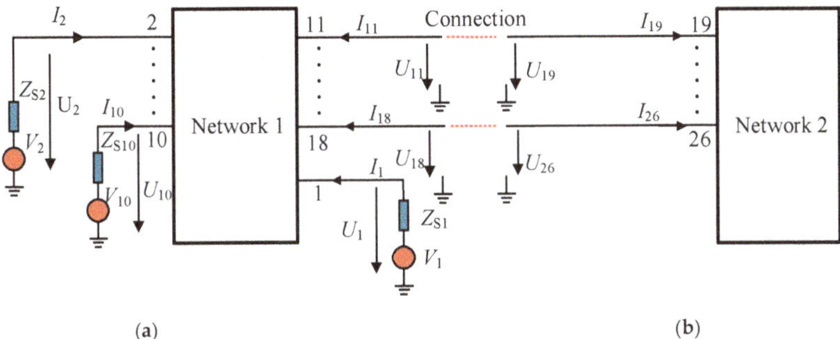

(a)   (b)

Figure 6. Equivalent multi-port network model. (**a**) Vehicle-level; (**b**) PDU.

In Figure 6, V_i and Z_{Si} ($i = 1, \cdots, 10$) were the equivalent voltage and internal impedance of components, U_i and I_i were the port voltage and current of the network. To simplify the representation, it should be noted that: (a) Both sensitive equipment and interference sources were modeled by the Thevenin's equivalent circuits and the voltage of sensitive ones was set to zero; (b) the measuring antenna was considered to be sensitive equipment. The following variables were defined to describe the networks:

$$
U = \begin{bmatrix} U_{vec} \\ U_{pdu} \end{bmatrix}, \; U_{vec} = \begin{bmatrix} U_1 \\ \vdots \\ U_{18} \end{bmatrix}, \; U_{pdu} = \begin{bmatrix} U_{19} \\ \vdots \\ U_{26} \end{bmatrix},
$$

$$
I = \begin{bmatrix} I_{vec} \\ I_{pdu} \end{bmatrix}, \; I_{vec} = \begin{bmatrix} I_1 \\ \vdots \\ I_{18} \end{bmatrix}, \; I_{pdu} = \begin{bmatrix} I_{19} \\ \vdots \\ I_{26} \end{bmatrix},
\tag{1}
$$

$$
V = \begin{bmatrix} V_2 \\ \vdots \\ V_{10} \end{bmatrix}, \; Z_S = \mathrm{diag}(Z_{S2}, \cdots, Z_{S10}), \; Z_L = [Z_{S1}],
\tag{2}
$$

$$
Z = \begin{bmatrix} Z_{vec} & \\ & Z_{pdu} \end{bmatrix}, \; Z_{vec} \in \mathbb{C}^{18 \times 18}, \; Z_{pdu} \in \mathbb{C}^{8 \times 8},
\tag{3}
$$

where Equation (1) defined the variables of the network ports, Equation (2) was the Thevenin's circuit parameters of components, and in Equation (3) Z_{vec} and Z_{pdu} were the network impedance parameters of Network 1 and Network 2, respectively, $\mathbb{C}$ was the set composed of complex matrices. The advantage of using network impedance parameter was that it was independent of port characteristics, which implies that the required parameters could be obtained separately by different technologies.

According to the connection of high-voltage system depicted in Figures 3 and 4b, the Network 1 ports indexed from 11 to 18 were connected to the Network 2 ports indexed from 19 to 26 one by one in sequence. To simplify the representation of the vehicle radiation model, the following topologies have been defined to describe the connection relationships:

$$
\begin{aligned}
\mathbf{G_U} \in \mathbb{R}^{8\times26}, \quad \mathbf{G_U}(i,j) &= \begin{cases} 1, & if\ i=1,\cdots,8\ and\ j=i+10 \\ -1, & if\ i=1,\cdots,8\ and\ j=i+18 \\ 0, & otherwise \end{cases}, \\
\mathbf{G_I} \in \mathbb{R}^{8\times26}, \quad \mathbf{G_I}(i,j) &= \begin{cases} 1, & if\ i=1,\cdots,8\ and\ j=i+10,\ i+18 \\ 0, & otherwise \end{cases}, \\
\mathbf{G_S} \in \mathbb{R}^{10\times26}, \quad \mathbf{G_S}(i,j) &= \begin{cases} 1, & if\ i=1,\cdots,9\ and\ j=i+1 \\ 0, & otherwise \end{cases}, \\
\mathbf{G_L} \in \mathbb{R}^{26}, \quad \mathbf{G_L}(i) &= \begin{cases} 1, & if\ i=1 \\ 0, & otherwise \end{cases}
\end{aligned}
\tag{4}
$$

where $\mathbb{R}$ was the set composed of matrices whose entity was selected from $\{-1, 0, 1\}$, $\mathbf{G_U}$ and $\mathbf{G_I}$ represented the connection relationships of voltage and current at the junctions of ports to satisfy Kirchhoff's law, $\mathbf{G_L}$ and $\mathbf{G_S}$ were used to combine all equivalent circuit parameters together.

Then the topological vehicle radiation model combining Network 1 and Network 2 became

$$
\mathbf{G_L}U = -Z_L\mathbf{G_L}I, \mathbf{G_S}U = V - Z_S\mathbf{G_S}I, U = ZI
\tag{5}
$$

At the junctions of ports, the following equations were established from Kirchhoff's law:

$$
\mathbf{G_U}U = 0, \mathbf{G_I}I = 0
\tag{6}
$$

The port voltage U was calculated by substituting Equation (6) into Equation (5) as

$$
U = Z \begin{bmatrix} \mathbf{G_L}Z + Z_L\mathbf{G_L} \\ \mathbf{G_S}Z + Z_S\mathbf{G_S} \\ \mathbf{G_U}Z \\ \mathbf{G_I} \end{bmatrix}^{-1} \begin{bmatrix} 0 \\ V \\ 0 \\ 0 \end{bmatrix}.
\tag{7}
$$

Then the port voltage U_1 of the measuring antenna was obtained. With the antenna factor, the strength of magnetic field could be calculated by

$$
|H_{ant}| = |U_1|\cdot AF_H
\tag{8}
$$

where $|H_{ant}|$ was the strength of magnetic field and AF_H was the antenna factor.

2.2. Acquiring Model Parameters

2.2.1. Network Parameter

The Z-parameter of the network was difficult to measure directly, because it was hard to construct the required open circuit condition in practice especially at the high frequency range. Instead, S-parameter was always used to describe the network characteristic, which can be measured by a network analyzer [24] or calculated by commercial three-dimensional electromagnetic field solving software, such as CST Microwave Studio (V2018, Paris, France), HFSS (v17.2, Pittsburgh, USA), FEKO (v2018, Wisconsin, USA), etc. [25]. In this study the model used to calculate the S-parameter of network by FEKO is shown in Figure 7.

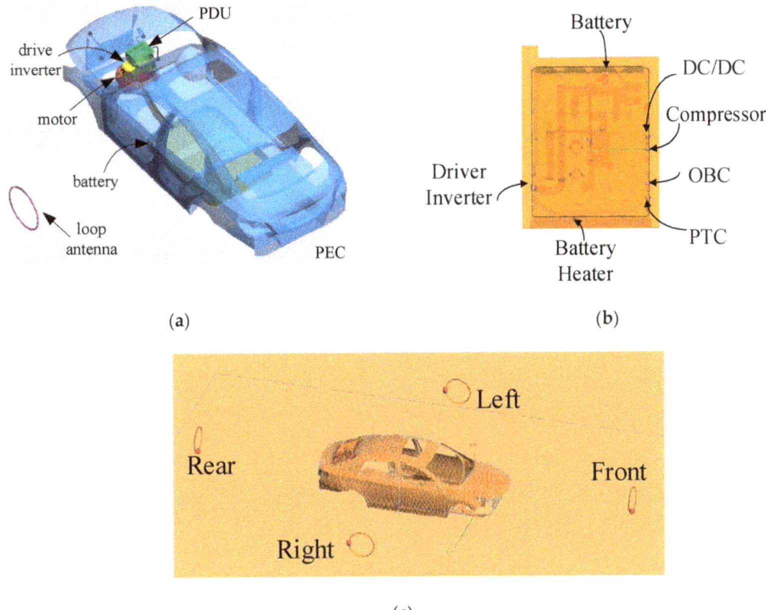

Figure 7. Models in FEKO. (**a**) electric vehicle (EV) model; (**b**) PDU model; (**c**) x polarization direction of loop antenna.

The FEKO models were solved by a workstation whose memory was 64G and CPU was E5-2650. When calculating the Network 1, the PDU was simplified to a metal box, whose internal coupling among high-voltage cables was solved by Network 2 separately. The calculated frequency range and the layout of loop antenna were defined according to SAE J551-5. The Z-parameter of network could be obtained from the S-parameter calculated by FEKO as the following and some network parameters are shown in Figure 8 as an example [18]:

$$Z = Z_0(E + S)(E - S)^{-1} \tag{9}$$

where Z, and S were the Z-parameter and S-parameter, Z_0 was the port impendence and E represented the identity matrix. Compared to the S-parameter, which varied in port impedance, the Z-parameter was independent of port characteristics. It was unnecessary to recalculate the network parameter to match the actual port impedance if the Z-parameter was used to predict the vehicle-level EMI.

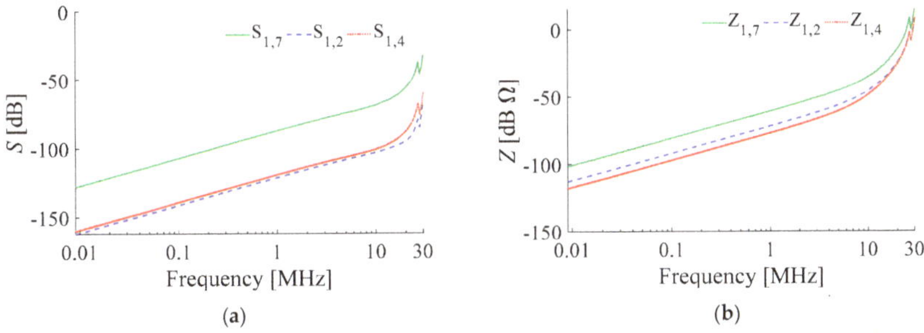

Figure 8. Network parameters. (**a**) S-parameter; (**b**) Z-parameter.

2.2.2. Equivalent Circuit Parameters

From the analysis in Section 2.1.1, the EMI was mainly induced by the CM interference, which was modeled by the Thevenin's equivalent circuits (shown in Figure 9) with the following steps: (a) Measure the CM current at the connector of each interference source in real vehicle or bench test; (b) measure the CM output impedance and terminating impedance of all components connected to the network; (c) the equivalent circuit parameters were obtained by calculating the equivalent voltage of interference source with the measured data as:

$$V_i = I_i \cdot (Z_{Si} + Z_{Li}) \tag{10}$$

where V_i and I_i were the CM voltage and current respectively, Z_{Si} was the output impedance and Z_{Li} was the terminating impedance. Here the CM current and impedance were measured by a broadband current clamp and a network analyzer, respectively.

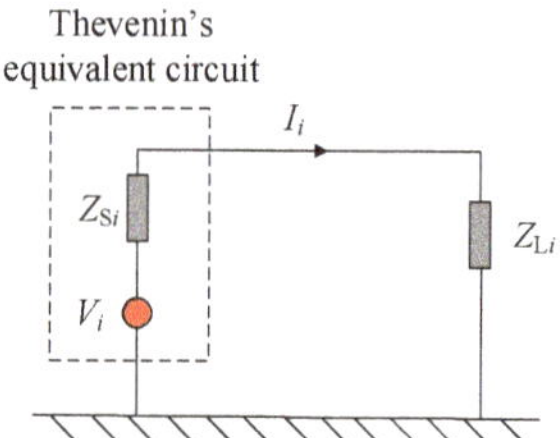

Figure 9. Equivalent model of component.

According to the analysis in Section 2.1, the ports indexed by 2, 4 and 7 were connected to the interference sources and others were considered to be loads whose impedances were directly measured by the network analyzer. Figure 10 shows the measured equivalent circuit parameters required to calculate the CM voltage by Equation (10), and the test condition was 40 km/h in accordance with SAE J551-5.

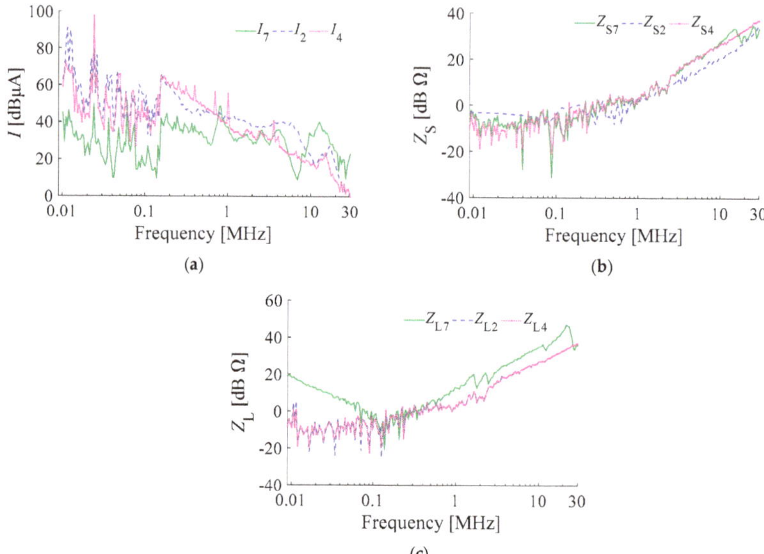

Figure 10. Equivalent circuit parameters. (**a**) Interference current; (**b**) equivalent output impedance; (**c**) equivalent terminating impedance.

3. Results and Analysis

3.1. Validation of EMI Prediction Model

With the model parameters obtained in Section 2.2, the radiated EMI was predicted by Equation (8), which was compared with the experimental results as shown in Figure 11. It shows that the trend of the predicted and measured values agree with each other well and the main peaks at about 25 KHz and 18.8 MHz could be predicted exactly. The overall prediction accuracy in the whole frequency range was about 71%. It indicates that the proposed approach could describe the vehicle-level EMI correctly and that this EMI model could be used to troubleshoot the EMI problem in the next section.

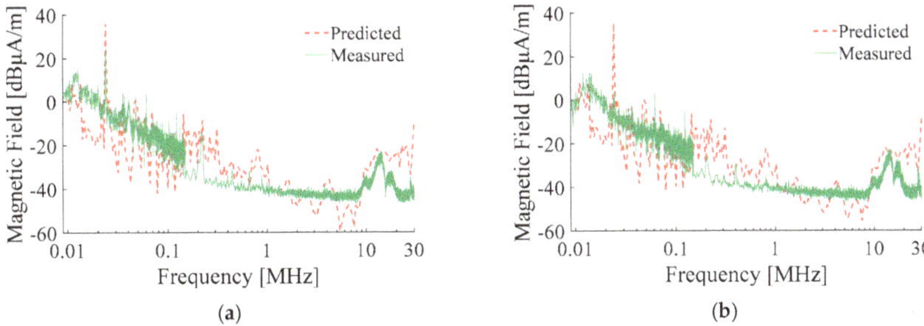

Figure 11. Comparing results of magnetic field with x polarization. (a) Left side; (b) right side.

3.2. Interference Source Diagnosis

At this late of a stage in the vehicle development process, it is impossible to change the vehicle layout design, so in this paper the diagnosis was used to find the main interference sources. One practical way to identify the source was by comparing the EMI under different combinational running states by switching on/off components, which was very resource-consuming and cannot provide an efficient improvement measure. In this section, a sensitivity-based approach has been presented to analyze and find out the main interference source using the EMI prediction model.

The magnetic field was proportional to the port voltage U_1 of the antenna, which was a sum of the products of the coupling coefficients and interferences, i.e., DC/DC and electric drive system. To find out which contributes most to the magnetic field, the sensitivity of the interferences to the induced voltage at the antenna was calculated and shown in Figure 12a. The results show that the magnetic field was more sensitive to the interference of DC/DC (V7). And to further validate this conclusion, the induced magnetic field components of different sources were calculated (see Figure 12b). The magnetic field generated by DC/DC was greater than that by motor by 15 dBμA/m and electric drive inverter by 47 dBμA/m at the frequency in excess of regulation, which was consistent with the sensitivity analysis results. As such, it is preferable to attenuate the interference of DC/DC rather than others and the EMI suppression was designed in the next section.

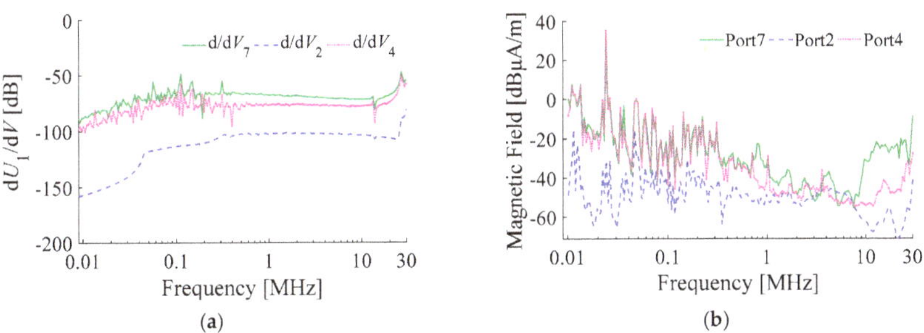

Figure 12. Diagnosis results. (a) Sensitive analysis; (b) magnetic field component of each port.

3.3. EMI Suppression

According to the previous analysis results, DC/DC was the main interference source. In this section, considering both the cost and filtering effect, a single-phase DC power filter (see Figure 13) was selected to prevent the disturbance of DC/DC from radiating by its connected high-voltage cables [26]. Its equivalent circuit model, including the parasitic effects is shown by Figure 13c, whose circuit parameters were provided by the manufacturer and listed as: LC = 0.45 mH, CY = 4.7 nF, CX = 470 uF, R = 0.47 MΩ, EPC = 67 nF, EPR = 16.8 kΩ, ESL = 7 nH, ESR = 0.05 mΩ.

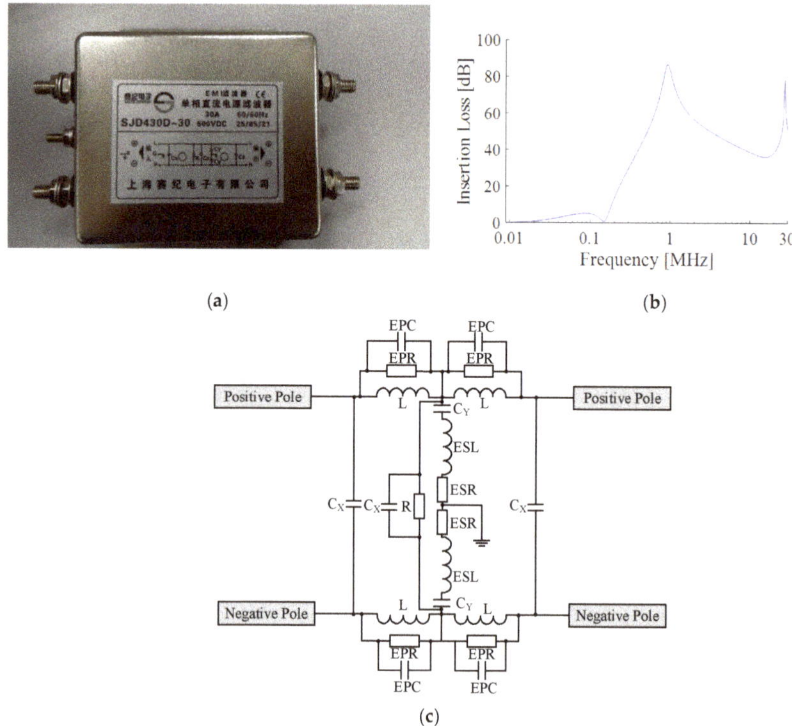

Figure 13. Single-phase DC power supply filter. (**a**) Product picture; (**b**) insertion loss; (**c**) schematic diagram.

Considering filter effectiveness, this filter should be installed closely to DC/DC. However, it was hard to carry out because of the compact size of DC/DC. Relatively, it was easier to install the filter in PDU, to which DC/DC was also connected. To further validate the effectiveness of the selected filter and choose its installation position, some further comparative simulations were conducted. The results in Figure 14 show that: (a) The selected filter could attenuate the EMI generated by DC/DC effectively at the exceeded frequency range; (b) generally it was better to install the filter at the port 7 of DC/DC (see Figure 3); (c) the radiated magnetic field was reduced by about 20 dBμA/m at around 18 MHz when the filter was installed at the port 22 of PDU (see Figure 4b), which was also acceptable compared with the exceeded test results shown in Figure 2.

Considering the difficulty of engineering, the filter was finally installed in PDU and the improvement test results of the magnetic field are shown in Figure 15, from which it was found that the presented filter and its installation position could attenuate the interference generated by DC/DC and the magnetic field emission passed the requirement of SAE J551-5.

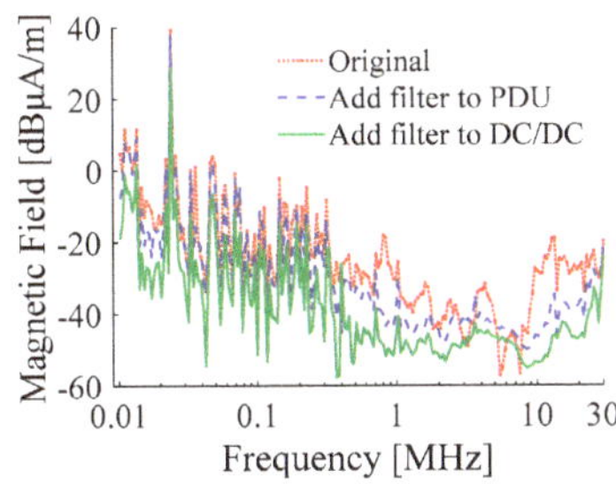

Figure 14. Comparison of radiated electromagnetic interference (EMI).

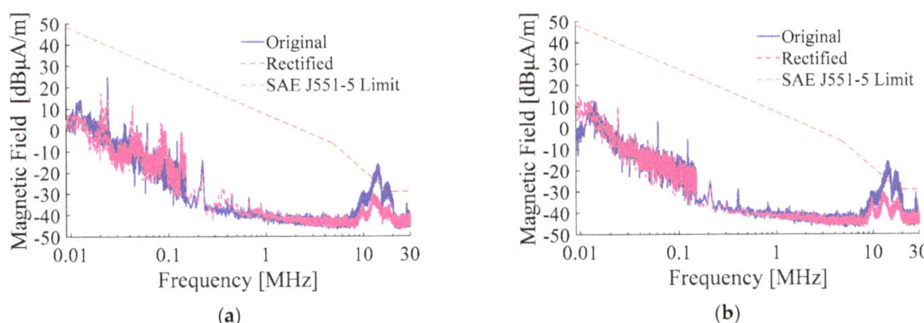

(a) (b)

Figure 15. Test results of x polarization direction. (a) Left side; (b) right side.

4. Discussion and Conclusions

This paper proposed a prediction method of vehicle-level EMI based on EMT and electric network theory to troubleshoot the EMC problem more efficiently. It was concluded from the analysis and application results that:

(1) The proposed EMT based method was an effective way to model the EMC problem of a complicated system with less engineering experience. For each subsystem, the parameter values could be derived independent of other parts with different technologies by adopting the network with a Z-parameter.

(2) The EMT based model for vehicle-level radiated emission at low frequency was accurate enough to predict and troubleshoot the EMC problem of the EV. It has the potential ability to be extended to higher frequency ranges and more EMC problems.

(3) The sensitive analysis method could be used to find out the main interference source, based on this EMI prediction model, by which the resource-consumed experimental diagnosis process was replaced successfully. However, further studies are needed on its theoretical base and on general application technologies.

(4) The presented EMI suppression approach could attenuate the interference generated by DC/DC sufficiently and the radiated emission at low frequency of studied EV finally met the requirements of SAE J551-5.

Author Contributions: Formal analysis, C.W.; writing-review and editing, F.G.; original draft preparation, H.D.; funding acquisition, Z.W.

Funding: This research was funded by the National Key R&D Program of China under grant 2017YFB0102504 and the Scientific Technological Plans of Chongqing under grant cstc2017zdcy-zdzx0042.

Acknowledgments: Thanks to the China Automotive Technology and Research Center, who provided the EV and the EMC laboratory.

Conflicts of Interest: The authors declare no conflict of interest.

References

1. Hamza, D.; Pahlevaninezhad, M. Integrated capacitor for common-mode EMI mitigation applicable to high frequency planar transformers used in electric vehicles DC/DC converters. In Proceedings of the IEEE Energy Conversion Congress and Exposition (ECCE), Pittsburgh, PA, USA, 14–18 September 2014; pp. 4809–4814.
2. Ivan, E.; Fernando, A.; Mateo, J.; Alvaro, P.; Javier, P.; Francisco, J.A. Common mode noise propagation and effects in a four-wheel driven electric vehicle. *IEEE Trans. Electromagn. Compat.* **2018**, *60*, 132–139.
3. De Miguel-Bilbao, S.; Blas, J.; Ramos, V. Effective analysis of human exposure conditions with body-worn dosimeters in the 2.4 GHz band. *J. Vis. Exp.* **2018**, *135*, e56525. [CrossRef] [PubMed]
4. De Miguel-Bilbao, S.; Ramos, V.; Blas, J. Responses to comments on assessment of polarization dependence of body shadow effect on dosimetry measurements in the 2.4 GHz band. *Bioelectromagnetics* **2017**, *38*, 650–652. [CrossRef] [PubMed]
5. Jose, H.; Artur, N.; Jose, S.; Magno, A. Proposal for a Brazilian regulation of electromagnetic compatibility applied to automotive vehicles. In Proceedings of the IEEE International Symposium on Electromagnetic Compatibility, Ottawa, ON, Canada, 25–29 July 2016; pp. 495–500.
6. Bernas, N.R. Electromagnetic compatibility design modification considerations in electronic equipment. In Proceedings of the IEEE Symposium on Product Compliance Engineering, San Diego, CA, USA, 10–12 October 2011; pp. 1–5.
7. Kim, T.H.; Kim, M.; Ahn, B.J.; Jung, S.Y. Study of EMC optimization of automotive electronic components using CAE. In Proceedings of the International Conference on Electrical Machines and Systems, Busan, Korea, 26–29 October 2013; pp. 26–29.
8. Li, P.; Jiang, L.J. Source reconstruction method-based radiated emission characterization for PCBs. *IEEE Trans. Electromagn. Compat.* **2013**, *55*, 933–940. [CrossRef]
9. Chun, H.T.; Han, S.H. Converter switching noise reduction for enhancing EMC performance in HEV and EV. In Proceedings of the International Exhibition and Conference for Power Electronics, Intelligent Motion, Renewable Energy and Energy Management, Nuremberg, Germany, 10–12 May 2016; pp. 1–8.
10. Piazza, M.C.; Luna, M.; Ragusa, A.; Vitale, G. An improved common mode active filter for EMI reduction in vehicular motor drives. In Proceedings of the IEEE Vehicle Power and Propulsion Conference, Chicago, IL, USA, 6–9 September 2011; pp. 1–8.
11. Gao, X.; Su, D. Suppression of a certain vehicle electrical field and magnetic field radiation resonance point. *IEEE Trans. Veh. Technol.* **2018**, *67*, 226–234. [CrossRef]
12. Yu, Z.; Mix, J.A.; Sajuyigbe, S.; Slattery, K.P.; Fan, J. An improved dipole-moment model based on near-field scanning for characterizing near-field coupling and far-field radiation from an IC. *IEEE Trans. Electromagn. Compat.* **2013**, *55*, 97–108. [CrossRef]
13. Yahyaoui, W.; Pichon, L.; Duval, F. A 3D PEEC method for the prediction of radiated fields from automotive cables. *IEEE Trans. Magn. Mag.* **2010**, *46*, 3053–3056. [CrossRef]
14. Li, P.; Yang, F.R.; Xu, W.Y. An efficient approach for analyzing shielding effectiveness of enclosure with connected accessory based on equivalent dipole modeling. *IEEE Trans. Electromagn. Compat.* **2016**, *58*, 103–110. [CrossRef]
15. Maeda, N.; Tanaka, K.; Imai, T. Estimation of inverter EM noise propagation to network communication through wire harness. *World Electr. Veh. Assoc. J.* **2007**, *1*, 121–125. [CrossRef]
16. Jeschke, S.; Hirsch, H.; Trautmann, M.; Maarleveld, M. EMI measurement on electric vehicle drive inverters using a passive motor impedance network. In Proceedings of the Asia-Pacific International Symposium on Electromagnetic Compatibility, Shenzhen, China, 18–21 May 2016; pp. 292–294.
17. Chen, S.; Nehl, T.W.; Lai, J.S.; Huang, X. Towards EMI prediction of a PM motor drive for automotive applications. In Proceedings of the Applied Power Electronics Conference and Exposition, New Orleans, LA, USA, 9–13 February 2003; pp. 14–22.
18. Gao, F.; Ye, C.; Wang, Z.; Li, X. Improvement of low frequency radiated emission in electric vehicle by numerical analysis. *J. Control Sci. Eng.* **2018**, *2018*, 5956973. [CrossRef]
19. Zhai, L.; Zhang, X.; Bondarenko, N.; Loken, D.; Doren, T.P.V.; Beetner, D.G. Mitigation emission strategy based on resonances from a power inverter system in electric vehicles. *Energies* **2016**, *9*, 419. [CrossRef]

20. Liu, G.; Chen, C.; Tu, Y. Anticipating full vehicle radiated EMI from module-level testing in automobiles. In Proceedings of the IEEE International Symposium on Electromagnetic Compatibility, Minneapolis, MN, USA, 19–23 August 2002; pp. 982–986.
21. Parmantier, J.P. Applications of EM topology on complex wiring systems. In Proceedings of the International Symposium on Electromagnetic Compatibility, Magdeburg, Germany, 2–6 August 1999; pp. 5–7.
22. Parmantier, J.P. Numerical coupling models for complex systems and results. *IEEE Trans. Electromagn. Compat.* **2004**, *46*, 359–367. [CrossRef]
23. Xiao, P.; Du, P.; Ren, D. A hybrid method for calculating the coupling to PCB inside a nested shielding enclosure based on electromagnetic topology. *IEEE Trans. Electromagn. Compat.* **2016**, *58*, 1701–1709. [CrossRef]
24. Chen, G.; Zhao, L.; Yu, W.H.; Yan, S.; Zhang, K.; Jin, J.-M. A general scheme for the discontinuous Galerkin time-domain modeling and S-parameter extraction of inhomogeneous waveports. *IEEE Trans. Microw. Theory Tech.* **2018**, *66*, 1701–1717. [CrossRef]
25. Cuffe, P.; Milano, F. Validating two novel equivalent impedance estimators. *IEEE Trans. Power Syst.* **2018**, *33*, 1151–1152. [CrossRef]
26. Varajao, D.; Araujo, R.E.; Miranda, L.M.; Lopes, J.A.P. EMI filter design for a single-stage bidirectional and isolated AC-DC matrix converter. *Electronics* **2018**, *7*, 318. [CrossRef]

Article

Robust ESD-Reliability Design of 300-V Power N-Channel LDMOSs with the Elliptical Cylinder Super-Junctions in the Drain Side

Shen-Li Chen [1,*], Pei-Lin Wu [1] and Yu-Jen Chen [2]

[1] Department of Electronic Engineering, National United University, Miaoli City 36063, Taiwan; wu09803026@gmail.com

[2] Department of Electrical Engineering, National Sun Yat-sen University, Kaohsiung City 80424, Taiwan; chenjen24@gmail.com

* Correspondence: jackchen@nuu.edu.tw or jackchen2100@gmail.com; Tel.: +886-37-382525

Received: 21 March 2020; Accepted: 28 April 2020; Published: 29 April 2020

Abstract: The weak ESD-immunity problem has been deeply persecuted in ultra high-voltage (UHV) metal-oxide-semiconductor field-effect transistors (MOSFETs) and urgently needs to be solved. In this paper, a UHV 300 V circular n-channel (n) lateral diffused MOSFET (nLDMOS) is taken as the benchmarked reference device for the electrostatic discharge (ESD) capability improvement. However, a super-junction (SJ) structure in the drain region will cause extra depletion zones in the long drain region and reduce the peak value of the channel electric field. Therefore, it may directly increase the resistance of the device to ESD. Then, in this reformation project for UHV nLDMOSs to ESD, two strengthening methods were used. Firstly, the SJ area ratio changed by the symmetric eight-zone elliptical-cylinder length (X) variance (i.e., X = 5, 10, 15 and 20 µm) is added into the drift region of drain side to explore the influence on ESD reliability. From the experimental results, it could be found that the breakdown voltages (V_{BK}) were changed slightly after adding this SJ structure. The V_{BK} values are filled between 391 and 393.5 V. Initially, the original reference sample is 393 V; the V_{BK} changing does not exceed 0.51%, which means that these components can be regarded as little changing in the conduction characteristic after adding these SJ structures under the normal operating conditions. In addition, in the ESD transient high-voltage bombardment situation, the human-body model (HBM) capability of the original reference device is 2500 V. Additionally, as SJs with the length X high-voltage P-type well (HVPW) are inserted into the drain-side drift region, the HBM robustness of these UHV nLDMOSs increases with the length X of the HVPW. When the length X (HVPW) is 20 µm, the HBM value can be upgraded to a maximum value of 5500 V, the ESD capability is increased by 120%. A linear relationship between the HBM immunity level and area ratio of SJs in the drains side in this work can be extracted. The second part revealed that, in the symmetric four-zone elliptical cylinder SJ modulation, the HBM robustness is generally promoted with the increase of HVPW SJ numbers (the highest HBM value (4500 V) of the M5 device improved by 80% as compared with the reference device under test (DUT)). Therefore, from this work, we can conclude that the addition of symmetric elliptical-cylinder SJ structures into the drain-side drift region of a UHV nLDMOS is a good strategy for improving the ESD immunity.

Keywords: electrostatic discharge (ESD); elliptical-cylinder type; human-body model (HBM); n-channel lateral-diffused MOSFET (nLDMOS); super-junction (SJ); ultra high-voltage (UHV)

---

1. Introduction

The lateral double-diffused MOSFET (LDMOS transistor) is the major dominant power component in the fabrication of power integrated circuits (PICs) because of many excellent electrical characteristics such as low on-resistance, high input-impedance, fast switching-speed and high breakdown-voltage. By the same token, due to the advantages of lower on-resistance and high voltage sustaining, UHV n-channel lateral diffused MOSFET (nLDMOS) devices have been commonly installed in many lighting, power-management systems, automotive systems, and 5G communication fields [1–16]. Even these UHV LDMOSs can be operated at very high voltage purposes, but compared with low-voltage (LV) and medium-voltage (MV) circuits, the electrostatic-discharge (ESD) immunity of UHV LDMOS related components is very feeble [17–28]. Nevertheless, a UHV nLDMOS component is mainly used in input/output (I/O) blocks and switching circuits, and it acts as a UHV device and at the same time as a self-protection ESD unit. Therefore, how to improve the ESD robustness of a UHV LDMOS is an important issue in these applications.

Moreover, in Figure 1, a new device-structure concept called the drift-region embedded super-junction (SJ) implemented in the vertical power device has been available commercially, which would break through the silicon device limit [29–49]. CoolMOS is registered by Infineon Technologies [50], these high-voltage SJ MOSFETs of CoolMOS components address applications for smart-phone chargers, notebook adapters, LED lighting as well as audio and TV power supplies. Here, the SJ idea is based upon achieving charge compensation in the off-state of a power MOSFET, in a set of alternating and heavily doped N- and P-pillars in the drift region of the high-voltage component. Meanwhile, provided that all of the pillars of SJs are fairly narrow and net dopants in both pillars are approximately equal, it is possible to deplete the pillars at relatively low voltage. Under the depletion situation, the N- and P-pillars appear to be an intrinsic (very low doped) layer and a near uniform electric field in the drain-side is achieved, therefore, resulting in a high breakdown voltage. However, in Figure 2, the SJ concept can be applied for lateral high-voltage devices because the charge interaction between the substrate (bulk) and SJ region exists, which is called the substrate-assisted depletion effect [32,40,47,49]. Of course, the breakdown-voltage behavior of this structure will be influenced.

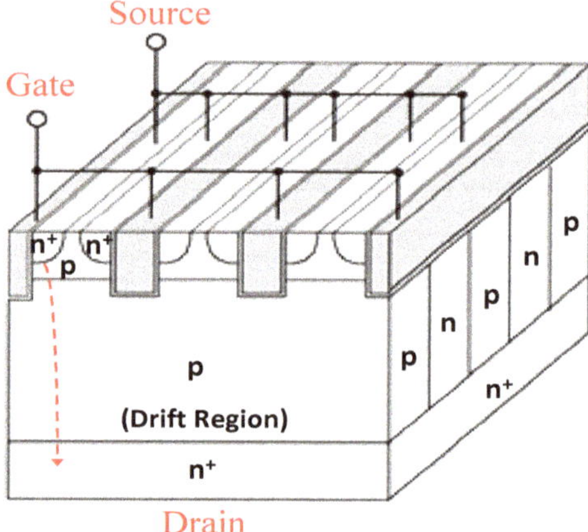

Figure 1. Vertical DMOS with an SJ (super-junction) structure in the drain side.

For improving the ESD capability of a UHV LDMOS, this study proposes a novel structure, where an elliptical cylinder SJ is added into the UHV circular nLDMOS. It [29–49] is known that an nLDMOS with these SJs in the drift region is quite a complicated structure and it will become quite different from a conventional UHV LDMOS. Generally, an nLDMOS-SJ composite device offers simultaneously high breakdown-voltage (V_{BK}) and low on-resistance (R_{on}) behavior [38,41,49,50]. Under normal circumstances, if there is an SJ structure (a P-pillar such as the HVPW layer) in the drain region, it will cause additional and extended extra depletion regions. These extended depletion regions will reduce the peak electric field along the conduction path, so it may cause the ability to resist ESD instant large electric fields. Then, how does this architecture change the HBM ESD capabilities of a UHV circular nLDMOS with elliptical cylinder super-junctions (SJs) structure? Moreover, in this article, we will choose two architectures (four-zone and eight-zone types) with a high degree of symmetry, and the layout is not too complicated (in order for the industrial practicality to be higher in the future). In order to realize these experimental samples, a TSMC 0.5 µm UHV bipolar-CMOS-DMOS (BCD) process is used in these UHV devices fabrications.

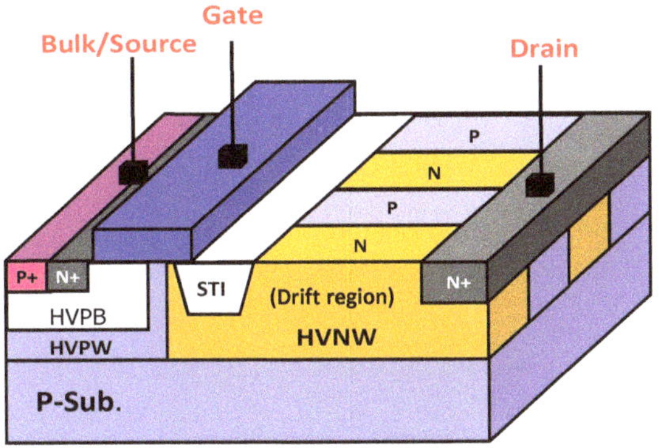

Figure 2. Lateral DMOS with an SJ structure in the drain side.

2. Device Layouts of UHV 300 V nLDMOS Related Devices

2.1. The Benchmarked Reference Sample (Pure nLDMOS)

The layout diagram and 3-D cross-sectional view (along the line AB of Figure 3a) of a UHV 300 V circular nLDMOS benchmarked reference device are shown in Figure 3a,b, respectively. In Figure 3a, due to the high operation voltage and high breakdown voltage need, a lightly n-doped HVNW layer is used in the drain-side drift region. The PBody and deep P-Well (DPW) layers will form a reduced surface-field (RESURF) structure, which causes the drift region to be completely depleted and increases the breakdown voltage of the device [30]. Similarly, the poly2 surround above the drift region is used to increase the breakdown voltage. In the rectangular-type layout, the current will flow through and concentrate at the electrode corner; therefore this component is easily damaged there. Instead, in Figure 3b, the layout of a UHV MOSFET adopts a circular mode due to the conduction current uniformly flowing along the shortest radial-outward path and a circular structure is used in order to prevent the current's uneven flowing. Meanwhile, a gate-grounded MOSFET (GGnMOS) structure is commonly used in the routing architecture of an ESD protection device, which discharges the instantaneous current of an external transient pulse mainly through the parasitic bipolar-junction transistor (BJT) conduction of this GGnMOS component. These experimental samples were fabricated by a TSMC 0.5 µm UHV BCD process with a channel length L of 2.5 µm and a channel width W of approximately 394.3 µm.

2.2. nLDMOS-SJs Samples with SJs Length Modulation

The Symmetrical Eight-Zone Elliptical Cylinder Type (M8) of LDMOS-SJs

Next, the cross-sectional view and layout diagram of a UHV 300 V nLDMOS with symmetrical eight-zone elliptical cylinder SJs (type-M8) in the drain region are presented in Figure 4. This novel arrangement is taken as the length of X (length X) modulation of the HVPW layer embedded in the drain-side drift region. Then, this arrangement is set to increase the length X of the HVPW zone shown in Figure 4b, and which are 5, 10, 15 and 20 µm, respectively. Meanwhile, the width (thickness) of these entire elliptical cylinder SJs are kept to 3 µm. Therefore, this study will mainly discuss the impact of HVPW distribution and how the HBM ability is affected by the HVPW length X.

Based on these devices and the reference sample (previous sub-section) structures, an equivalent circuit of the nLDMOS-SJ devices with SJ length modulation is presented in Figure 5; R_{bulk} is the parasitic resistances of the source-to-bulk; the BJT is a parasitic device of the gate-grounded nLDMOS device; the parasitic resistance $(R_{drift})_{SJ}$ in the drift region is sufficiently high and varied with the SJs length modulation; and R_{drain} is the parasitic resistances of the drain electrode.

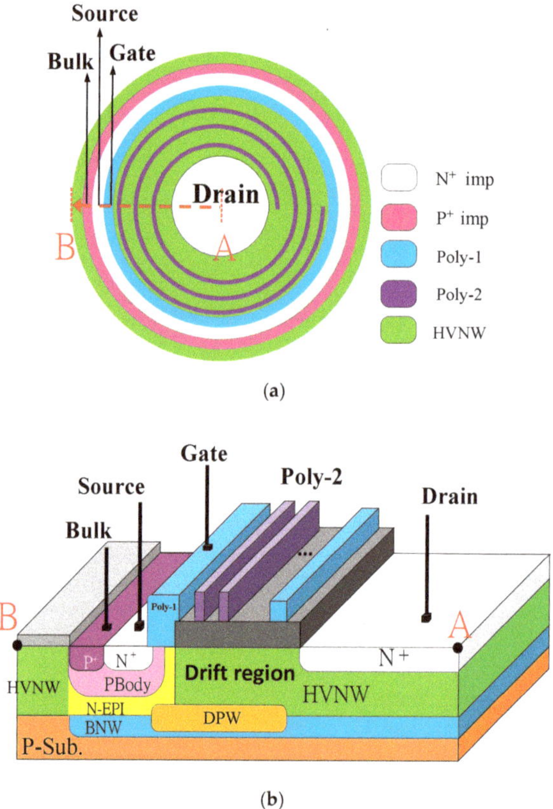

Figure 3. (**a**) Device layout diagrams, and (**b**) 3-D structure view of the UHV 300 V circular LDMOS ref. device.

2.3. SJs Number Modulation: Six LDMOS-SJs Devices with the Symmetrical Four-Zone Elliptical Cylinder Type

In Figure 6, the following items are the SJs' number modulation of symmetrical four-zone elliptical HVPW cylinders, which are classified as the group A (M1, M5, M9) and the group B (M2, M6, M10), respectively. In these six LDMOS-SJ devices, each has a common feature, that is, the SJ on each side is perpendicular to the SJs on the adjacent sides. Additionally the length X of the HVPWs shown in Figure 6 are 20 µm and 5 µm for the group A (M1, M5 and M9) and B (M2, M6 and M10), respectively. Meanwhile, the thicknesses of these six elliptical cylinder SJs are kept to 3 µm. In the case of increasing the HVPWs in this item, we first set the modulation without affecting the breakdown voltage of the electrical characteristics so much, and it is desirable to understand how the modulation impacted on HBM ESD reliability.

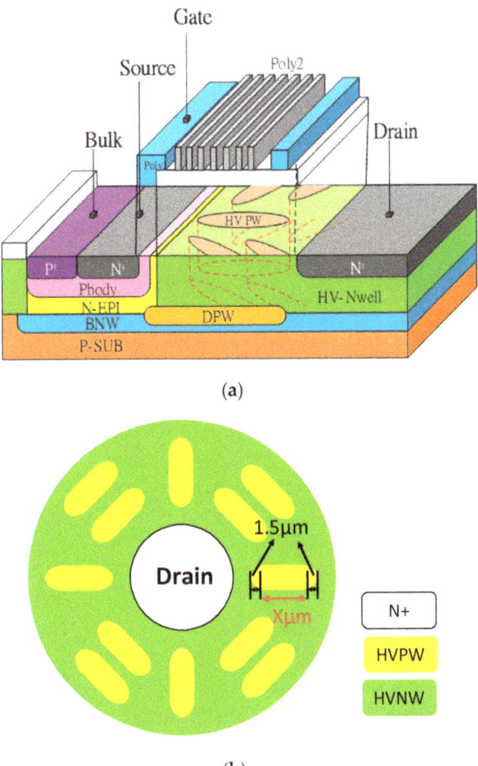

Figure 4. (**a**) 3-D structure and (**b**) drain region magnified diagrams of the UHV 300 V LDMOS (type-M8) with symmetrical eight-zone SJs in the drain region.

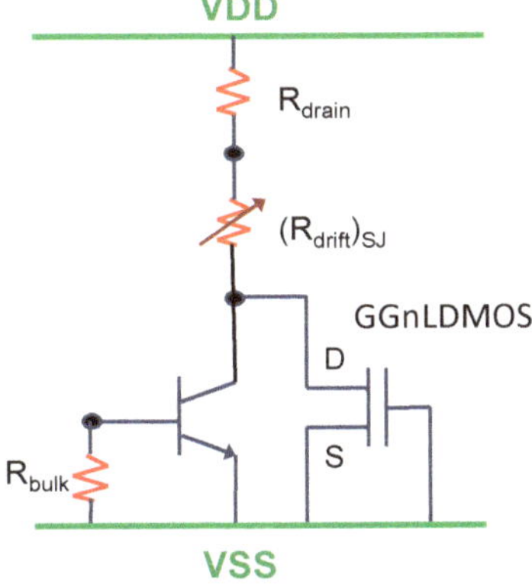

Figure 5. An equivalent circuit of a GGnLDMOS device with symmetrical SJs in the drain region.

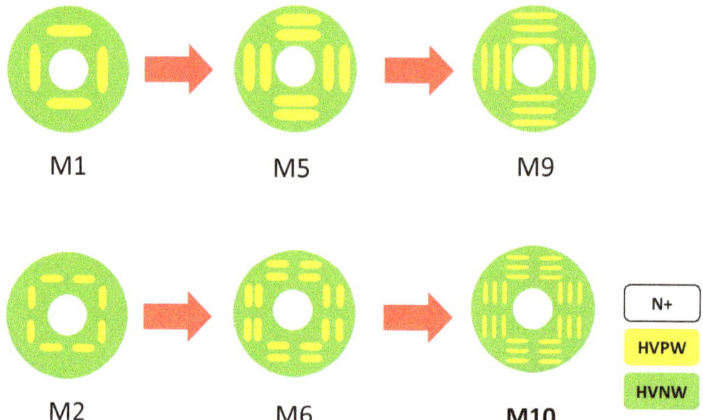

Figure 6. Device layout diagrams of six UHV 300 V LDMOSs with symmetrical four-zone SJs in the drain region.

3. HBM Testing System

In the HBM ESD-reliability level testing, a KeyTek MK2 ESD testing machine was used. The ESD testing mode was human-body model (HBM) and the maximum zapping-voltage can be zapped up to ±8000 V. The testing waveforms were confirmed by the MIL-STD-883 EOS/ESD and ANSI/ESDA/JEDEC JS-001 test standards [51,52]. Then, we used the positive-to-VSS (PS) mode (applied a zapping voltage to the drain terminal of this device) for these testing samples. Furthermore, the testing arrangement is shown in Figure 7. The testing voltages for HBM testing were in the range of 500 to 8000 V and changed by 250 V per interval in the step-1 HV zapping. A quantity of I-V curve was selected as the failure judge criterion in the step-2 leakage-current measurement. Therefore, the leakage-current was fixed at 1 µA for sensing before/post HBM zapping and compared for I-V curve variations. If the voltage-shift changed by more than ±30% ($\Delta V_{shift} > \pm 30\%$), then this pin was decided as not qualified at this HBM zapped level.

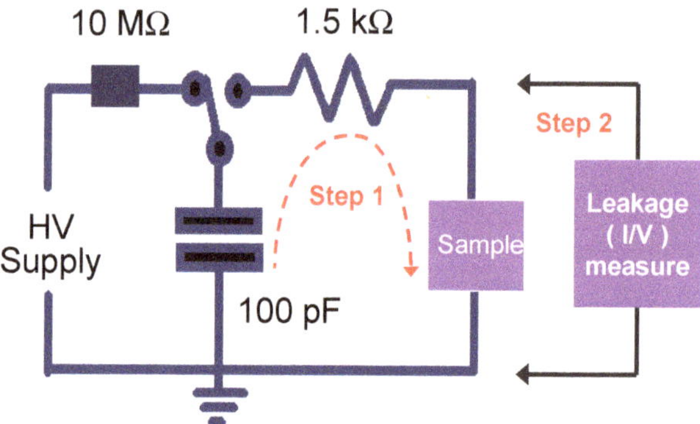

Figure 7. Testing procedures of an HBM ESD testing for UHV nLDMOS related devices.

4. Experimental Data and Discussion

Even LDMOSs with the super-junction (SJ) structures were recently developed to obtain the trade-off targets between the breakdown-voltage (V_{BK}) capability and the on-resistance behavior which are always the key issues in the design of power electronics. The SJ architecture is chiefly to gain charge compensation in the off-state (an ESD device with GGnMOS configuration is operated at the normally-off state) of a UHV device, by alternating doped N- and P-pillars comprising the drift region near the device drain side.

Firstly, the experimental results of normal output I-V characteristics of the UHV 300 V reference nLDMOS and nLDMOS-SJs type-M8 with X = 10 μm are shown in Figure 8. Obviously, these output I-V behaviors changed very little after adding elliptical cylinder SJs. For example, V_g = 5 V and V_d = 40 V, and the drain currents were equal to 2.551 mA and 2.5 mA, respectively. The output I-V variation does not exceed 1.88%. Meanwhile, on the other hand, the breakdown voltages of these nLDMOS related devices with symmetrical eight-zone elliptical cylinder SJs in the drain region are presented in Figure 9. From these experimental results, there is not much change in the breakdown-voltage (V_{BK}) test after adding the SJ structure, which fall between 391 and 393.5 V (the original benchmarked device under test (DUT) is 393 V). The V_{BK}'s value variation does not exceed 0.51%, it means that the basic electrical properties of the nLDMOS components are carefully controlled so that they do not change much. That is, in this study, we want to enhance the ESD-reliability capability without affecting the original electrical behavior afterwards by painstakingly adding these SJ structures.

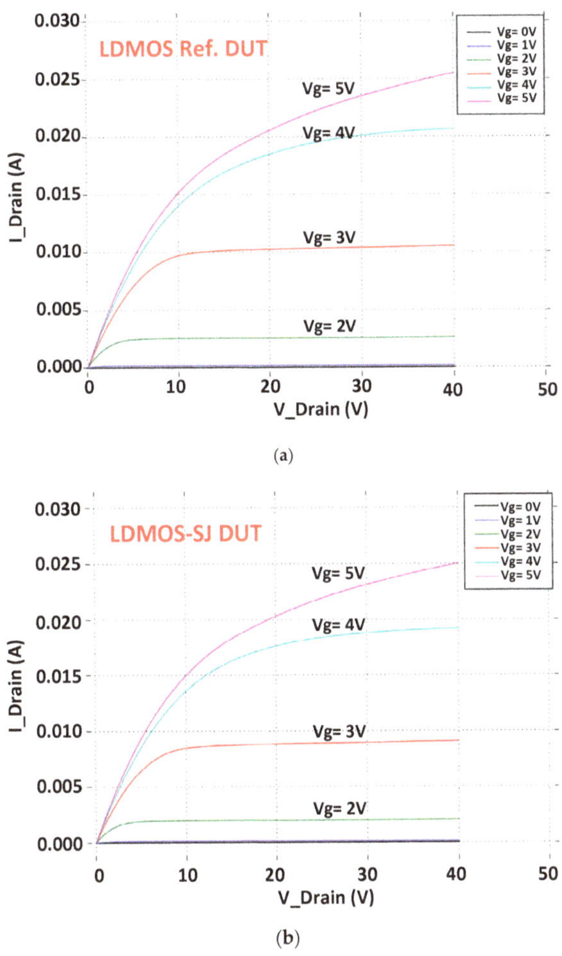

Figure 8. The normal output I-V characteristics of (**a**) UHV 300 V ref. nLDMOS, and (**b**) nLDMOS-SJs type-M8 device with X = 10 μm.

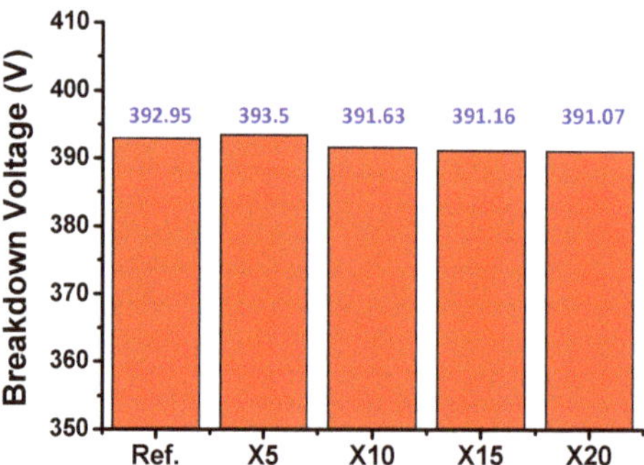

Figure 9. V_{BK} values (shown by the corresponding numbers) comparison of the UHV 300 V GGnLDMOS type-M8 devices with different SJs' length Xs.

In Figure 10, the HBM immunity level of the benchmarked GGnLDMOS reference DUT is 2500 V. As the symmetrical eight-zone elliptical cylinder SJs (type-M8) are embedded in the drain-side drift region of the nLDMOS and the length X of the HVPW layer is modulated, it is found that the HBM capability of these nLDMOS-SJ samples increases as the length X of the HVPW layer (also called the HV PW/NW area ratio) becomes longer and larger. When the HVPW length X is 20 μm, the HBM capability value can be promoted up to 5500 V, so its ESD capability is improved by 120% as compared with the original benchmarked GGnLDMOS device. Due to the previous description of the equivalent circuit of Figure 5, an nLDMOS device embedded SJ in the drift region can be equivalently regarded as a variable linear resistor [53]. Therefore, by the linear-regression technique, a strongly linear relationship for the HBM immunity levels of these M8s versus the HVPW/HVNW area ratio can be found in Equation (1),

$$V_{HBM} = (V_{HBM})_0 + C \times (Area \quad Ratio\%)_{HVPW/HVNW}$$
$$\approx 2240V + 5157.1V \times (Area \quad Ratio\%)_{HVPW/HVNW} \tag{1}$$

where the $(V_{HBM})_0$ is the HBM fitting level for the nLDMOS reference sample and C is a linear-proportional slope constant, such as here for these type-M8 samples the $(V_{HBM})_0$ is 2240 V and C is 5157.1 V, respectively.

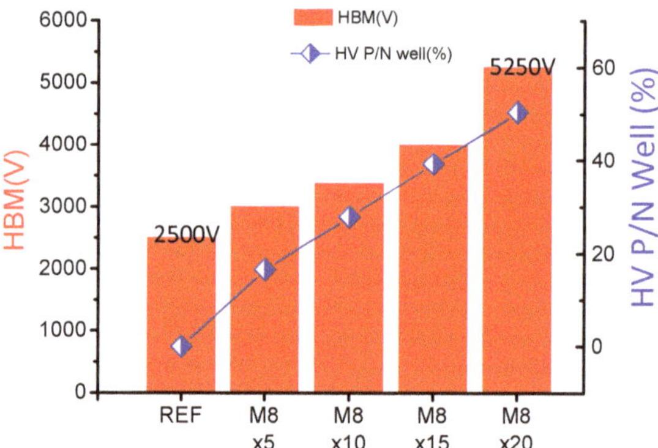

Figure 10. HBM values and SJ (HVPW) area % comparisons of UHV 300 V GGnLDMOSs type-M8 devices with different SJ length Xs.

By using the Silvaco EDA software for electric field verifications, Figure 11a,b are the 3-D side view diagrams of a thin-slice structure for the nLDMOS reference device and the nLDMOS-SJ M8_X20 device, respectively. Then, a high voltage of 400 V was applied at the drain terminals of these two DUTs for comparing high electric-field distributions. From Figure 11c, the electric field of the circular nLDMOS reference device was concentrated at the depletion area of the LOCOS edge under this high voltage biase, and the peak value of the electric field is about 1.0247×10^6 V/cm. However, in Figure 11d for the nLDMOS-SJ M8_X20 device, the electric field is concentrated between HVPW and N-EPI, and the depletion region by the SJ (HVPW and HVNW). The maximum electric field of the nLDMOS-SJ M8_X20 device is downgraded to approximately 1.9192×10^5 V/cm, and the peak value of this electric field decreases to the ratio of 18.73% compared with the nLDMOS reference device. Obviously, it reveals that the nLDMOS-SJ structure can indeed reduce the peak electric field of the device.

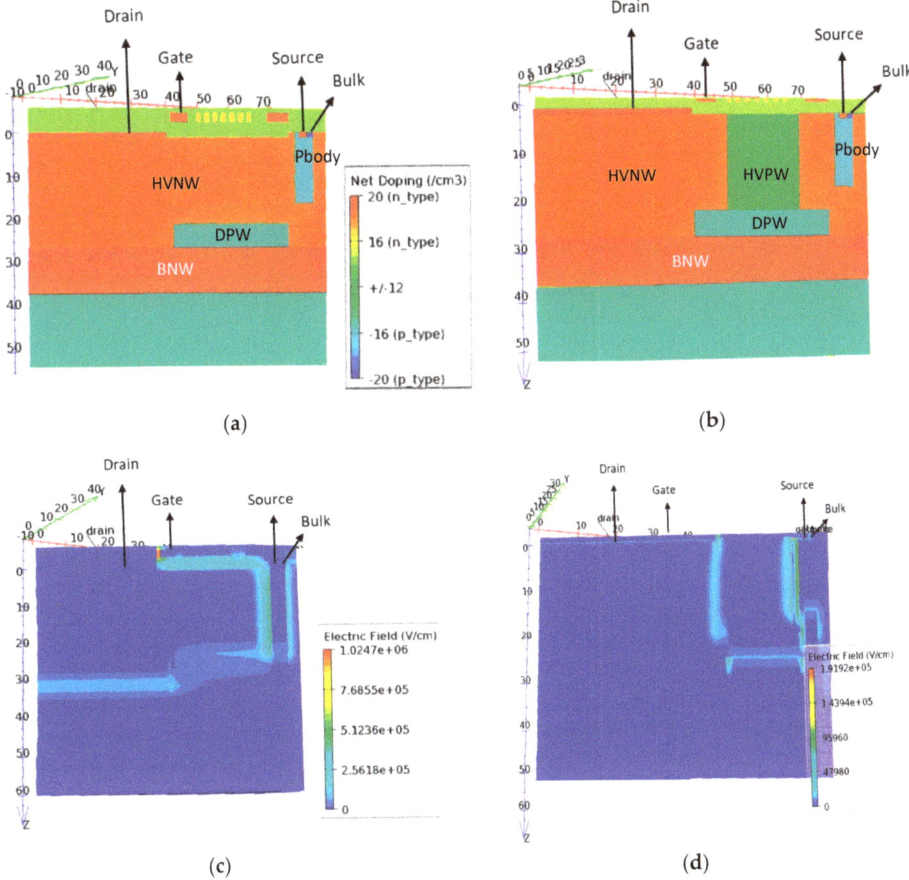

Figure 11. (**a**) 3-D thin-slice structure of the nLDMOS ref. device, (**b**) 3-D thin-slice structure of the nLDMOS-SJ M8_X20 device, (**c**) electric field distribution of the nLDMOS ref. device, and (**d**) electric field distribution of the nLDMOS-SJ M8_X20 device.

Next, for proof again by the symmetric four-zone elliptical cylinder SJ modulation, the HVPW structures (M1, M5 and M9, and M2, M6 and M10) are embedded in the drain side, and the HBM values are presented in Figure 12. Because they are not as highly symmetrical as the type-M8 devices, the HBM capability versus the HVPW/HVNW area ratio cannot be satisfactorily linearly related, but can be with another high-order proportional relationship for these group A and B samples. Nevertheless, the HBM ESD capability is somewhat higher with the increase in the number of HVPW SJs. The highest HBM value of M5 (4500 V), its (the M5) ESD capability, is improved by 80% as compared with the original benchmarked GGnLDMOS DUT. Meanwhile, it is found that the HVPW/HVNW area ratio of the group A samples is generally higher than that of the corresponding group B devices, so their ESD HBM abilities are also relatively high. In the case of M9 and M10 (three HVPW SJ arrays in the outward direction), HBM values are downgraded than that of the M5 and M6 (two HVPW SJ arrays in the outward direction). This is caused by the thin-oxide definition (OD) distance from the outermost HVPW to the gate and the heat dissipation cross-sectional area being too small, resulting in a decreasing data value in the HBM robustness.

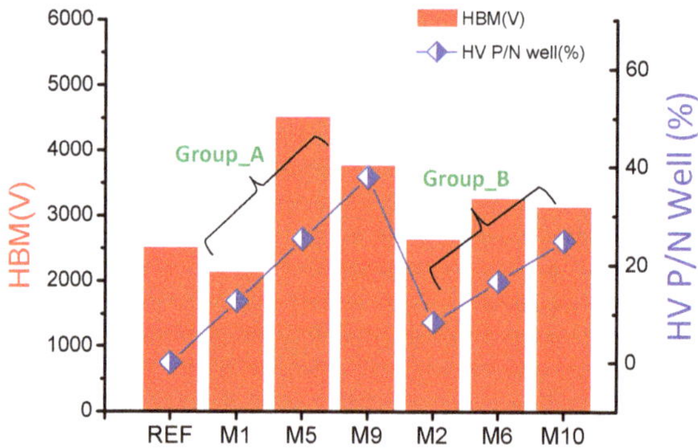

Figure 12. HBM values and SJ (HVPW) area % comparison of UHV 300 V GGnLDMOSs with different symmetrical four-zone SJs.

5. Conclusions

This paper is focused on the enhancements of HBM-immunity levels for UHV 300 V nLDMOS devices being added with 1) symmetrical eight-zone elliptical cylinder SJs (type-M8) in the drain region, in which the length Xs of SJs are 5, 10, 15 and 20 μm, respectively; and 2) symmetrical four-zone elliptical cylinder SJs in the drain region. The first part focuses on the symmetrical eight-zone nLDMOS-SJ experimental samples; we can find that the breakdown-voltages of these samples do change a little due to the addition of this SJ structure. The highest test data does not change by more than 0.51% compared to the original benchmarked GGnLDMOS sample, which means that the basic normal electrical characteristics of nLDMOS-SJ components change very little. In addition, the HBM testing values of these samples are optimal for the X20 sample, and it can be upgraded to 5500 V (120% higher than that of the benchmarked DUT). Meanwhile, a strongly linear relationship between the HBM immunity levels versus the HVPW/HVNW area ratio could be found. That is a very useful indicator of how to improve the ESD immunity in UHV nLDMOS transistors. The second part revealed that, for the symmetric four-zone elliptical cylinder SJs in the drain side, the HBM capability is generally higher with the increase in the number of HVPW SJs, with the highest HBM value of M5 (4500 V) improved by 80%. Then, it can be concluded that a UHV nLDMOS device embedded with symmetric elliptical cylinder SJs in the drift region is a good strategy and the HBM capability of these UHV nLDMOS transistors could be effectively improved without changing the basic electrical properties and adding any extra cell area. Therefore, it is a very positive method for the ESD-reliability enhancement in UHV LDMOS components.

Author Contributions: Conceptualization, S.-L.C. and P.-L.W.; Data curation, P.-L.W. and Y.-J.C.; Formal analysis, P.-L.W. and S.-L.C.; Funding acquisition, S.-L.C.; Methodology, S.-L.C. and P.-L.W.; Project administration, P.-L.W. and S.-L.C.; Resources, Y.-J.C.; Supervision, S.-L.C.; Validation, S.-L.C. and P.-L.W.; Visualization, Y.-J.C.; Writing—original draft, S.-L.C.; Writing—review & editing, S.-L.C. All authors have read and agreed to the published version of the manuscript.

Funding: This research received no external funding.

Acknowledgments: In this work, authors would like to thank the Taiwan Semiconductor Research Institute in Taiwan for providing the process information and fabrication platform. Authors would also like to acknowledge the financial support of the Ministry of Science and Technology of Taiwan.

Conflicts of Interest: The authors declare no conflict of interest

Nomenclature

BCD	Bipolar-CMOS-DMOS
BJT	Bipolar-junction transistor
BNW	Buried N-type well
DPW	Deep P-type well
DUT	Device under test
ESD	Electrostatic discharge
GGnMOS	Gate-grounded nMOSFET
nLDMOS	N-channel lateral diffused MOSFET
HBM	Human-body model
HVNW	High-voltage N-type well
HVPB	High-voltage P-type Base
HVPW	High-voltage P-type well
IC	Integrated circuit
I/O	Input/output
Length X	The length of X (the variable length of elliptical cylinders)
LV	Low voltage
MV	Medium voltage
N-EPI	N-type epitaxy layer
OD	Thin-oxide definition
PBody	P-type body layer
RESURF	Reduced surface field
R_{on}	On-resistance
SJ	Super junction
STI	Shallow trench isolation
UHV	Ultra high-voltage
V_{bk}	Breakdown voltage

References

1. Aoike, N.; Hoshino, M.; Iwabuchi, A. Automotive HID headlamps producing compact electronic ballasts using power ICs. *IEEE Ind. Appl. Mag.* **2002**, *8*, 37–41. [CrossRef]
2. Tee, E.K.C.; Julai, N.; Hai, H.Y.; Pal, D.K.; Hua, T.S. Design of 0.18um high voltage LDMOS for automotive application. In Proceedings of the IEEE International Conference on Semiconductor Electronics, Johor Bahru, Malaysia, 27 November 2008; pp. 6–10.
3. DeVincentis, M.; Hollingsworth, G.; Gilliard, R. Long life solid-state RF powered light sources for projection display and general lighting applications. In Proceedings of the IEEE MTT-S International Microwave Symposium Digest, Atlanta, GA, USA, 15–20 June 2008; pp. 1497–1500.
4. Shirai, K.; Yonemura, K.; Watanabe, K.; Kimura, K. Ultra-low on-resistance LDMOS implementation in 0.13μm CD and BiCD process technologies for analog power IC's. In Proceedings of the 21st International Symposium on Power Semiconductor Devices & ICs (ISPSD), Barcelona, Spain, 14–18 June 2009; pp. 77–79.
5. Wang, Y.; Zhang, D.; Lv, Y.; Gong, D.; Shao, K.; Wang, Z.; He, D.; Cheng, X. A simulation study of SOI RESURF junctions for HV LDMOS (>600V). In Proceedings of the International Workshop on Junction Technology, Shanghai, China, 10–11 May 2010; pp. 1–4.
6. Zhu, J.; Qian, Q.; Sun, W.; Lu, S.; Lin, Y. A novel compact isolated structure for 600V Gate Drive IC. In Proceedings of the IEEE International Conference of Electron Devices and Solid-State Circuits, Tianjin, China, 17–18 November 2011; pp. 1–2.

7. Liu, S.; Sun, W.; Huang, T.; Zhang, C. Novel 200V power devices with large current capability and high reliability by inverted HV-well SOI technology. In Proceedings of the 25th International Symposium on Power Semiconductor Devices & ICs (ISPSD), Kanazawa, Japan, 26–30 May 2013; pp. 115–118.

8. Huang, Y.-J.; Ker, M.-D.; Huang, Y.-J.; Tsai, C.-C.; Jou, Y.-N.; Lin, G.-L. ESD protection design with latchup-free immunity in 120V SOI process. In Proceedings of the IEEE SOI-3D-Subthreshold Microelectronics Technology Unified Conference (S3S), Rohnert Park, CA, USA, 5–8 October 2015; pp. 1–2.

9. Ko, K.; Park, J.; Eum, J.; Lee, K.; Lee, S.; Lee, J. Proposal of 0.13um new structure LDMOS for automotive PMIC. In Proceedings of the 73rd Annual Device Research Conference (DRC), Columbus, OH, USA, 21–24 June 2015; pp. 119–120.

10. Zierak, M.; Feilchenfeld, N.; Li, C.; Letavic, T. Fully-isolated silicon RF LDMOS for high-efficiency mobile power conversion and RF amplification. In Proceedings of the IEEE 27th International Symposium on Power Semiconductor Devices & ICs (ISPSD), Hong Kong, China, 10–14 May 2015; pp. 337–340.

11. Chen, Y.; Chang, C.; Yang, P. A Novel Primary-Side Controlled Universal-Input AC–DC LED Driver Based on a Source-Driving Control Scheme. *IEEE Trans. Power Electron.* **2015**, *30*, 4327–4335. [CrossRef]

12. Chang, C.; Jiang, T.; Yang, P.; Xu, Y.; Chen, Y. Adaptive line voltage compensation scheme for a source-driving controlled AC–DC LED driver. *IET Circuitsdevices Syst.* **2017**, *11*, 21–28. [CrossRef]

13. Matsuda, J.-I.; Kuwana, A.; Kojima, J.-Y.; Tsukiji, N.; Kobayashi, H. Wide SOA and High Reliability 60-100 V LDMOS Transistors with Low Switching Loss and Low Specific On-Resistance. In Proceedings of the IEEE International Conference on Solid-State and Integrated Circuit Technology (ICSSICT), Qingdao, China, 31 October–3 November 2018; pp. 1–3.

14. Wu, C.-Y.; You, H.-C. Effect of SOI LDMOS Epitaxial Layer Thickness on Breakdown Voltage. In Proceedings of the International Symposium on Computer, Consumer and Control (IS3C), Taichung, Taiwan, 6–8 December 2018; pp. 80–83.

15. Theeuwen, S.J.C.H.; Mollee, H.; Heeres, R.; Van Rijs, F. LDMOS Technology for Power Amplifiers Up to 12 GHz. In Proceedings of the 13th European Microwave Integrated Circuits Conference (EuMIC), Madrid, Spain, 23–25 September 2018; pp. 162–165.

16. Malik, W.A.; Abdulkawi, W.M.; Sheta, A.A.; Alzakari, N.F. A 6 Watts LDMOS Wideband Balanced Power Amplifier. In Proceedings of the 16th International Bhurban Conference on Applied Sciences and Technology (IBCAST), Islamabad, Pakistan, 8–12 January 2019; pp. 1053–1056.

17. van Zwol, J.; van den Berg, A.; Smedes, T. ESD Protection by Keep-On Design for a 550 V Fluorescent Lamp Control IC with Integrated LDMOS Power Stage. In Proceedings of the Electrical Overstress/Electrostatic Discharge Symposium, Orlando, FL, USA, 10–12 September 2002; pp. 270–276.

18. Vashchenko, V.-A.; Gallerano, A.; Shibkov, A. On chip ESD protection of 600V voltage node. In Proceedings of the IEEE 23rd International Symposium on Power Semiconductor Devices and ICs (ISPSD), San Diego, CA, USA, 23–26 May 2011; pp. 128–131.

19. Lee, J.-H.; Kao, T.-C.; Chan, C.-L.; Su, J.-L.; Su, H.-D.; Chang, K.-C. The ESD failure mechanism of ultra-HV 700V LDMOS. In Proceedings of the IEEE 23rd International Symposium on Power Semiconductor Devices and ICs (ISPSD), San Diego, CA, USA, 23–26 May 2011; pp. 188–191.

20. Kanawati, R.; Parvin, A.; Rotshtein, I.; Quon, D.; Yankelevich, A.; Aloni, E.; Eyal, A.; Shapira, S. ESD protection schemes and devices embedded in a 700V integrated power management platform. In Proceedings of the IEEE International Conference on Microwaves Communications, Antennas and Electronics Systems (COMCAS), Tel Aviv, Israel, 7–9 November 2011; pp. 1–4.

21. Tsai, J.-R.; Lee, Y.-M.; Tsai, M.-C.; Sheu, G.; Yang, S.-M. Development of ESD robustness enhancement of a novel 800V LDMOS multiple RESURF with linear P-top rings. In Proceedings of the TENCON 2011—IEEE Region 10 Conference, Bali, Indonesia, 21–24 November 2011; pp. 760–763.

22. Kim, S.-B.; Geum, J.; Kyoung, S.; Sung, M.-Y. On-chip stacked Punchthrough diode design for 900V power MOSFET gate ESD protection. In Proceedings of the 12th IEEE International Conference on Solid-State and Integrated Circuit Technology (ICSICT), Guilin, China, 28–31 October 2014; pp. 1–3.

23. Wu, C.-H.; Lee, J.-H.; Lien, C. A Novel Drain Design for ESD Improvement of UHV-LDMOS. *IEEE Trans. Electron Devices* **2015**, *62*, 4135–4138. [CrossRef]

24. Chen, Z.; Salman, A.; Mathur, G.; Boselli, G. Design and optimization on ESD self-protection schemes for 700V LDMOS in high voltage power IC. In Proceedings of the 37th Electrical Overstress/Electrostatic Discharge Symposium (EOS/ESD), Reno, NV, USA, 27 September–2 October 2015; pp. 1–6.

25. Sheu, G.; Li, Y.-H. An innovated 500V UHV ESD self-protection device structure. In Proceedings of the International Conference on Electron Devices and Solid-State Circuits (EDSSC), Hsinchu, Taiwan, 18–20 October 2017; pp. 1–2.

26. Kim, S.; LaFonteese, D.; Zhu, D.; Sridhar, D.S.; Pendharkar, S.; Endoh, H.; Boku, K. A new ESD self-protection structure for 700V high side gate drive IC. In Proceedings of the International Symposium on Power Semiconductor Devices and ICs (ISPSD), Sapporo, Japan, 28 May–1 June 2017; pp. 467–470.

27. Pjenčák, J.; Agam, M.; Šeliga, L.; Yao, T.; Suwhanov, A. Novel approach for NLDMOS performance enhancement by critical electric field engineering. In Proceedings of the IEEE 30th International Symposium on Power Semiconductor Devices and ICs (ISPSD), Chicago, IL, USA, 13–17 May 2018; pp. 307–310.

28. Bandi, L.; Hemanth, A.; Kumar, B.H.; Wijaya, A.C.; Sheu, G. Fully Ion-Implanted 1200V LDMOS with Linear P-Top Technology. In Proceedings of the IEEE International Conference on Consumer Electronics—Taiwan (ICCE-TW), Yilan, Taiwan, 20–22 May 2019; pp. 1–2.

29. Amberetu, M.A.; Salama, C.A.T. 150-V class superjunction power LDMOS transistor switch on SOI. In Proceedings of the 14th International Symposium on Power Semiconductor Devices and ICs (ISPSD), Sante Fe, NM, USA, 7 June 2002; pp. 101–104.

30. Permthammasin, K.; Wachutka, G.; Schmitt, M.; Kapels, H. Performance Analysis of Novel 600V Super-Junction Power LDMOS Transistors with Embedded P-Type Round Pillars. In Proceedings of the International Conference on Simulation of Semiconductor Processes and Devices, Tokyo, Japan, 1–3 September 2005; pp. 179–182.

31. Permthammasin, K.; Wachutka, G.; Schmitt, M.; Kapels, H. New 600V Lateral Superjunction Power MOSFETs Based on Embedded Non-Uniform Column Structure. In Proceedings of the International Conference on Advanced Semiconductor Devices and Microsystems, Smolenice Castle, Slovakia, 16–18 October 2006; pp. 263–266.

32. Park, I.-Y.C.; Salama, C.A.T. Super Junction LDMOS Transistors-Implementing super junction LDMOS transistors to overcome substrate depletion effects. *IEEE Circuits Devices Mag.* **2006**, *22*, 10–15. [CrossRef]

33. Duan, B.; Yang, Y.; Zhang, B. New Superjunction LDMOS with N-Type Charges' Compensation Layer. *IEEE Electron Device Lett.* **2009**, *30*, 305–307. [CrossRef]

34. Park, I.-Y.; Choi, Y.-K.; Ko, K.-Y.; Yoon, C.-J.; Kim, Y.-S.; Kim, M.-Y.; Kim, H.-T.; Lim, H.-C.; Kim, N.-J.; Yoo, K.-D. Implementation of buffered Super-Junction LDMOS in a 0.18 um BCD process. In Proceedings of the 21st International Symposium on Power Semiconductor Devices & ICs (ISPSD), Barcelona, Spain, 14–18 June 2009; pp. 192–195.

35. Qiao, M.; Wang, W.-L.; Li, Z.-J.; Zhang, B. Super junction LDMOS technologies for power integrated circuits. In Proceedings of the IEEE 11th International Conference on Solid-State and Integrated Circuit Technology (ICSICT), Xi'an, China, 29 October–1 November 2012; pp. 1–4.

36. Panigrahi, S.K.; Gogineni, U.; Baghini, M.S.; Iravani, F. 120 V super junction LDMOS transistor. In Proceedings of the IEEE International Conference of Electron Devices and Solid-state Circuits (EDSSC), Hong Kong, China, 3–5 June 2013; pp. 1–2.

37. Duan, B.; Yuan, S.; Cao, Z.; Yang, Y. New Superjunction LDMOS with the Complete Charge Compensation by the Electric Field Modulation. *IEEE Electron Device Lett.* **2014**, *35*, 1115–1117. [CrossRef]

38. Duan, B.; Cao, Z.; Yuan, S.; Yang, Y. Complete 3D-Reduced Surface Field Superjunction Lateral Double-Diffused MOSFET Breaking Silicon Limit. *IEEE Electron Device Lett.* **2015**, *36*, 1348–1350. [CrossRef]

39. Jagannathan, R.; Hoenes, H.-P.; Duggal, T. Impacts of package, layout and freewheeling diode on switching characteristics of Super Junction MOSFET in automotive DC-DC applications. In Proceedings of the IEEE International Conference on Power Electronics, Drives and Energy Systems (PEDES), Kerala, India, 14–17 December 2016; pp. 1–6.

40. Cao, Z.; Duan, B.; Cai, H.; Yuan, S.; Yang, Y. Theoretical Analyses of Complete 3-D Reduced Surface Field LDMOS With Folded-Substrate Breaking Limit of Superjunction LDMOS. *IEEE Trans. Electron Devices* **2016**, *63*, 4865–4872. [CrossRef]

41. Zhang, W.; Zhang, B.; Qiao, M.; Li, Z.; Luo, X.; Li, Z. Optimization of Lateral Superjunction Based on the Minimum Specific ON-Resistance. *IEEE Trans. Electron Devices* **2016**, *63*, 1984–1990. [CrossRef]

42. Tsai, J.-Y.; Hu, H.-H. New Super-Junction LDMOS Based on Poly-Si Thin-Film Transistors. *IEEE J. Electron Devices Soc.* **2016**, *4*, 430–435. [CrossRef]

43. Park, J.; Lee, J.-H. A 650 V Super-Junction MOSFETw ith Novel Hexagonal Structure for Superior Static Performance and High BV Resilience to Charge Imbalance: A TCAD Simulation Study. *IEEE Electron Device Lett.* **2017**, *38*, 111–114. [CrossRef]

44. Cao, Z.; Duan, B.; Yuan, S.; Guo, H.; Lv, J.; Shi, T.; Yang, Y. Novel superjunction LDMOS with multi-floating buried layers. In Proceedings of the 29th International Symposium on Power Semiconductor Devices and ICs (ISPSD), Sapporo, Japan, 28 May–1 June 2017; pp. 283–286.

45. Gui, H.; Zhang, Z.; Ren, R.; Chen, R.; Niu, J.; Tolbert, L.M.; Wang, F.; Blalock, B.J.; Costinett, D.J.; Choi, B.B. SiC MOSFET versus Si Super Junction MOSFET-Switching Loss Comparison in Different Switching Cell Configurations. In Proceedings of the IEEE Energy Conversion Congress and Exposition (ECCE), Portland, OR, USA, 23–27 September 2018; pp. 6146–6151.

46. Cheng, J.; Chen, W.; Li, P. Improvement of Deep-Trench LDMOS With Variation Vertical Doping for Charge-Balance Super-Junction. *IEEE Trans. Electron Devices* **2018**, *65*, 1404–1410. [CrossRef]

47. Cao, Z.; Duan, B.; Shi, T.; Dong, Z.; Guo, H.; Yang, Y. Theory Analyses of SJ-LDMOS With Multiple Floating Buried Layers Based on Bulk Electric Field Modulation. *IEEE Trans. Electron Devices* **2018**, *65*, 2565–2572. [CrossRef]

48. Tang, P.-P.; Wang, Y.; Bao, M.-T.; Luo, X.; Cao, F.; Yu, C.-H. Improving breakdown performance for SOI LDMOS with sidewall field plate. *Micro Nano Lett.* **2019**, *14*, 420–423. [CrossRef]

49. Duan, B.; Li, M.; Dong, Z.; Wang, Y.; Yang, Y. New Super-Junction LDMOS Breaking Silicon Limit by Multi-Ring Assisted Depletion Substrate. *IEEE Trans. Electron Devices* **2019**, *66*, 4836–4841. [CrossRef]

50. *CoolMOS™ SJ MOSFETs Selection Guide*; Infineon Technologies: Neubiberg, Germany, 2019; pp. 1–32.

51. MIL-STD-883K, Method 3015. 2016. Available online: https://infostore.saiglobal.com/en-us/Standards/MIL-STD-883-K-733707_SAIG_MIL_MIL_1702986 (accessed on 5 March 2018).

52. ANSI/ESDA/JEDEC JS-001. 2017. Available online: https://www.jedec.org/document_search?search_api_views_fulltext=JS-001-2017 (accessed on 21 March 2020).

53. Quddus, M.-T.; Tu, L.; Ishiguro, T. Drain voltage dependence of on resistance in 700V super junction LDMOS transistor. In Proceedings of the International Symposium on Power Semiconductor Devices & ICs (ISPSD), Kitakyushu, Japan, 24–27 May 2004; pp. 201–204.

Article

An Experimental Study of the Failure Mode of ZnO Varistors Under Multiple Lightning Strokes

Chunlong Zhang [1,2,3,4], Hongyan Xing [1,2,3,*], Pengfei Li [3,4], Chunying Li [3,4], Dongbo Lv [3,4] and Shaojie Yang [4]

[1]   Collaborative Innovation Center on Forecast and Evaluation of Meteorological Disasters, Nanjing University of Information Science & Technology, Nanjing 210044, China; 20151114051@nuist.edu.cn
[2]   Jiangsu Key Laboratory of Meteorological Observation and Information Processing, Nanjing University of Information Science & Technology, Nanjing 210044, China
[3]   School of Atmospheric Physics, Nanjing University of Information Science & Technology, Nanjing 210044, China; 20131214195@nuist.edu.cn (P.L.); 20156282032@nuist.edu.cn (C.L.); 20156282031@nuist.edu.cn (D.L.)
[4]   Meteorological Disaster Prevention Technology Center of Heilongjiang Province, Harbin, 150030, China; sjyang@grmc.gov.cn
*   Correspondence: xinghy@nuist.edu.cn; Tel.: +86-025-58731370

Received: 15 January 2019; Accepted: 31 January 2019; Published: 2 February 2019

Abstract: In this study, in order to explore the failure mode of ZnO varistors under multiple lightning strokes, a five-pulse 8/20 µs nominal lightning current with pulse intervals of 50 ms was applied to ZnO varistors. Scanning electron microscopy (SEM) and X-ray diffractometry (XRD) were used to analyze the microstructure of the material. The failure processes of ZnO varistors caused by multiple lightning impulse currents were described. The performance changes of ZnO varistors after multiple lightning impulses were analyzed from both macro and micro perspectives. According to the results of this study's experiments, the macroscopic failure mode of ZnO varistors after multiple lightning impulses involved the rapid deterioration of the electrical parameters with the increase of the number of impulse groups, until destruction occurred by side-corner cracking. The microstructural examination indicated that, after the multiple lightning strokes, the proportion of Bi in the crystal phases was altered, the grain size of the ZnO varistors became smaller, and the white intergranular phase (Bi-rich grain boundary layer) increased significantly. The failure mechanism was thermal damage and grain boundary structure damage caused by temperature gradient thermal stress, generated by multiple lightning currents.

Keywords: failure mode; impulse current; microstructure; multiple lightning; ZnO varistors

---

1. Introduction

The installation of surge protective devices is one of the most economical and effective means to avoid or reduce damage caused by lightning impulses in power distribution systems. The core component of a surge protective device is a ZnO varistor, which has many advantages, including a good nonlinear property, small normal leakage current, low level of residual pressure, and no follow current. ZnO varistors have been widely applied for the protection of electronic and electrical equipment from lightning [1,2]. However, after a natural lightning strike, ZnO varistors often have varying degrees of failure.

An impulse test is the most direct means to study the performance of ZnO varistors. Currently, an 8/20 µs single pulse waveform had been adopted for such tests [3,4]. A host of studies have been presented regarding the performance changes that occur in ZnO varistors under single pulse lightning impulses [5–8]. However, modern lightning observations and artificially triggered lightning acquisition data have shown that two-thirds of the lightning events in natural settings are a multi-pulse process. The statistics show that nearly 70% of cloud-to-ground lightning strikes involve from 2 and up to 20 strikes, with an average number of between 3 and 5, and a time duration between strikes of 15 ms to 150 ms. There is a significant difference between the single-pulse lightning waveform and the multiple lightning strikes of natural lightning [9,10], and the total time and energy of a multi-pulse are several times that of a single pulse. When a low-voltage distribution line is struck by a lightning, the lightning current flowing through the surge protective device generates heat in the ZnO varistor due to power loss, causing the body to heat up. When a multi-pulse lightning current and single-pulse lightning current flow through a ZnO varistor, there is a huge difference in duration and energy absorption, which inevitably leads to a difference in the temperature increase of the ZnO varistor and the final electrical performance parameters. Therefore, lightning single pulse test methods have been unable to properly simulate the damage caused by lightning [11]. Darveniza et al. [12] pointed out that multiple lightning impulse currents could potentially cause severe damage to lightning protection devices protecting the equipment in power distribution systems. Previous studies have shown that six-pulse lightning current impulses have resulted in serious damage to lightning protection components, and confident conclusions cannot currently be drawn regarding their impact according to the lightning protection test standard. Lee et al. [13,14] found that ZnO varistors degraded when subjected to multiple impulse currents, and their life mainly depended on the amplitude of the lightning surge. Haryono et al. [15,16] analyzed the damage effects to ZnO varistors undergoing multiple lightning impulse currents from the perspective of energy absorption. At the present time, the research regarding the performance of ZnO varistors undergoing multiple lightning impulse current has been mainly based on the analysis of the macroscopic electrical properties [17–19]. Qingheng Chen et al. [20] used the simulation network model to analyze the current, temperature, and thermal stress in zinc oxide varistors. The results showed that reducing the average size of the ZnO grains can significantly reduce the temperature difference inside the ZnO varistor. Thermal stress increases the impact energy absorption capacity of zinc oxide varistors. Pengfei Li et al. [21] analyzed the damage form of metal oxide under multiple lightning impulses and found that it was mainly the edge that was cracking. However, few research studies have been conducted to investigate the damage failure mechanisms under multi-pulse continuous impulses, especially regarding the changes in microstructures. Tsuboi et al. [22] believed that damage could occur in the internal components of ZnO when subjected to multiple lightning impulses, resulting in the failure of the ZnO varistor, which provided a theoretical basis for the modeling of this study.

Therefore, in this study, a five-pulse current with a time interval of 50 ms was applied to the samples. The actual lightning strike to the ZnO varistors in practical applications was simulated as realistically as possible, and the macroscopic damage form and static parameter variation characteristics were monitored. At the same time, the ZnO varistors were analyzed before and after the impulse by scanning electron microscopy (SEM) and X-ray diffractometry (XRD). It was expected that we would investigate the failure mode of the ZnO varistors during the natural lightning strikes from the perspective of macroscopic and microscopic bonding. Indeed, some experimental data regarding the performance of ZnO varistors under multiple lightning impulse currents are shown here. The failure modes of the ZnO varistors undergoing multiple lightning impulse currents are discussed on the basis of the results of thermal effect. This is believed to be particularly important for improving lightning protection and safety performances of ZnO varistors.

2. Experiment

2.1. Impulse Test and Waveform

The multiple lightning impulse equipment used in this experiment was a 20-pulse lightning impulse current generator, as shown in Figure 1. Multi-channel discharge technology was adopted to simulate the multiple lightning stroke processes. The waveform generator circuit diagram is shown in Figure 2, in which C is the capacitor, G is the impact gap, Rs is the resistor, and Ls is the inductor. With the impulse test end as the axis, there are 10 trigger channels at each end. When the primary current is fully triggered, 20 high-voltage pulses can be generated, and the time interval can be changed from 1 ms to 999 ms.

Figure 1. 20-pulse lightning impulse current generator.

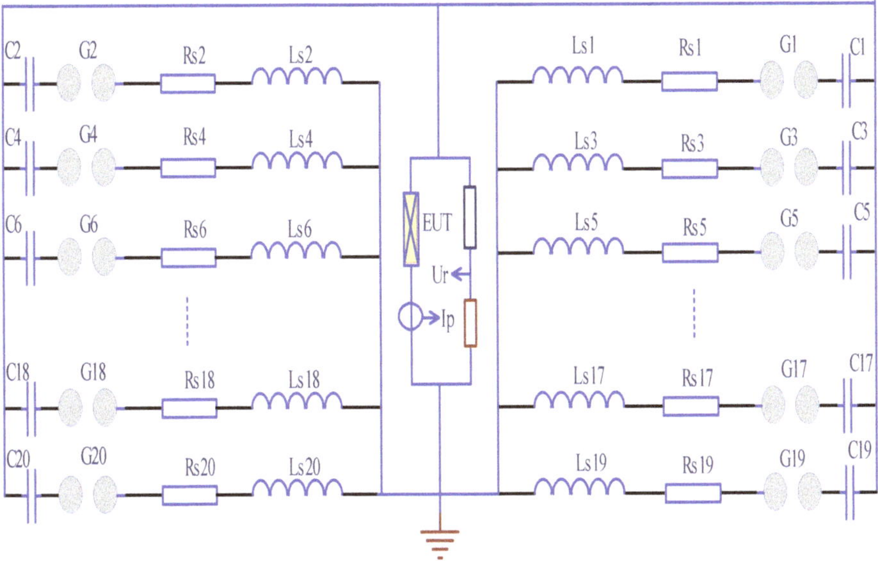

Figure 2. Multiple lightning impulse current generator circuit diagram.

In this study, Section 1 of IEC 62305-1 Lightning Protection: General Rule [23] provides a special definition for multiple lightning. Lightning with an average of three to four impulses, and intervals of approximately 50 ms was defined as multiple lightning. Therefore, the multiple lightning strikes in this test were represented by a group of multiple impulse currents, which included five consecutive impulse currents, each of which had 8/20 μs waveforms. The time between two consecutive pulses was 50 ms, and the pulse amplitudes were the nominal discharge currents of the selected ZnO varistors. The waveform diagram is shown in Figure 3, where the five yellow vertical lines represent the five pulses, and the following 8/20 μs waveform denotes the full waveform figure of a single pulse.

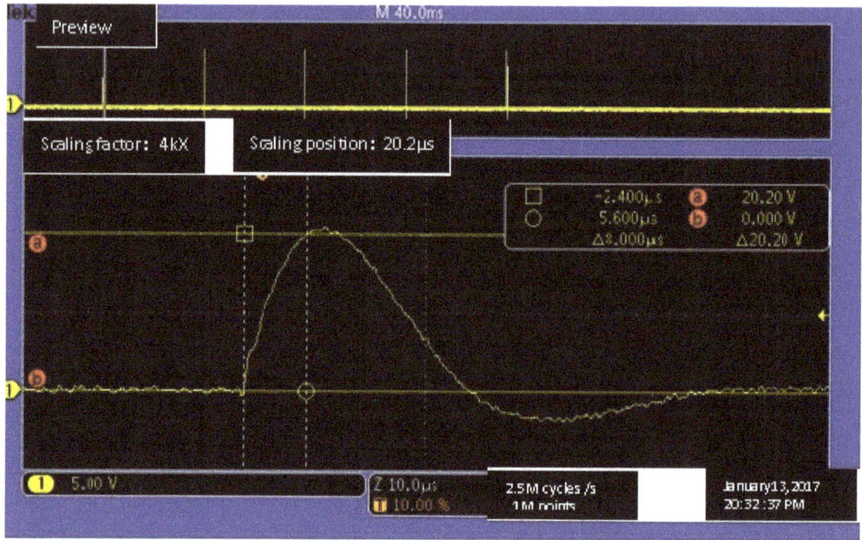

Figure 3. Five-pulse lightning waveform diagram.

2.2. Sample Preparation

In this study, the various ZnO varistors that were used were provided by the same manufacturer. The basic parameters of these samples were nominal discharge current I_n = 20 kA and the maximum continuous operation voltage U_c = 385 V, and the static parameters (varistor voltage and leakage current) were as shown in Table 1. The varistor voltage is a voltage corresponding to a current of the varistor of 1 mA and is a standard of a voltage whose current rapidly rises with voltage, expressed by U_{1mA}. The leakage current refers to the current flowing through the varistor at a specified temperature and maximum DC voltage, generally expressed by I_{ie}. Under the premise of approaching U_{1mA}, the smaller the I_{ie} was, the better the voltage limiting the performance of the varistor would be. In this study, seven ZnO varistor blocks, with the majority approaching the static parameters, were selected as the samples and were denoted as A1 to A7.

Table 1. Changes in the electrical parameters before and after the impulse in the ZnO varistors.

No.	Initial U_{1mA} (V)	Initial I_{ie} (μA)	U_{1mA} When Failure (V)	I_{ie} When Failure (μA)	T_{max} When Failure (°C)	Groups of Impulse
A1	690	0	602	8.2	224	17
A2	689	0	600	8.1	225	16
A3	690	0	604	7.5	222	15
A4	691	0.1	615	9.4	221	16
A5	688	0.1	605	8.6	216	16
A6	690	0.1	603	8.5	228	15
A7	689	0.1	602	7.8	218	16

2.3. Experimental Procedure

The flowchart of the experimental procedure is shown in Figure 4.

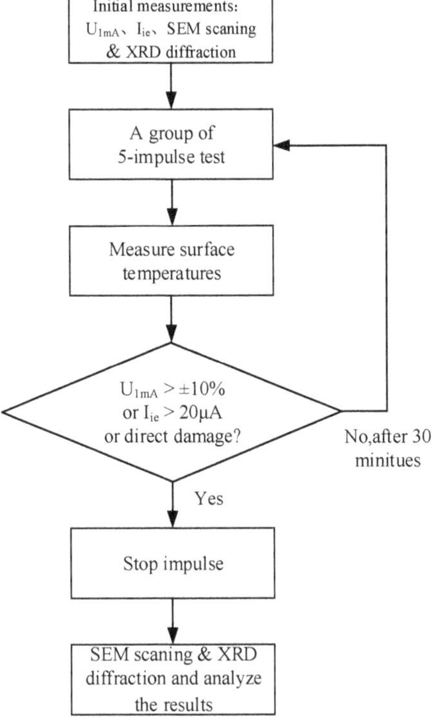

Figure 4. Flowchart of the experimental procedure.

(1) Initial measurements: The samples are characterized with the U_{1mA}, I_{ie}, and by photographs taken at the beginning of the test. We used scanning electron microscopy (SEM) and an X-ray diffractometer (XRD) on the ZnO varistor blocks before the impulse test.

(2) Impulse test: we adjusted the charging voltage of the generator to output the demand impulse currents. Then, multiple lightning impulse currents were applied to the ZnO varistors. The time between the application of one group of impulse currents to a ZnO varistor block, and that of the next group of impulse currents was 30 minutes; with such long duration of time, we were able to return to the original conditions. The process of the impulses is shown in Figure 5.

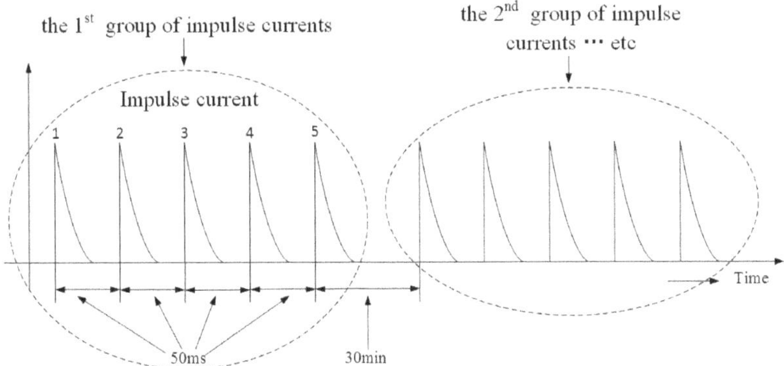

Figure 5. Groups of impulse currents applied to the ZnO varistors.

(3) We measured the surface temperatures, varistor voltage U_{1mA}, and leakage current I_{ie} of the ZnO varistors after each impulse. Then, we checked the surface of the ZnO varistors for flashover or puncture and took photographs of the damaged ZnO varistors. When the change in amplitude of the U_{1mA} reached beyond $\pm 10\%$ of the original, the I_{ie} exceeded 20 µA, or direct damage occurred, the ZnO varistors were judged as having failed. Subsequently, the impulse test was ceased, and the data were recorded.

(4) We used scanning electron microscopy (SEM) and an X-ray diffractometer (XRD) on the ZnO varistor blocks after the impulse test in order to observe the microstructural changes of the ZnO varistors.

3. Results and Discussion

3.1. Macroscopic Properties

The average level change directions of the U_{1mA} and I_{ie} after the increases in the impulse groups for the ZnO varistors under multiple lightning impulse currents can be seen in Figure 6. It can be observed that the U_{1mA} showed a trend of decreasing-stable-decreasing with the increase of impulse groups, and was observed to drop after the first group of impulses and the last group of impulses, with an average decline rate as high as 4%. The I_{ie} all showed increasing trends with larger rising rates, with an average growth rate of 0.66 µA/a group. Following the 16 groups of impulses, it was observed that the U_{1mA} fell sharply, with an average drop rate of more than 10% of the original U_{1mA}. This resulted in failures occurring in the ZnO varistors. Then, after another group of impulses, the ZnO varistor blocks became damaged.

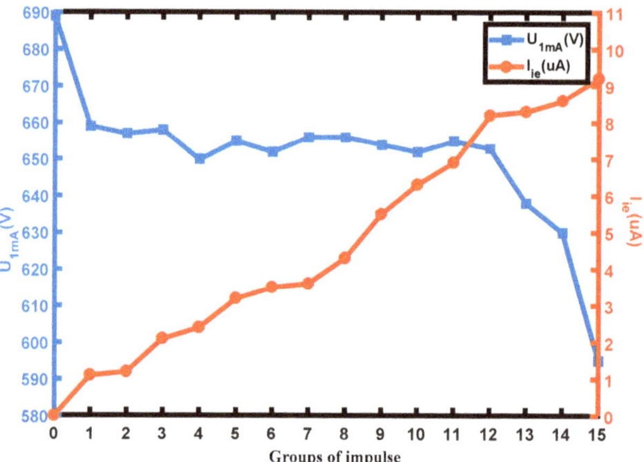

Figure 6. Varistor voltage U_{1mA} and leakage current I_{ie} variation diagrams of the ZnO varistors under multiple lightning impulse currents.

As shown in Table 1, under the five-pulse lightning currents, the ZnO varistor blocks A1 to A7 was able to withstand an average of 16 groups of impulses. When the U_{1mA} had fallen one to two groups of impulses. Figure 7b shows the damage forms of the ZnO varistor block A2 during its failure, which presented as a side-corner cracking along with patch collapse [24]. As shown in Figure 7c, the damage forms of the ZnO varistor blocks A1 to A4 in the different codes were observed to be highly consistent.

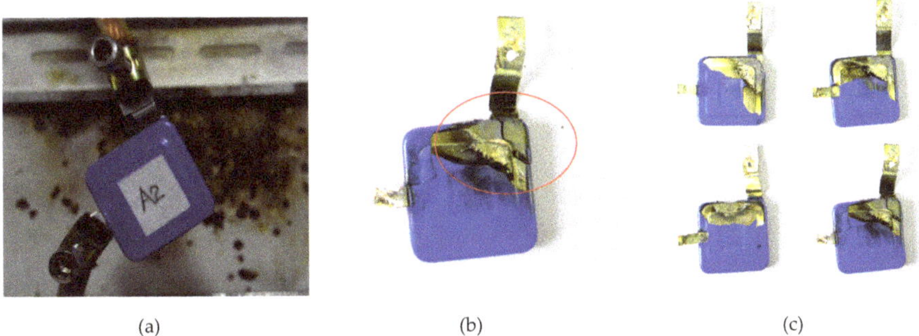

(a)   (b)   (c)

Figure 7. The appearance state diagram of the ZnO varistors before and after the multiple lightning impulse currents. (**a**) The state diagram of Sample A2 before impulses; (**b**) Damage state diagram of Sample A2; (**c**) Damage diagrams of Samples A1 to A4.

3.2. Microscopic Properties

Figure 8 details the SEM images of the ZnO varistors prior to the impulses, with a resolution of 10 µm. It can be seen from the figure that the internal structure of the ZnO varistor was mainly composed of four types of gray matter, marked as +1, +2, +3, and +4. Through this study's EDS analysis, it was determined that the main components of +1 in the white area were O, Zn, and Bi; those of +2 in the gray area were O and Zn; those of +3 at the white-gray junction were O, Zn, and Bi, and those of +4 in the deep black region were O and Zn. The proportion of elements in different regions is shown in Figure 8b. It can be seen that the main components of the ZnO varistor are zinc oxide and small amounts of Bi compounds. The gray area in the figure indicates the zinc oxide grain, and the white area indicates the rich-Bi phase around the zinc oxide grain. It was found that the cell distributions of the ZnO varistor were not uniform, and the rich-Bi phase was concentrated in a certain region. Map analysis at the +1 position of the ZnO varistors is shown in Figure 8c.

(a)

Figure 8. *Cont.*

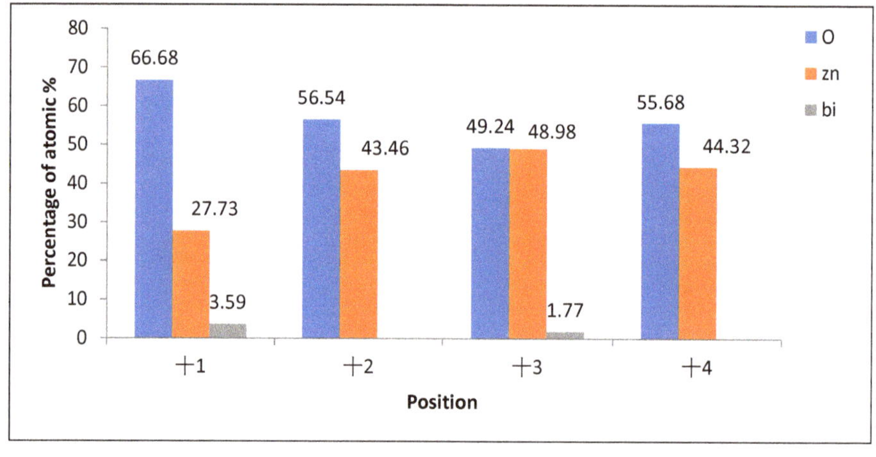

(b)

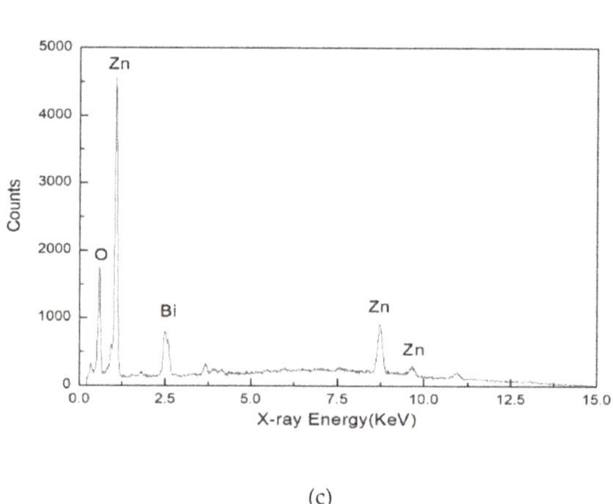

(c)

Figure 8. (**a**) SEM images of the ZnO varistors (resolution of 10 μm); (**b**) EDS analysis results of the ZnO varistors. (+1, +2, +3, +4); (**c**) Map analysis at the +1 position of the ZnO varistors.

Figure 9a,b shows the XRD diffraction pattern of the ZnO varistors, before and after the impulses. It can be seen in the figure that the crystal phase compositions mainly include four parts: zincite, syn; manganese oxide; bismuth oxide; and bismite, syn. Through the analysis of the figures, it was determined that the proportion of Bi in the crystal phases before the impulse test was 0.9%, 0.8%, and 0.6%, respectively. The proportion after the impulse test was 0.0%, 1.0%, and 1.3%, respectively. The proportion of Bi in the crystal phases has been converted.

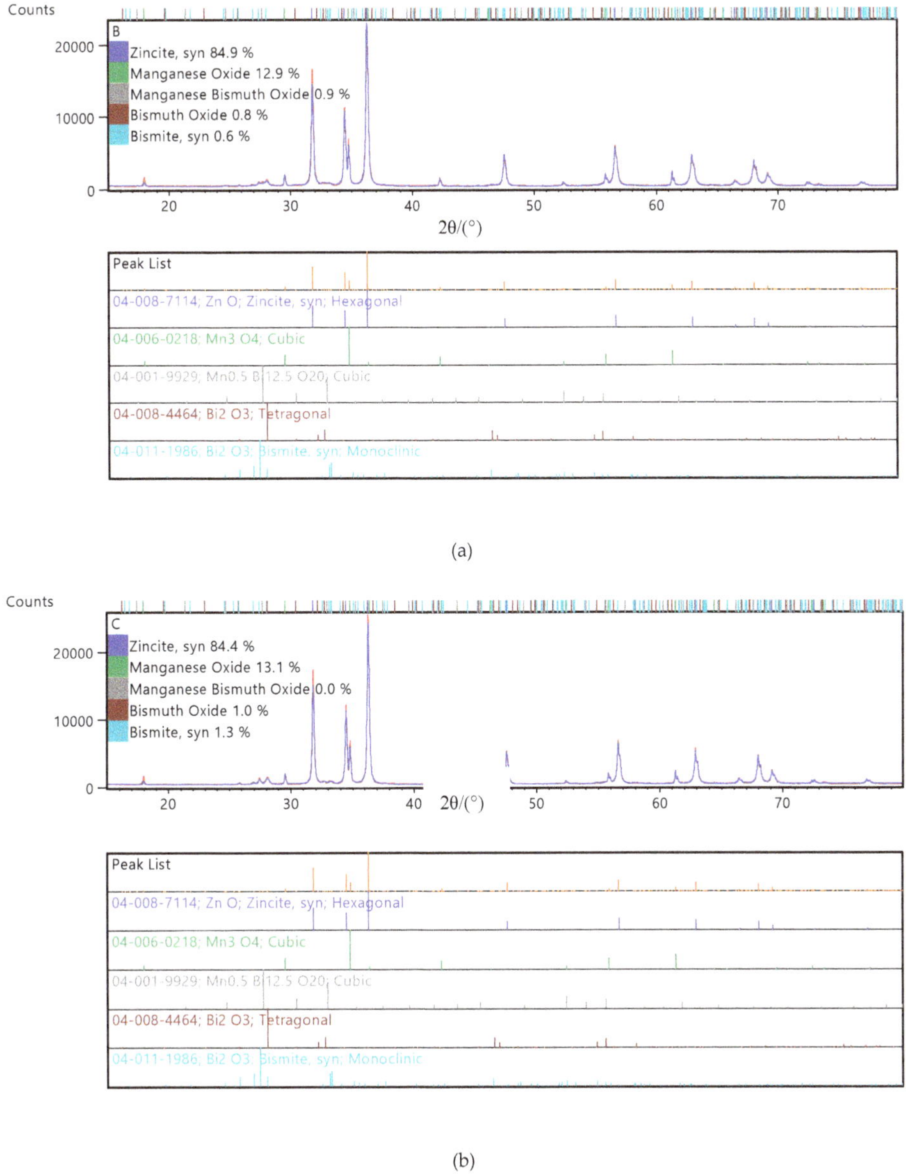

(a)

(b)

Figure 9. XRD diffraction results of the ZnO varistors before (**a**) and after (**b**) the impulses.

Figure 10a,b illustrate the SEM images of the ZnO varistors, before and after the impulses, with a resolution of 80 μm. It was observed that after the impulses, the grain sizes became smaller, while the white areas increased. The differences in the microstructures of the materials in the ZnO varistors indicate that the multiple lighting impulse currents led to the ZnO grains becoming smaller, along with the grain boundary growth.

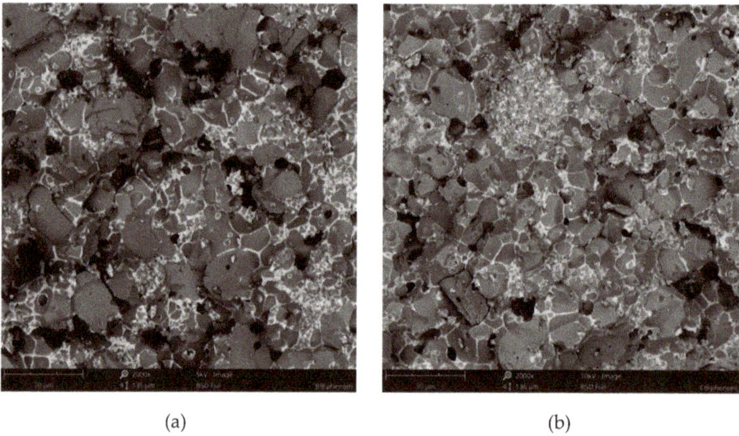

(a)   (b)

Figure 10. SEM images of the ZnO varistors (resolution of 80 μm). (**a**) Before the impulses. (**b**) After the impulses.

3.3. Failure Mechanism

The ZnO varistors have the appearance of layers of insulation coating. When impulse current passes through the ZnO varistors, the pulse intervals are ms in length, and the internal heat of the ZnO varistors will be almost instantaneously concentrated, accompanied by a sharp rise in temperature. In terms of the energy convection and exchange, this process can be seen as an adiabatic temperature rise. At the time that the failures occurred in the ZnO varistors, the surface temperature had risen to over 200 °C, and the average temperature was 223 °C, as shown in Table 1. According to Equation (1), in the cases of the injected energy equivalent, the temperature increases of the ZnO varistors units with the intake of energy could be directly determined by the thermal physical property parameters ρ and Cp. However, during the production process of the ZnO varistors, absolute material mixing uniformity could not be achieved. As can be seen from the SEM images of the ZnO varistor blocks, the microstructures show inhomogeneities of the internal material distribution, which were observed to have led to significant differences in the thermal physical properties of different parts of the same ZnO varistors. For example, inhomogeneities existed in the thermal physical properties of the ZnO varistor blocks. In this study, t was determined that the ρ and Cp differences in different parts led to different adiabatic temperature increases. Furthermore, the infrared imaging measurement results showed different temperature increases in different locations of the ZnO varistor blocks after absorbing the impulse energy. Figure 11 shows the temperature distribution diagram of the ZnO varistor blocks A1 and A4 following the impulses. It can be seen in the figure that there were considerable differences observed in the temperature increases in different parts. The energy at the local hot spot occurred too late to be passed around, which led to a large temperature gradient between the hot spot and the surrounding area. Thermal stress to the temperature gradient occurred in the interior of the ZnO varistor block. When the thermal stress reached a certain value, it caused burst damage to the ZnO varistor block [25].

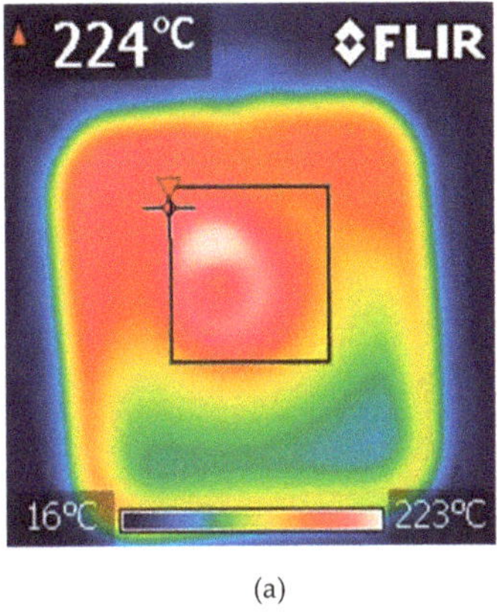

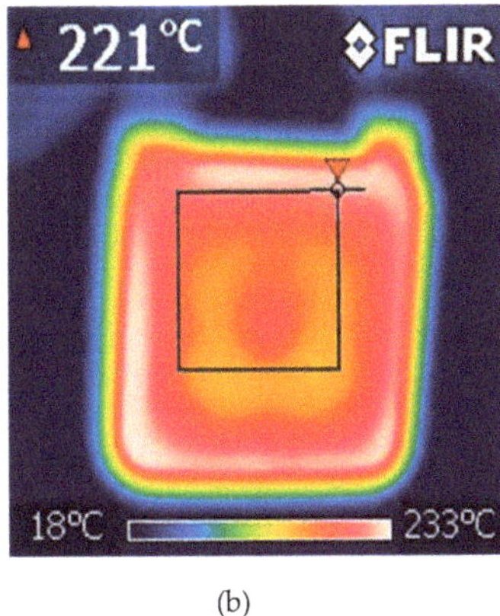

(a)   (b)

Figure 11. Temperature distribution map of the ZnO varistor after impulses. (**a**) Varistor block A1. (**b**) Varistor block A4.

The adiabatic temperature increases of the ZnO varistor units after the absorption of energy W (unit: J) were as follows [25]:

$$\Delta T = \frac{\Delta W}{\Delta V \rho c_p} \tag{1}$$

In Figure 11, ΔV is the volume of the ZnO varistor units; ρ is the proportion of the ZnO varistors, with an average of 5600 kg/m^3; and C_p is the constant-pressure heat capacity of the ZnO varistor units, which was approximately 500 J/kg °C at 20 °C.

If the temperature increases of the adjacent two units were ΔT_1 and ΔT_2, respectively, then the thermal stress f under the effects of the temperature gradient was as follows [25]:

$$f = \frac{Ea}{1-\mu}(\Delta T_1 - \Delta T_2) \tag{2}$$

where E is the elastic modulus; a denotes the thermal expansion coefficient; and μ is the Poisson's ratio. When the thermal stress f produced under the action of the temperature gradient exceeded the thermal stress threshold during the rupture failure of the ZnO varistors material, it could potentially lead to the destruction of the ZnO varistors.

The experimental phenomenon in which the varistor voltage presented a declining trend could be explained by the microstructure characteristics of the ZnO varistors. As shown in Figure 10, the SEM images confirmed that the interiors of the ZnO varistors were composed of many zinc-oxide grains, and the distances between various grains were fixed. Meanwhile, transverse and longitudinal capacitances were formed between the grains, which displayed a vertical distribution.

The microanalysis showed that the grain sizes of the ZnO varistors became smaller, while the area of the grain boundary layer increased, which led to the increases in the capacitance of the grain boundary layer as follows:

$$C = \frac{\varepsilon S}{4\pi kd} = \frac{Q}{V} \tag{3}$$

In Equation (3) [26], C represents the capacitance; ε is the dielectric constant; S is the cross-sectional area; k denotes the electrostatic force constant; d is the anode-to-cathode distance; Q represents the charge quantity; and V is the voltage.

It can be seen from the above equation that, as the number of impulse groups increased, the capacitance of the grain boundary layer increased, while the varistor voltage gradually decreased.

In the ZnO varistors under the multiple lightning impulse current, the leakage current value gradually increased. This was due to the fact that after several impulse groups, the ZnO varistors absorbed the impulse energy, and the temperatures rose. This sped up the rates of ion migration, resulting in the leakage current values displaying a growing trend.

4. Conclusions

In this study, experiments were conducted to investigate the microstructure and macroscopic properties of ZnO varistors under multiple lightning impulse currents.

(1) The macroscopic failure mode of ZnO varistors under multiple lightning impulses could be described as the rapid deterioration of the electrical parameters with the increase of the number of impulse groups before destruction occurred by side-corner bursting. Under a five-pulse lightning current, the ZnO varistors were able to withstand an average of 16 groups of impulses.

(2) The microstructural examination indicated that, after the multiple lightning strokes, the proportion of Bi in the crystal phases was converted, the grain size of the ZnO varistors became smaller, and the white intergranular phase (Bi-rich grain boundary layer) increased significantly.

(3) The ZnO varistors undergoing multiple lightning impulse currents presented adiabatic temperature increases. It was observed that, due to the unevenness of the material, the ZnO varistors displayed local temperature increases after absorbing impulse heat. The failure mechanism was thermal damage and grain boundary structure damage caused by the temperature gradient thermal stress generated by multiple lightning currents.

Author Contributions: Data curation, C.L.; Investigation, H.X.; Methodology, D.L. and S.Y.; Writing—original draft, C.Z. and P.L.

Funding: This research was funded by the National Nature Science Foundation of China, NSFC Grants No. 61671248, Open Foundation of Jiangsu Key Laboratory of Meteorological Observation and Information Processing, KDXS1603, and A Project Funded by the Priority Academic Program Development of Jiangsu Higher Education Institutions (PAPD) and The APC was funded by Heilongjiang Provincial Meteorological Bureau Scientific Research Project, HQ2015005.

Acknowledgments: The authors are thankful to the anonymous reviewers for their valuable comments.

References

1. He, J.; Liu, J. Development of ZnO varistors in metal oxide arrestors utilized in ultra high voltage systems. *High Volt. Eng.* **2011**, *37*, 634–643.

2. He, J.; Yuan, Z. Effective protection distances of low-voltage SPD with different voltage protection levels. *IEEE Trans. Power Delivery* **2010**, *25*, 187–195.

3. British Standards Document. *Low-Voltage Surge Protective Devices. Surge Protective Devices Connected to Low-Voltage Power Systems. Requirements and Test Methods*; IEC61643-11; International Electrotechnical Commission (IEC): London, UK, 2011.

4. Ringler, K.G.; Kirkby, P. The energy absorption capability and time-to-failure of varistors used in station-class metal-oxide surge arresters. *IEEE Trans. Power Delivery* **1997**, *12*, 203–212. [CrossRef]

5. Takada, M.; Yoshikado, S. Effect of SnO_2 addition on electrical degradation of ZnO varistors. *Key Eng. Mater.* **2007**, *350*, 213–216. [CrossRef]

6. De Salles, C.; Martinez, M.L.B.; de Queiroz, A.A.A. Ageing of metal oxide varistors due to surges. In Proceedings of the 2011 International Symposium on Lightning Protection, Fortaleza, Brazil, 3–7 October 2011.

7. Tsukamoto, N. Study of degradation by impulse having 4/10µs and 8/20µs waveform for MOVs (metal oxide varistors). In Proceedings of the 2014 International Conference on Lightning Protection, Shanghai, China, 11–18 October 2014; pp. 620–623.

8. Bassi, W.; Tatizawa, H. Early prediction of surge arrester failures by dielectric characterization. *IEEE Electr. Insul. Mag.* **2016**, *32*, 35–42. [CrossRef]

9. Matsumoto, Y.; Sakuma, O. Measurement of lightning surges on test transmission line equipped with arresters struck by natural and triggered lightning. *IEEE Trans. Power Deliv.* **1996**, *11*, 996–1002. [CrossRef]

10. Zhang, Q.; Qie, X.; Kong, X. Comparative analysis on return stroke current of triggered and natural lightning flashes. *Proc. CSEE* **2007**, *27*, 67–71.

11. Heidler, F.; Zischank, W.; Flisowski, Z. Parameters of lightning current given in IEC 62305-background, experience and outlook. In Proceedings of the 29th International Conference on Lightning Protection, Uppsala, Sweden, 23–26 June 2008.

12. Darveniza, M.; Tumma, L.R. Multipulse lightning currents and metal-oxide arresters. *IEEE Trans. Power Deliv.* **1997**, *12*, 1168–1175. [CrossRef]

13. Lee, B.-H.; Kang, S.-M. Properties of ZnO varistor blocks under multiple lightning impulse voltages. *Curr. Appl. Phys.* **2006**, *6*, 844–851. [CrossRef]

14. Lee, B.-H.; Kang, S.-M.; Pak, K.-Y. Effect of multiple lightning impulse currents on Zinc oxide arrester blocks. In Proceedings of the 27th International Conference on Lightning Protection, Avignon, France, 13–16 September 2004; pp. 634–639.

15. Haryono, T.; Sirait, K.T. The damage of ZnO arrester block due to multiple impulse currents. *TELKOMNIKA* **2011**, *9*, 171. [CrossRef]

16. Haryono, T.; Sirait, K.T. Effect of multiple lightning strikes on the performance of ZnO lightning arrester block. *High Volt. Eng.* **2011**, *37*, 2763–2771.

17. Vahidi, B.; Cornick, K.; Greaves, D. Effects of multiple stroke on ZnO surge arresters. In Proceedings of the 24th International Conference on Lightning Protection, Binningam, UK, 14–18 September 1998.

18. Heinrich, C.; Wagner, S.; Richter, B. Multipulse tests on surge arresters for medium and low voltage systems. In Proceedings of the 24th International Conference on Lightning Protection, Birmingham, UK, 14–18 September 1998; pp. 790–794.

19. Munukutla, K.; Vittal, V. A practical evaluation of surge arrester placement for transmission line lightning protection. *IEEE Trans. Power Deliv.* **2010**, *25*, 1742–1748. [CrossRef]

20. Chen, Q.; He, J. Effects of Grain Size on Temperature and Thermal Stress in Zinc Oxide Nonlinear Resistor Pieces. *Sci. China Ser. E* **2002**, *3*, 323–330.

21. Li, P.; Zhang, C. Destructive forms of metal oxides under multiple lightning impulses. *High Volt. Eng.* **2017**, *43*, 3792–3799.

22. Tsuboi, T.; Takami, J. Energy absorption capacity of a 500k V surge arrester for direct and multiple lightning strokes. *IEEE Trans. Dielectr. Electr. Insul.* **2014**, *22*, 916–924. [CrossRef]

23. International Electrotechnical Commission (IEC). *Protection against Lightning—Part 1: General Priciples*; BS EN/IEC 62305-1; IEC: London, UK, 2006.

24. Darveniza, M.; Mercer, D.R. Laboratory studies of the effects of multipulse lightning currents on distribution surge arresters. *IEEE Trans. Power Deliv.* **1993**, *8*, 1035–1044. [CrossRef]

25. Wu, W.; He, J. Impact Damage Principle of metal Oxide Valve Plates. In *Properties and Applications of Nonlinear Metal Oxide, Varistors*, 5th ed.; Tsinghua University Club: Beijing, China, 1998; pp. 147–149.

26. Zou, R. Discussion on Dielectric Constants in Secondary School Physics Textbooks. *Tech. Phys. Teach.* **2000**, *4*, 22–23.

 *electronics*

Article

An Experimental Study on the Effect of Multiple Lightning Waveform Parameters on the Aging Characteristics of ZnO Varistors

Chunlong Zhang [1,2,3,4], Chunying Li [3,4], Dongbo Lv [3,4], Hao Zhu [3,5] and Hongyan Xing [1,2,3,*]

1   Collaborative Innovation Center on Forecast and Evaluation of Meteorological Disasters, Nanjing University of Information Science & Technology, Nanjing 210044, China; 20151114051@nuist.edu.cn
2   Jiangsu Key Laboratory of Meteorological Observation and Information Processing, Nanjing University of Information Science & Technology, Nanjing 210044, China
3   School of Atmospheric Physics, Nanjing University of Information Science & Technology, Nanjing 210044, China; 20156282032@nuist.edu.cn (C.L.); 20156282031@nuist.edu.cn (D.L.); 20156282035@nuist.edu.cn (H.Z.)
4   Heilongjiang Meteorological Disaster Prevention Technology Center, Harbin 150030, China
5   Anhui Meteorological Disaster Prevention Center, Hefei 230061, China
*   Correspondence: xinghy@nuist.edu.cn; Tel.: +86-025-58731370

Received: 7 April 2020; Accepted: 1 June 2020; Published: 3 June 2020

Abstract: In this study, in order to study the effect of multi-pulse waveform parameters on the aging characteristics of ZnO varistor, the aging rate and surface temperature rise of ZnO varistor under the impact of multi-pulse current were analyzed. The number of pulses and the pulse interval under multiple pulses play a decisive role in the aging rate of ZnO varistor. The greater the number of pulses and the smaller the pulse interval, the higher the temperature rise of the ZnO varistor and the faster the aging rate, the more likely to be failure and damage. The surface temperature distribution of the ZnO varistor under multi-pulse is not uniform, and the more pulses, the more uneven the temperature distribution, but the surface temperature rise has a nonlinear relationship with the number of pulses. The relationship between pulse interval, impact times and average surface temperature rise is established. The aging mechanism of the ZnO varistor under a multi-pulse lightning stroke was revealed from the perspective of energy absorption and heat transfer modelling. The energy sustained by the ZnO varistor under multiple pulses have a nonlinear multiple relationship with the energy of the single pulse current wave at the same amplitude. The superimposed cumulative energy of the impact under multiple pulses accelerates the aging process of the ZnO varistor, and eventually produces an irreversible structural destruction.

Keywords: aging characteristics; impulse current; multiple lightning; ZnO varistors; energy absorption

---

1. Introduction

The installation of a surge protector (SPD) is one of the most economical and effective protection methods for reducing the damage of lightning strikes in low-voltage power distribution systems. As the core component of the ZnO varistor, it has good nonlinearity, low residual voltage and rapid response. Its advantages include achieving lightning protection for electronic and electrical equipment by limiting overvoltage and discharge surge current [1–3]. At present, SPDs produced in various countries around the world are basically developed and produced in accordance with the technical standards of single-pulse products, and are tested by the lightning high-voltage laboratory, using 10/350 µs or 8/20 µs single-pulse waveforms. Although many lightning protection products of low-voltage power distribution systems have passed tests according to technical standards, protection failure often occurs when lightning strikes, and fire accidents may even occur. The data collected by modern lightning observations and collected by artificial lightning shows that 80% of the lightning in nature is multi-pulse. The traditional single-pulse test cannot effectively reflect the performance change of SPD subjected to natural multi-pulse lightning strikes, which leads to the lightning test deviating from the real lightning discharge process.

The problem of lightning strike failure of ZnO varistor has always been one of the core problems of lightning protection in low-voltage distribution systems. It is an important means of improving the safety performance of ZnO varistors by studying the lightning strike failure characteristics and failure mechanism of ZnO varistors through laboratory simulation of natural lightning strikes [4,5]. The damage characteristics and aging mechanism of ZnO varistor under multi-pulse is very important to study the failure mode of SPD. Compared with single pulse, ZnO varistor is aggravated by the thermal effect under multi-pulse current impact. Changes, impact resistance and other aspects will face more severe tests.

In recent years, with the deepening of people's understanding of the physical characteristics of multi-pulse lightning, domestic and foreign scholars have made some achievements in the study of the damage and aging mechanism of ZnO varistor under multi-pulse, but they are not systematic and in-depth.

Darveniza et al. [6,7] carried out a series of studies on the destruction effect of multi-pulse lightning strikes on ZnO varistor in power distribution system in 1993, and judged the ZnO varistor by analyzing the rate of change of varistor voltage change ratio before and after impact. At the same time, it proposed that multi-pulse test methods should be added to the lightning test technical standards. High-speed camera and microscopic electron microscope scanning can be used to analyze the electrical properties of ZnO varistor. Through high-speed camera shooting, it was found that the ZnO varistor will experience flashover after being subjected to multi-pulse lightning strikes. The electrical parameters of the ZnO varistor will change significantly. Microscopic electron microscopy scanning can find changes in the internal structure.

Electrical parameters and changes in the microstructure and performance of the multi-pulse ZnO varistor were analyzed in two ways, but the damage and aging mechanism of the ZnO varistor was not analyzed [8,9]. Rousseau A et al. [10] pointed out in a report on the 2014 International Conference on Lightning Protection (ICLP), held in Shanghai, that the design and production process of ZnO varistors have an important impact on their ability to withstand multi-pulse shocks, but the report did not give details on the material and structural factors of impact capability. By changing the temperature and humidity of ZnO varistors, some scholars found that the aging rate of a wet ZnO varistor under multi-pulse impact is significantly faster than that under dry conditions. Increased humidity will accelerate the aging rate, indicating that the environmental conditions of temperature and humidity are important factors affecting the aging of a ZnO varistor [11–13].

In 2011, C. de Salles [14] applied a series of shock pulses to ZnO varistors at different temperatures, and compared the leakage current and power loss of each pulse when the ZnO varistor reached room temperature. After statistical analysis, it was proposed that aging is related to the number of pulses applied, but there is no quantitative analysis of the relationship between aging and temperature and pulse energy absorption. B. Vahidi [15] and C. Heinrich [16], respectively, explained the destructive effects of multi-pulse lightning strikes on ZnO arresters in the power distribution system at the 24th International Conference on Lightning Protection. Haryono T et al. [17,18] conducted an impact experiment on a ZnO varistor using an 8/20 μs impulse current with five pulses and a pulse interval of 35 ms. This increases the probability of damage. This phenomenon also exists under single pulse impact, because the amplitude of the injected lightning current is proportional to the impact energy. A large current amplitude will inevitably make the ZnO varistor have a higher temperature rise. However, the damage result of the ZnO varistor is not related to the thermal effect of multi-pulse lightning current.

In the above study of the aging law and mechanism, the influence of waveform parameters, such as pulse interval, on the temperature rise and aging rate of the ZnO varistor was not analyzed. For the energy absorption capacity of the ZnO varistor under a single pulse, only the amplitude of the current needs to be considered, that is, the ultimate current strength that can be withstood [19]. Compared with the single pulse, the multi-pulse has different waveform parameters, such as different pulse number and pulse interval. The change of each factor may change the energy value applied to the ZnO varistor. The resulting temperature rise and the degree of damage are also different, so it is necessary to study the effect of different waveform parameters on the aging characteristics of the ZnO varistor under pulse. In this paper, the controlled variable method is used to study the effect of multi-pulse waveform parameters, such as the number of pulses, pulse interval and number of impacts on the aging rate of the ZnO varistor, and its relationship with the surface temperature rise. It further reveals, from the perspective of energy absorption, the aging mechanism of the ZnO varistor under pulsed lightning.

2. Experiment

2.1. Impulse Test and Waveform

The experimental platform built in this paper can produce a lightning strike process similar to or similar to the natural lightning in the laboratory environment. It is used in multi-pulse lightning shock test experiments. The impact test platform is shown in Figure 1.

Figure 1. A 20-pulse lightning impulse current generator.

The main technical parameter design of the multi-pulse lightning test device is shown in Table 1. The high-voltage ignition device is a key device for realizing pulse triggering, and is composed of a pulse controllable timer, pulse polarity control, a pulse silicon stack trigger, a pulse spark plug and a pulse booster. The synchronous trigger (0 μs) or asynchronous trigger (1 μs–999 ms) time command is sent to the pulse controllable timer through the 232 communication protocol. The pulse controllable timer sends out 20 pulse trigger signals. After photoelectric conversion, it is connected to the trigger box through the optical fiber line with the transmitting head and the receiving head, and the booster generates a high voltage. The pulse silicon stack trigger triggers the conducting silicon stack respectively, and the current after the output triggers is boosted. The device reaches a high voltage capable of ionizing air and conducts to the spark plug. The high voltage is discharged through the discharge electrode after breaking through a certain thickness of air, thereby completing a multi-pulse discharge ignition.

Table 1. The main technical parameters of the experimental setup.

No.	Technical Parameters	Technical Index
1	Charging device input power	220 V/40 kV/80 kV
2	Charging voltage	80 kVDC ± 10%
3	Maximum charging current	1 A
4	Pulse capacitor	40 kV/16 μF
5	Total pulse energy	240 kJ
6	8/20 μs output maximum current	150 kA
7	Trigger gap voltage range	5 kV–120 kV
8	Trigger time controllable accuracy	±1 μs
9	Trigger delay range	0–999 ms

The main body of the multi-pulse lightning experiment platform is an impulse current generator. The basic principle is to connect multiple capacitors in series and parallel, charge them through a high-voltage DC device through rectified voltage or constant current, and then generate a large current to ZnO through gap discharge. Figure 2 shows the basic circuit of the inrush current generator. L and R are the total inductance and resistance of all devices; C is the total capacitance of multiple capacitors connected in parallel; D is the silicon stack; G is the ignition discharge ball gap; CRO is the measurement oscilloscope; T is the charging transformer; O is test terminal sample; r is protection resistance; S is shunt for measuring sample current; $C1$ and $C2$ are voltage dividers for measuring sample voltage. When the circuit is running, the rectifier device is charged and transported, according to the voltage required by the capacitor bank, and, at the same time, sends a trigger pulse to penetrate the ball gap G, and the capacitor bank can be discharged on L, R and the sample.

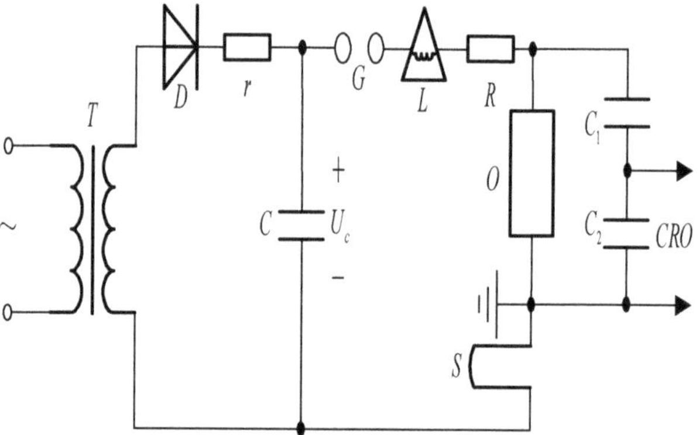

Figure 2. Basic circuit of an inrush current generator.

At present, in the international multi-pulse lightning observation record, the maximum number of pulses does not exceed 20, so, in this paper, the lightning impulse test platform is developed according to the maximum number of pulses in the research process. The 20-pulse lightning current waveform collected by the oscilloscope is shown in Figure 3, which can clearly display the waveform shape of the multi-pulse. The number of pulses is 20.

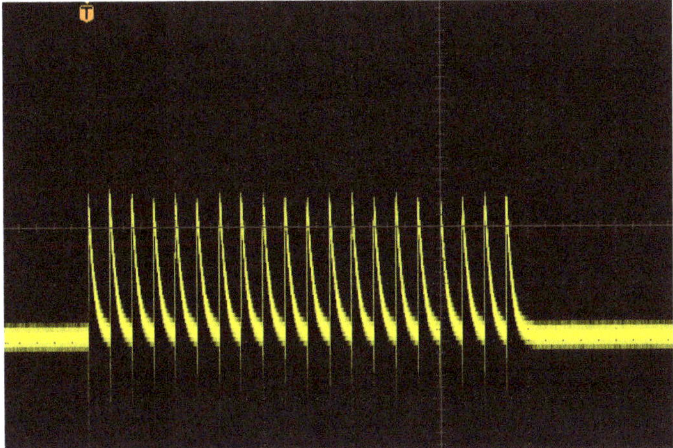

Figure 3. A 20 pulse lightning waveform.

2.2. Sample Preparation

In the experiment, the same batch of ZnO varistor square sheets from a certain manufacturer was selected. The nominal discharge current of the sample was 20 kA, the maximum discharge current was 40 kA, and the maximum continuous operating voltage was 385 V. The static parameters of the varistor voltage and leakage current were measured and selected A sample with the closest static parameters and electrical characteristics was used as the test product. The varistor voltage $U_{1\,mA}$ is between 680 V and 690 V, and the leakage current I_{ie} is 0.1 μA. The samples were numbered P1–P18, every three samples are a group and different numbers correspond to different test waveforms, as shown in Table 2. P1–P12 are the experimental samples used for the aging of ZnO varistors with different pulse time intervals. The number of pulses is 5, and the pulse interval Δt is 20 ms, 50 ms, 100 ms and 500 ms, respectively. The P6 sample corresponds to a pulse interval of 50 ms, the P7–P9 sample corresponds to a pulse interval of 100 ms and the P10–P12 sample corresponds to a pulse interval of 500 ms. P13–P18 are the experimental samples used for the aging of ZnO varistors with different pulse numbers. The pulse interval is 50 ms, P13–P15 samples correspond to 3 pulses, P16–P18 samples correspond to 4 pulses.

Table 2. The static parameters of different samples and the corresponding impulse current waveform parameters.

No.	Initial $U_{1\,mA}$ (V)	Initial I_{ie} (μA)	n	Δt (ms)
P1	682	0.1		
P2	680	0.1	5	20
P3	687	0.1		
P4	686	0.1		
P5	688	0.2	5	50
P6	686	0.1		
P7	687	0.1		
P8	687	0.1	5	100
P9	682	0.1		
P10	684	0.1		
P11	683	0.1	5	500
P12	665	0.1		
P13	690	0.1		
P14	690	0.1	3	50
P15	690	0.1		
P16	688	0.1		
P17	689	0.1	4	50
P18	689	0.1		

2.3. Experimental Procedure

The test procedure is carried out according to the limited voltage impulse test item of IEC61643. The waveform parameters are designed according to the lightning multi-pulse waveform parameters specified in IEC62305-1. The flowchart of the experimental procedure is shown in Figure 4.

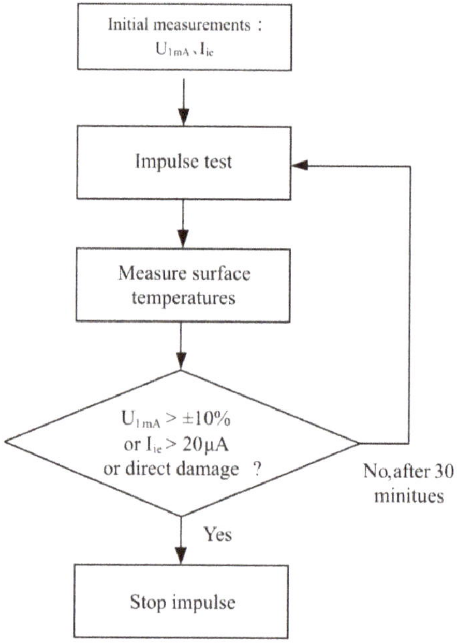

Figure 4. Flowchart of the experimental procedure.

(1) Initial measurements: the samples are characterized with the $U_{1\,mA}$, I_{ie}, and by photographs taken at the beginning of the test.

(2) Impulse test: adjust the charging voltage of the impulse current generator, so that it can output the current amplitude that meets the experiment. A multi-pulse lightning current is used to perform an impact test on a ZnO varistor. Taking a 5-pulse impact with a pulse interval of 50 ms as an example, every 5 pulse currents are recorded as a group of impulse current. After cooling for 30 min, the next group of impulse current was performed. The impact time interval between each two adjacent groups is 30 min. This time interval is sufficient to allow the ZnO varistor to cool to room temperature before the next impact. Until the static parameter of the ZnO varistor changes in $U_{1\,mA}$, by more than $\pm 10\%$, or I_{ie} exceeds 20 μA or the appearance damage occurs directly, the ZnO varistor is judged to be invalid, the impact test is stopped, and the relevant data is recorded. Finally, the static parameters of the sample after different multi-pulse intervals Δt and n are obtained.

(3) Measure the temperature distribution of the sample surface after each impact. Figure 5 is the auxiliary experimental equipment. Figure 5a is an infrared thermometer, used to measure whether the sample reaches the cooling temperature, model is FLUKE63; Figure 5b is an infrared thermal imager, used to measure the surface temperature distribution of the sample after impact, model FLUKE TiS2; Figure 5c is the varistor tester, used to measure the static parameters of the sample before and after impact, the model is FC-2GA; Figure 5d is the LCR tester, used to measure the resistance of the sample before and after impact, the model is HIOKIIM3523.

(4) Change the pulse interval and the number of pulses, repeat the impact experiment process in (2), and record the relevant data. The specific pulse number and pulse interval are shown in Table 1.

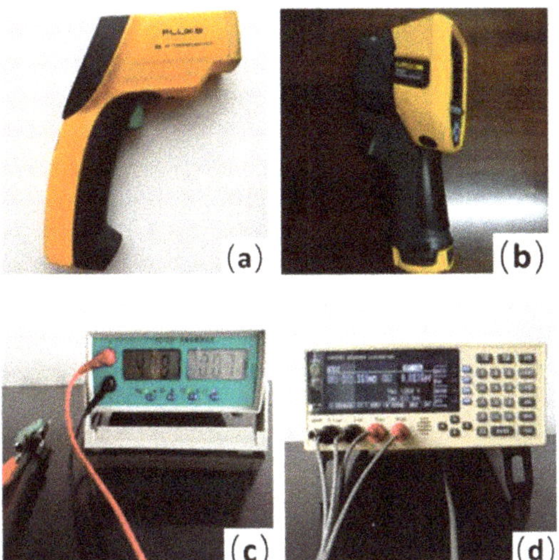

Figure 5. Experimental auxiliary measuring equipment. (**a**) Infrared thermometer, (**b**) Infrared thermal imager, (**c**) Varistor tester, (**d**) LCR tester.

3. Results and Discussion

3.1. Effect of Multi-Pulse Waveform Parameters on Electrical Performance Parameters of ZnO Varistor

The static parameters of the ZnO varistor after the pulse is impacted at different pulse intervals under five pulses are shown in Table 3. The P1–P3 sample can withstand 14 shocks after a pulse interval of 20 ms, the minimum varistor voltage is 615 V, and the maximum leakage current is 12.6 μA. The P4–P6 sample can withstand 15 shocks after a pulse interval of 50 ms, the minimum varistor voltage is 620 V, and the maximum leakage current is 9.9 μA. The P7–P9 samples can withstand 16 shocks after a pulse interval of 100 ms, the minimum varistor voltage is 625 V, and the maximum leakage current is 7.4 μA. The P10–P12 sample can withstand 17 shocks after a pulse interval of 500 ms, the minimum varistor voltage is 628 V, and the maximum leakage current is 6.2 μA.

Table 3. The electrical parameters after impulse at different pulse intervals.

No.	n	$U_{1\,\text{mA}}$ (V)	I_{ie} (μA)	No.	n	$U_{1\,\text{mA}}$ (V)	I_{ie} (μA)
P1	14	610	12.6	P7	15	625	7.0
P2	14	611	11.9	P8	15	624	7.2
P3	14	610	12.3	P9	16	625	7.4
P4	15	615	8.9	P10	17	628	6.2
P5	15	614	9.9	P11	17	628	6.1
P6	16	612	9.3	P12	17	629	5.8

As shown in Table 2, it can be found that, under the impact of five pulses and amplitude of 20 kA, the pulse interval length has an effect on the number of ZnO varistors' impact resistance. It shows that different pulse time intervals have an effect on the performance of a ZnO varistor.

Taking the leakage current parameter in the static parameter as an example, the influence of the pulse interval and the number of pulses on the electrical performance of the ZnO varistor under multi-pulse impact is analyzed. Studies have shown that the leakage current of ZnO varistors increases slowly under a single pulse impact test. Figure 6 shows the change trend of leakage current, with the number of impacts under four pulse intervals of 20 ms, 50 ms, 100 ms and 500 ms. It can be seen that under multi-pulse impact, the leakage current of ZnO varistor increases, until it deteriorates. The trend of the change curve at different intervals tends to be the same, but the shorter the pulse time interval, the faster the leakage current changes. This shows that the pulse interval is an important indicator that affects the aging rate.

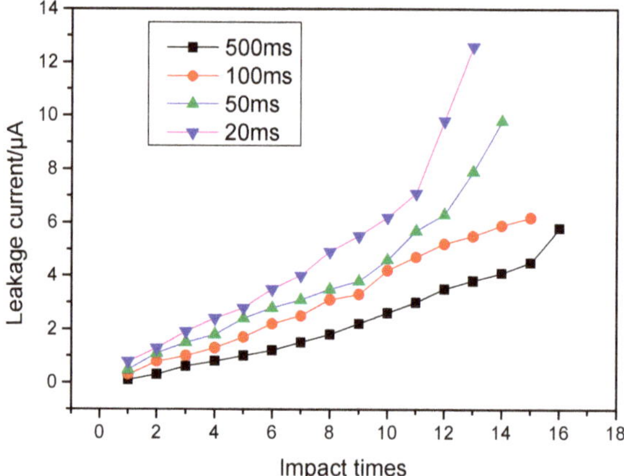

Figure 6. The relationship between the leakage current of ZnO varistors and the number of impacts at different pulse intervals.

Through the impulse test, the values of the varistor voltage and leakage current, measured before the sample was damaged by the pulse under three pulses and four pulses, are shown in Table 4. The ZnO varistor sample can withstand 23 shocks when impacted under three pulses. The minimum varistor voltage before damage is 634 V, and the maximum leakage current is 6.9 µA. The ZnO varistor sample can withstand 20 shocks when impacted under four pulses. The minimum varistor voltage before damage is 620 V, and the maximum leakage current is 7.9 µA.

Table 4. The electrical parameters after the impulse with different number of pulses.

No.	n	$U_{1\,mA}$ (V)	I_{ie} (µA)
P13	23	636	6.8
P14	23	634	6.6
P15	23	634	6.9
P16	20	623	7.5
P17	20	621	7.8
P18	20	620	7.9

Comparing the experimental data analysis, it can be seen that due to the reduction of the number of pulses, the energy injected by the varistor is significantly reduced, and the number of shock resistances under three pulses and four pulses is significantly higher than that of five pulses. The relationship between the leakage current of ZnO varistors with different pulse numbers and the number of impacts is shown in Figure 7. As can be seen from the figure, the leakage current under three pulses, four pulses and five pulses all increase with the number of impacts. Under the same conditions, the greater the number of pulses, the faster the rate of change of the leakage current of the ZnO varistor. Under the same number of impacts, the greater the number of pulses, the greater the leakage current. The failure is the first to be reached under five pulses, and the corresponding leakage current value is also the largest. This shows that the aging rate is positively correlated with the number of pulses under multiple pulses.

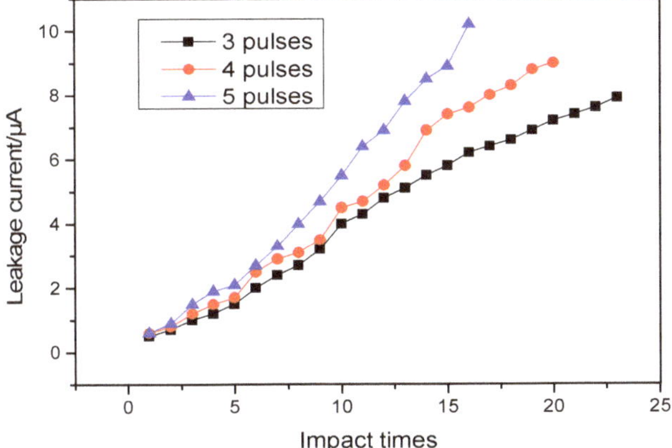

Figure 7. Relationship between the leakage current of ZnO varistors and the impact times under different pulse numbers.

3.2. Relationship between Different Waveform Parameters and Temperature Rise of ZnO Varistor

The three-dimensional distributions of the average temperature rise of ZnO varistor under three pulses, four pulses and five pulses are shown in Figures 8–10. The left horizontal axis is the number of impacts, the right horizontal axis is the pulse interval and the vertical axis is the average temperature rise of the surface of the ZnO varistor. Each point on the curved surface in the figure corresponds to the average surface temperature rise at different impact times and pulse intervals.

The average temperature rise of the surface of ZnO varistor under three pulses varies with the number of impacts and the pulse interval, as shown in Figure 8. The average temperature rise can reach up to 135 °C. After fitting, the relationship between the temperature under three pulses and the number of impacts and the pulse interval is as follows:

$$Z_{3p} = 1.579n - 0.1132\Delta t - 0.05041n^2 + 117.5 \tag{1}$$

where Z_{3p} is the average surface temperature rise of the ZnO varistor under three pulses, n is the number of impacts and Δt is the pulse interval.

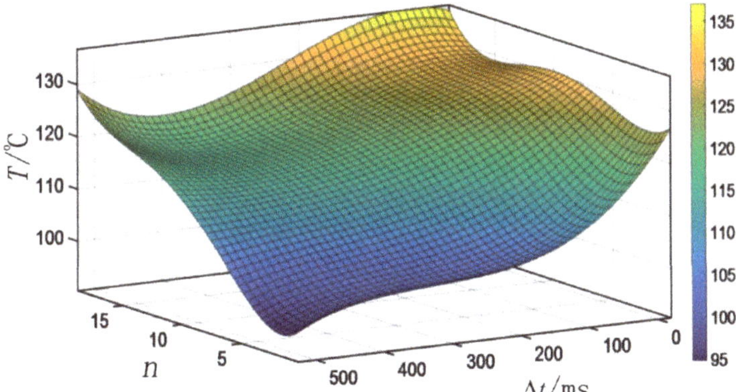

Figure 8. Average temperature rise of the ZnO varistor surface under three pulses.

The change of the average temperature increase of the surface of the ZnO varistor under four pulses with the number of impacts, and the pulse interval is shown in Figure 9. The average temperature rise can be up to 140 °C. After fitting, the relationship between the temperature under four pulses and the number of impacts and the pulse interval is as follows:

$$Z_{4p} = 3.619n - 0.07714\Delta t - 0.1113n^2 + 115.2 \tag{2}$$

where Z_{4p} is the average surface temperature rise of the ZnO varistor under four pulses, n is the number of impacts and Δt is the pulse interval.

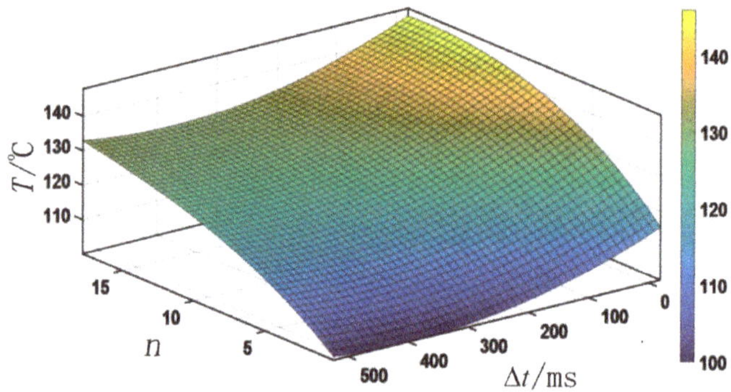

Figure 9. Average temperature rise of the ZnO varistor surface under four pulses.

The change of the average temperature rise of the surface of the ZnO varistor under five pulses with the number of impacts and pulse intervals is shown in Figure 10, and the average temperature rise can be up to 155 °C. After fitting, the relationship between the temperature and the number of impacts and pulse intervals under five pulses is:

$$Z_{5p} = 3.73n - 0.07701\Delta t - 0.1196n^2 + 129.9 \tag{3}$$

where Z_{5p} is the average surface temperature rise of the ZnO varistor under five pulses, n is the number of impacts and Δt is the pulse interval.

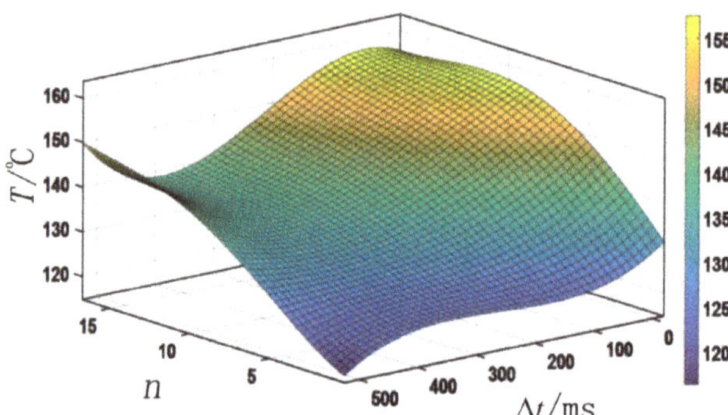

Figure 10. Average temperature rise of the ZnO varistor surface under five pulses.

It can be seen from the temperature rise distribution graph that, no matter what the number of pulses is, when the pulse interval is fixed, the surface temperature of the sample gradually increases with the increase of the number of impacts, but the rate of temperature increase in the late stage of impact is lower than that in the early stage of impact. Under the same conditions, the smaller the pulse interval, the greater the temperature rise value, and the pulse interval has a great influence on the temperature rise. Comparing the overall temperature rise of three pulse, four pulse and five pulse, it can be seen that the more pulses, the higher the average temperature rise. As the number of pulses increases, the average temperature rise of the surface of the ZnO varistor has a nonlinear relationship with the number of pulses. This is because heat will be dissipated in the pulse interval, and the accumulation of energy is not a simple summation.

When the temperature rise of the ZnO varistor reaches a certain level, serious thermal instability will occur. If the heat accumulation is too large in a short time, or the accumulated heat cannot be dissipated in time, it will cause the ZnO varistor to heat damage. Therefore, the greater the number of pulses and the smaller the pulse interval, the more likely that the ZnO varistor will be thermally damaged.

An infrared imager was used to scan the surface of the ZnO varistor sheet immediately after the impact, to obtain a thermographic image of the surface temperature distribution of the ZnO varistor. Figure 11 shows the surface temperature distribution of the ZnO varistor at different pulse numbers at 50 ms intervals. The thermal imaging images use different color reaction temperature values. The white bright area represents the high temperature area. It can be seen that, no matter how many pulses, ZnO The temperature distribution on the surface of the varistor is non-uniform, and the greater the number of pulses, the more non-uniform the temperature distribution. The temperature gradient is caused by the non-uniform material of the ZnO varistor.

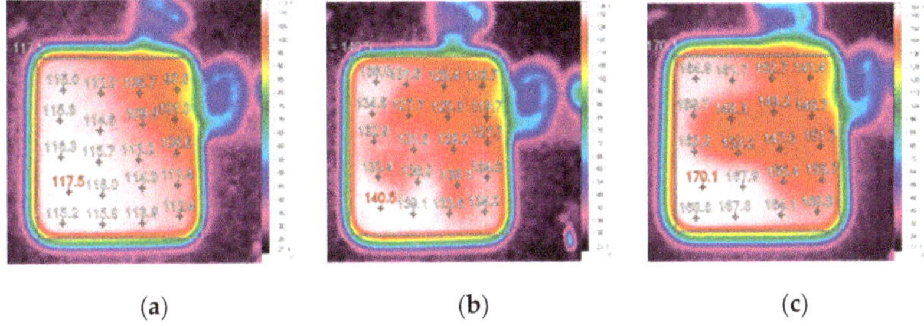

(a)   (b)   (c)

Figure 11. Surface temperature distribution of ZnO varistor under different pulse numbers. (**a**)Three pulses; (**b**) four pulses; (**c**) five pulses.

Statistical analysis of the sample surface temperature value, the sample surface temperature comparison under different pulse numbers is shown in Figure 12, the blue curve represents the maximum temperature T_{max}, the black curve represents the minimum temperature T_{min}, the red curve represents the average temperature Tav. It can be seen from the figure that as the number of pulses increases, the values of T_{max}, T_{min}, and T_{av} all increase. Under three pulses, the maximum temperature of the sample surface is around 117 °C. Under four pulses, the maximum temperature of the sample surface is around 140 °C. Under five pulses, the maximum surface temperature of the sample is around 170 °C. As the number of pulses increases, the maximum temperature T_{max} increases the most. The average temperature has a linear relationship with the number of pulses, but the relationship between the maximum temperature T_{max} and the minimum temperature T_{min} and the number of pulses is nonlinear. It can be seen that the injected energy has a linear relationship with the number of pulses. The more pulses, the higher the average temperature. This is precisely due to the uniformity of the varistor, and the greater the number of pulses, the greater the effect of structural non-uniformity on the non-uniform temperature rise distribution.

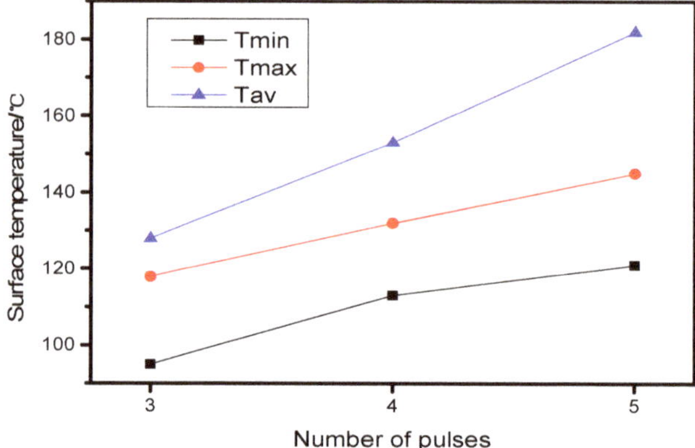

Figure 12. Comparison of the sample surface temperature under different pulse numbers.

3.3. Aging Mechanism Based on Energy Absorption

When the lightning current passes through the ZnO varistor, it must absorb energy itself when discharging the lightning current and suppressing transient overvoltage, that is, the injected energy is greater than the absorbed energy. The residual voltage is an important parameter that reflects the overvoltage protection level of the ZnO varistor, which is the maximum peak voltage between the two ends when the nominal current flows through the ZnO varistor. Therefore, the current waveform and residual voltage waveform after the lightning current impact should be collected, and the current and residual voltage data should be used as the calculation basis for energy absorption.

The impulse current and residual voltage waveform of the ZnO varistor under the action of a single 8/20 μs pulse waveform are shown in Figure 13. The red waveform is the current waveform, and the blue waveform is the residual voltage waveform when the applied impulse current is 20 kA. The value of residual voltage after impact is 1.32 kV. In the main part of the impulse current flow, the 8/20 μs current wave is a nonlinear wave, and the residual voltage wave can be approximately linearly decreasing.

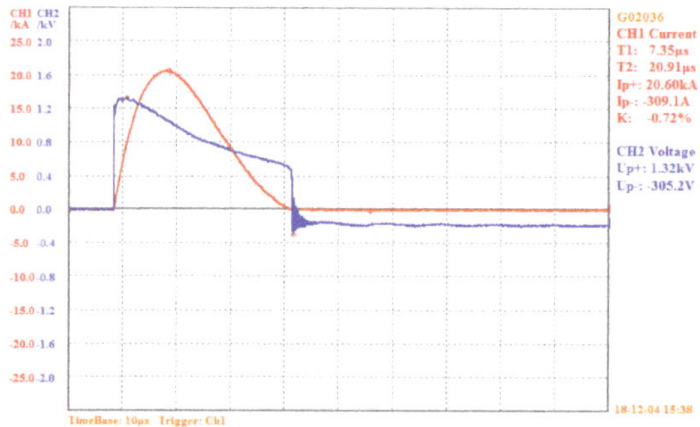

Figure 13. Waveform of current and residual voltage of ZnO varistor under a single 8/20 μs pulse.

The 8/20 μs current wave can be expressed by power exponent:

$$i(t) = AI_\mathrm{m}t^3 e^{-\frac{1}{\tau}} \tag{4}$$

where I_m is the peak current, unit kA, A and τ is the fitting constant. A = 0.1243 μs^{-3}, τ = 3.91 μs.

The residual pressure wave can be approximated as a linear decreasing function expressed as:

$$v(t) = \left(\frac{V_{min} - V_{max}}{T}\right)t + V_{max} \tag{5}$$

where V_{max} is the maximum residual voltage in the figure, the unit is kV, V_{min} is the minimum residual voltage in the figure, the unit is kV and T is the time when the positive residual voltage in the figure is minimum.

The energy W received by the ZnO varistor during the entire current wave time is usually determined by the actual voltage and current of the ZnO varistor through the lightning current, that is:

$$W = \int_0^T v(t)i(t)\mathrm{d}t \tag{6}$$

Substituting Equations (4)–(6), we get:

$$W = \int_0^T \left[AI_\mathrm{m}\left(\frac{V_{min} - V_{max}}{T}\right)t^4 e^{-\frac{1}{\tau}} + AI_\mathrm{m}V_{max}t^3\right]\mathrm{d}t \tag{7}$$

$$a = AI_\mathrm{m}\left(\frac{V_{min} - V_{max}}{T}\right), \; b = AI_\mathrm{m}V_{max} \tag{8}$$

when $T = 30$, we get:

$$W = 19402a + 1330b \tag{9}$$

Under multi-pulse, the pulse interval of 50 ms can be approximated as an adiabatic process. In an ideal state, if the current amplitude and residual voltage under each pulse are the same, the energy sustained by the ZnO varistor under multi-pulse current is the same amplitude The corresponding multiple of the energy of the single pulse current wave is linear. However, the experimental results show that in a complete multi-pulse shock process, the impact current amplitude and residual voltage value of each pulse displayed on the oscilloscope are different. Therefore, the energy sustained by the ZnO varistor under multi-pulse current energy of a single pulse current wave at the same amplitude is not linear. This is because, during the actual impact, the energy accumulation causes the starting temperature of the next pulse to be higher than the temperature at the impact of the previous pulse. Figure 14 shows the current waveform and residual voltage waveform collected on the oscilloscope under five pulses, respectively. The five vertical lines in the figure above represent five pulses. The green box is the corresponding pulse sequence, and the green box is the first pulse, and the red ellipses are the other four pulses in a group of shocks. The lower waveform in the figure is an enlarged display of the waveform in the green box, where the yellow waveform is the current waveform and the blue waveform is the residual voltage waveform.

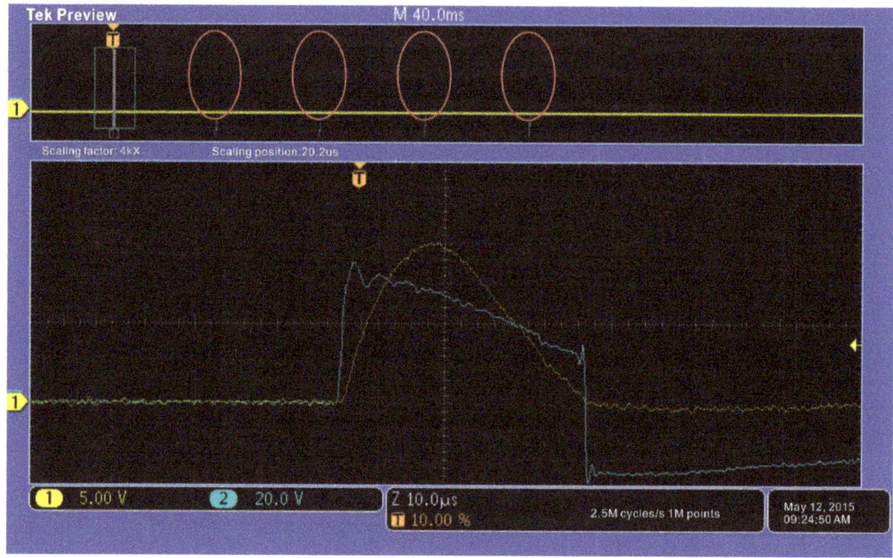

Figure 14. Current and voltage waveforms under five pulses.

Taking the current amplitude and residual voltage value of each pulse waveform acting on the ZnO varistor as energy calculation parameters, calculating the energy value under the action of each pulse, adding up the energy of five single pulses, according to the obtained impulse current and residual voltage waveforms, the five pulse current impulse results under 20 kA conditions can be obtained. As shown in Table 5, the energy absorbed by the ZnO varistor under the action of the 5-pulse current is:

$$W_{5p} = 19402(a_1 + a_2 + a_3 + a_4 + a_5) + 1330(b_1 + b_2 + b_3 + b_4 + b_5) \tag{10}$$

Substituting the impact result of the five pulse current in the table into Equation (9), the five pulse absorption energy is 1664.94 J.

Table 5. Impact energy of the sample under five pulses.

n	V_{max}	V_{min}	I_m	a	b	W/J
1st pulse	1.35	0.71	19.48 kA	−0.0051	0.3268	329.60
2nd pulse	1.37	0.68	19.62 kA	−0.0052	0.3219	327.22
3rd pulse	1.38	0.65	19.70 kA	−0.0059	0.3379	334.93
4th pulse	1.40	0.63	19.83 kA	−0.0063	0.3450	336.62
5th pulse	1.42	0.60	19.88 kA	−0.0067	0.3508	336.57

Similarly, the impact energy values of the three pulse current and the four pulse current are also calculated according to the above energy accumulation method. The energy absorbed by three pulses is 991.75 J, and the energy absorbed by four pulses is 1328.37 J.

Taking five pulses as an example, the impact energy accumulation of ZnO varistor under different pulse sequences is shown in Figure 15. The number on the horizontal axis in Figure 15 represents the pulse sequence number within a shock period. For example, 1 represents the first pulse. The energy accumulation from the first pulse to the fifth pulse can be obtained. After each pulse is applied, the energy gradually increases and there is a cumulative effect of energy superposition. Except for the first pulse, the temperature of each subsequent pulse current injection is close to the temperature after the previous end. At this time, the ZnO varistor is subjected to 5 different temperature gradient thermal stresses.It can be seen that the temperature of the ZnO varistor gradually rises during the entire impact cycle and is subjected to multiple "short strikes" of thermal stress in consecutive short intervals. At this time, the Schottky barrier height must fall faster than the single pulse. As a result, the breakdown voltage of the grain boundary drops faster. In the later stage of the impact, the electrical properties of the sample are mainly controlled by the thermal destruction of the grain and the grain boundary. The varistor characteristic of the sample disappears, causing the local temperature of the sample to rise sharply, causing the sample to burst and damage. To sum up, the energy superimposed cumulative effect of multi-pulse impact accelerates the aging process of the ZnO varistor, and eventually produces irreversible structural damage.

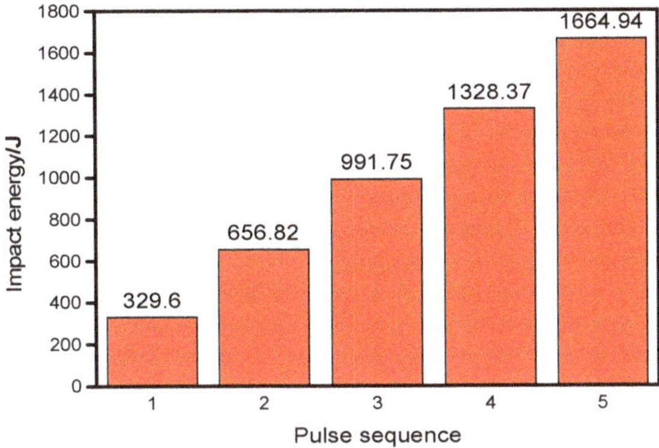

Figure 15. Current and voltage waveforms under five pulses.

3.4. Heat Transfer Modelling

In the impact process of the 8/20 single pulse current wave, since the action time is very short, it can be considered that the interior of the varistor is insulated. However, in the case of multiple pulses, the problem of heat transfer should be considered, because there is heat conduction during the intermittent time of two adjacent pulses. The heat generated by the first pulse after the heat exchange between the heat dissipation, and the surrounding air during the intermittent time, is lower than that at the end of the first pulse, and this temperature is used as the initial temperature of the resistance during the second pulse. The temperature can be calculated by analogy.

According to the temperature distribution of the varistor thermal imaging, there are always weak points with relatively large current density and relatively high temperature inside the varistor. The destruction of the resistor starts from the weak point, assuming there is a small cylinder inside the varistor. Its equivalent resistance under high current is lower than the rest of the varistor. Next, the temperature variation of the varistor under a certain number of impacts will be simulated and calculated.

In the interval time of two adjacent pulse currents, the heat conduction inside the varistor can be described by the following equation in one-dimensional cylindrical coordinates:

$$\frac{\partial^2 T}{\partial r^2} + \frac{1}{r}\frac{\partial T}{\partial t} = \frac{c_p \rho}{k}\frac{\partial T}{\partial t} \tag{11}$$

where k is the thermal conductivity, c_p is the heat capacity under constant pressure, ρ is the specific gravity and r is the radial distance from the center. The surface heat transfer at the side surface can be expressed as:

Boundary conditions:

$$-k\frac{\partial T}{\partial r} = \alpha(T - T_\alpha)\, r = R \tag{12}$$

where α is the comprehensive heat transfer coefficient of the surface. At room temperature of 20 °C, $\alpha = 7.0$, T_a is the ambient temperature, and R is the width of the varistor. Here, the finite difference method is used for calculation.

$$\frac{T_i^n - 2T_{i+1}^n + T_{i+2}^n}{\Delta r^2} + \frac{T_{i+1}^n - T_i^n}{i\Delta r\Delta r} = \frac{c_p\rho}{k}\cdot\frac{T_{i+1}^n - T_i^n}{\Delta t} \quad i = 1 \tag{13}$$

$$\frac{T_{i-1}^n - 2T_i^n + T_{i+1}^n}{\Delta r^2} + \frac{T_{i+1}^n - T_{i-1}^n}{i\Delta r\Delta r} = \frac{c_p\rho}{k}\cdot\frac{T_{i+1}^{n+1} - T_i^n}{\Delta t} \quad i = 1,2\ldots M-1 \tag{14}$$

At the node on the boundary, T_{i+1} does not exist, but the finite difference equation of the node at the boundary can be obtained by the boundary condition. Set on the boundary to extend the heat conduction area by a distance Δr, so that there is a hypothetical node of temperature $T_{-(i-1)}$, then the heat conduction equation at the boundary node is:

$$\frac{T_{i-1}^n - 2T_i^n + T_{-(i-1)}^n}{\Delta r^2} + \frac{T_{-(i-1)}^n - T_{i-1}^n}{R2\Delta r} = \frac{c_p\rho}{k}\cdot\frac{T_i^{n+1} - T_i^n}{\Delta t} \tag{15}$$

Using the central difference formula, the finite difference form of the boundary conditions is:

$$-k\frac{T_{-(i-1)}^n - T_{i-1}^n}{2\Delta r} = \alpha\left(T_i^n - T_a\right) \tag{16}$$

Combine the two equations above and get:

$$\frac{T_{i-1}^n - 2T_i^n + \frac{2\Delta t}{k}\alpha\left(T_i^n - T_a\right)}{\Delta r^2} + \frac{\alpha\left(T_i^n - T_a\right)}{kR} = \frac{c_p\rho}{k}\cdot\frac{T_i^{n+1} - T_i^n}{\Delta t} \quad i = M \tag{17}$$

This is the heat conduction equation of the node located on the convection boundary. Combining Formulas (15)–(17), the temperature distribution of the varistor before the second pulse comes after the end of the first pulse can be obtained. The finite difference expression given above is to find the temperature T_i^{n+1} at the (n + 1)th time step (i = 1, 2, ...), based on the known temperature T_i^n at the nth time step at the previous moment. (i = 1, 2, ... , M).

Discrete 8/20 us current waveform into units with time step $\Delta t = 0.4$ μs, and in the interval time of 10 ms–500 ms, the time step $\Delta t = 0.5$ ms used for heat conduction. According to the model description calculation, the results corresponding to the multi-pulse current shocks with interval times of 20 ms, 50 ms, 100 ms and 500 ms are shown in Figure 16 as a, b, c and d, to indicate the relationship between the temperature and the impact time. The horizontal axis in the figure is the impact time, where period A is the time from the beginning of the first pulse to the end of the first pulse, and the temperature rises from the initial room temperature at this time; period B is the time from the end of the first pulse to the start of the second pulse. During this time, the varistor heat dissipates and the temperature drops; period C is when the second pulse begins to function until the end time, at which time the varistor temperature rises sharply. It can be seen from the figure that the shorter the intermittent time, the higher the initial temperature at the second pulse current. Due to the sensitivity of the current-voltage characteristics to temperature, the current is more concentrated, and the temperature at the end of the entire impact test is higher. This is also the reason why the varistor is most prone to aging when there is an impact with an intermittent time of 20 ms.

The shorter the pulse interval, the higher the temperature rise of the varistor, which is consistent with the results obtained in the experiment.

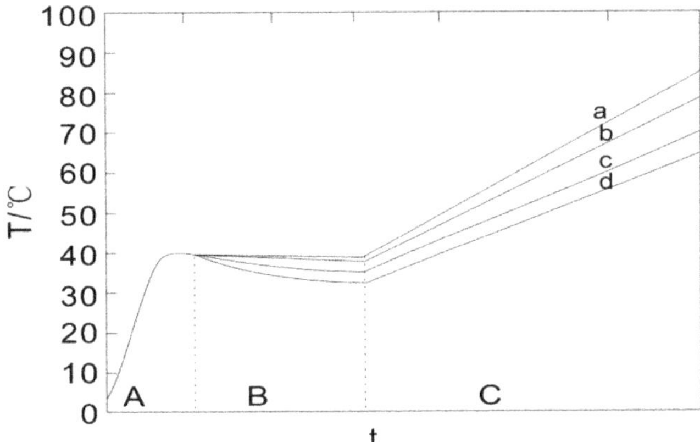

Figure 16. Variation of varistor temperature with impact time.

Taking the five pulse 50 ms interval as an example, the simulation calculation of temperature over time is performed. The calculation result is shown in Figure 17. The temperature rise after the action of each pulse can be obtained from the figure. It can be seen that the temperature and the number of pulses are non-linear.

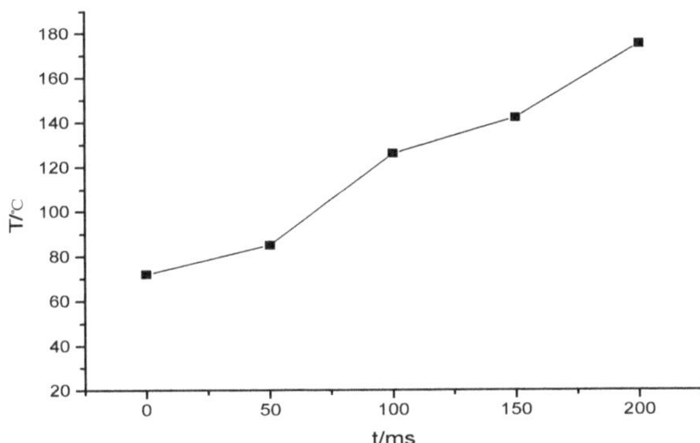

Figure 17. Variation of varistor temperature with impact time under the five pulse.

4. Conclusions

In this study, experiments were conducted to investigate the properties of ZnO varistors under multiple lightning impulse currents.

(1) The number of pulses and the pulse interval under multiple pulses play a decisive role in the aging rate. We analyzed the impact of pulse waveform parameters on aging. The more pulses and the smaller the pulse interval, the faster the aging speed of ZnO varistor, and the more likely there will be failure and damage.

(2) The temperature distribution on the surface of the ZnO varistor is non-uniform, and the greater the number of pulses, the more uneven the temperature distribution. The existing temperature gradient is caused by the uneven microscopic material of the ZnO varistor. The relationship between the average temperature rise and the number of shocks and pulse interval under different pulse numbers was established. The more the number of shocks and the smaller the pulse interval, the greater the temperature rise value. Comparing the overall temperature rise of three pulse,

four pulse and five pulse, it can be seen that the more pulses, the higher the average temperature rise. As the number of pulses increases, the average temperature rise on the surface of the ZnO varistor has a nonlinear relationship with the number of pulses.

(3) The current amplitude and residual voltage value of each pulse waveform acting on the ZnO varistor are used as energy calculation parameters to calculate the energy value under the action of each pulse. The energy sustained by the ZnO varistor under multi-pulse current is not linearly related to the energy of the single-pulse current wave at the same amplitude. The superimposed cumulative energy of the impact under multi-pulse accelerates the aging process of the ZnO varistor and eventually produces an irreversible structural destruction.

(4) Heat transfer simulation results show that the shorter the pulse interval, the higher the temperature rise of the varistor, which is consistent with the results obtained in the experiment.

Author Contributions: Data curation, D.L.; Formal analysis, C.Z.; Investigation, C.L.; Methodology, D.L.; Project administration, C.Z. and H.X.; Software, C.L.; Supervision, H.Z. All authors have read and agreed to the published version of the manuscript.

Funding: This research was funded by the National Nature Science Foundation of China grant number 61671248 and The APC was funded by Heilongjiang Natural Fund Joint Guidance Project (LH2019E120).

Conflicts of Interest: The authors declare no conflict of interest.

References

1. Zhang, C. *Research on the Aging and Deterioration of Surge Protectors in Low-Voltage Power Systems*; Nanjing University of Information Science and Technology: Nanjing, China, 2013.
2. Su, Y.; Sun, L.; Yang, Z. Research progress of low voltage varistor. *Electron. Compon. Mater.* **2010**, *29*, 74–78.
3. Yang, Z.; Chen, L. Analysis and application of capacitance change during the deterioration of zinc oxide varistor. *High Volt. Technol.* **2010**, *36*, 2167–2172.
4. Li, X.; Zhang, J.; Chen, L. Research on the design method of combined surge protector. *Electr. Porcelain Light. Arrester* **2017**, *4*, 16–22.
5. Yang, D.; Zhang, X.; Xu, Y. The failure mode and failure reason of SPD in low-voltage power supply system. *Electr. Porcelain Arrester* **2007**, *4*, 43–46.
6. Darveniza, M.; Tumma, L.R. Multipulse lightning currents and metal-oxide arresters. *IEEE Trans. Power Deliv.* **1997**, *12*, 1168–1175. [CrossRef]
7. Darveniza, M.; Mercer, D.R. Laboratory studies of the effects of multipulse lightning currents on distribution surge arresters. *IEEE Trans. Power Deliv.* **1993**, *8*, 1035–1044. [CrossRef]
8. Lee, B.-H.; Kang, S.-M. Properties of ZnO varistor blocks under multiple lightning impulse voltages. *Curr. Appl. Phys.* **2006**, *6*, 844–851. [CrossRef]
9. Lee, B.-H.; Kang, S.-M.; Pak, K.-Y.; Choi, H.S. Effect of multiple lightning impulse currents on Zinc oxide arrester blocks. In Proceedings of the 27th International Conference on Lightning Protection, Avignon, France, 13–16 September 2004; pp. 634–639.
10. Rousseau, A.; Zang, X.; Tao, M. Multiple shots on SPDs—Additional tests. In Proceedings of the 2014 International Conference on Lightning Protection (ICLP), Shanghai, China, 11–18 October 2014; pp. 997–1001.
11. Li, P.; Yang, Z.; Cao, H. Research on the aging performance of MOA under multi-pulse impact. *China Electr. Power* **2016**, *49*, 69–74.
12. Li, P.; Zhang, C.; Lv, D. The destruction form of metal oxide under multi-pulse lightning strike. *High Volt. Technol.* **2017**, *43*, 3792–3799.
13. Xiao, Y.; Yang, Z.; Liu, J. MOV performance of single-chip and double-chip in parallel under high-voltage multi-pulse impact. *China Electr. Power* **2016**, *49*, 55–59.
14. de Salles, C.; Martinez, M.L.B.; de Queiroz, Á.A.A. Ageing of metal oxide varistors due to surges. In Proceedings of the 2011 International Symposium on Lightning Protection (XI SIPDA), Fortaleza, Brazil, 3–7 October 2011; pp. 171–176. [CrossRef]
15. Vahidi, B.; Cornick, K.; Greaves, D. Effects of Multiple Stroke on ZnO Surge Arresters. In Proceedings of the 24th International Conference on Lightning Protection, Birmingham, UK, 14–18 September 1998; pp. 960–963.

16. Heinrich, C.; Wagner, S.; Richter, B.; Kalkner, W. Multipulse tests on surge arresters for medium and low voltage systems. In Proceedings of the 24th International Conference on Lightning Protection, Birmingham, UK, 14–18 September 1998; pp. 790–794.

17. Haryono, T.; Sirait, K.T. The damage of ZnO arrester block due to multiple impulse currents. *Telkomnika* **2011**, *9*, 171. [CrossRef]

18. Haryono, T.; Sirait, K.T. Effect of multiple lightning strikes on the performance of ZnO lightning arrester block. *High Volt. Eng.* **2011**, *37*, 2763–2771.

19. Zhang, C.; Zhang, S. Zinc oxide nonlinear resistance impact aging failure mechanism. *High Volt. Technol.* **2000**, *5*, 48–49.

MDPI
St. Alban-Anlage 66
4052 Basel
Switzerland
Tel. +41 61 683 77 34
Fax +41 61 302 89 18
www.mdpi.com

Electronics Editorial Office
E-mail: electronics@mdpi.com
www.mdpi.com/journal/electronics